Implementation of Geographic Information Systems

Implementation of Geographic Information Systems

Edited by **Marina De Lima**

$\mathcal{CL}$ LANRYE
INTERNATIONAL

New Jersey

Published by Clanrye International,
55 Van Reypen Street,
Jersey City, NJ 07306, USA
www.clanryeinternational.com

Implementation of Geographic Information Systems
Edited by Marina De Lima

International Standard Book Number: 978-1-63240-301-8 (Hardback)

Contents

Permissions

List of Contributors

Preface

This book was inspired by the evolution of our times; to answer the curiosity of inquisitive minds. Many developments have occurred across the globe in the recent past which has transformed the progress in the field.

The book provides advanced information regarding the implementation of geographic information systems. In theoretical and professional spheres, Geographic Information Systems (GIS) hold special and crucial significance. The use of GIS has been on a steep rise in various professional fields. From civil engineers to spatial economists; from sociologists to urban planners, and criminal justice officers, the applications of GIS have found relevance in numerous professions. Hence, the understanding of Geographic Information Systems has become critical in academics and professional work. This book contains research findings that explain GIS's applications in different subsets of social sciences. Along with compilations of numerous case studies conducted around the world, the book amalgamates the theories of GIS and their practical implementations in varied conditions. This book sheds light on the applications and usage of GIS in a broad array of fields like geospatial analysis and modeling, land use analysis, and infrastructure network analysis like transportation and water distribution. For the purpose of teaching, application and research, this book is extremely handy. The lucidity of this book adds substantially to the understanding of the expansive subject of GIS.

This book was developed from a mere concept to drafts to chapters and finally compiled together as a complete text to benefit the readers across all nations. To ensure the quality of the content we instilled two significant steps in our procedure. The first was to appoint an editorial team that would verify the data and statistics provided in the book and also select the most appropriate and valuable contributions from the plentiful contributions we received from authors worldwide. The next step was to appoint an expert of the topic as the Editor-in-Chief, who would head the project and finally make the necessary amendments and modifications to make the text reader-friendly. I was then commissioned to examine all the material to present the topics in the most comprehensible and productive format.

I would like to take this opportunity to thank all the contributing authors who were supportive enough to contribute their time and knowledge to this project. I also wish to convey my regards to my family who have been extremely supportive during the entire project.

Editor

Multi-Scale GIS Data-Driven Method for Early Assessment of Wetlands Impacted by Transportation Corridors

Rodrigo Nobrega, Colin Brooks, Charles O'Hara and Bethany Stich

Additional information is available at the end of the chapter

1. Introduction

The correlation between transportation systems and adverse impacts on the natural environment have been investigated at different scales of observation (Kuitunen et al., 1998; Bouman et al., 1999; Corrales et al., 2000; Formann et al., 2003, Wheeler et al., 2005, Fletcher and Hutto, 2008). There is a growing body of literature reporting and quantifying the effects caused by transportation infrastructure on the proximate biophysical setting as shown in (Keller & Largiardèr, 2003) as well as on the socio-economic setting as shown in (Boarnet & Chalermpong, 2001). The environmental consequences of landscape fragmentation in different phases of transportation project development have been investigated and tabulated by (Corrales et al., 2000). However, the disparity of definitions for the biophysical landscape can make it difficult to communicate clearly and even more difficulty to establish consistent management policies. Landscape invariably comprises an area of land containing a mosaic of patches or land elements (McGarigal & Marks, 1995; Hilty et al., 2006). The overall knowledge-base of transportation systems and methods to consider, minimize, and mitigate adverse impacts on natural systems and biophysical settings have gradually been absorbed and adopted by transportation and Environmental Impact Assessment (EIA) practitioners to design balanced engineering solutions and deliver transportation infrastructure in an environmentally responsible manner. The body of science and knowledge supporting practitioners has grown through in-depth reviews about transportation and ecological effects (Spellerberg, 1998; and Formann et al., 2003) Similarly, the knowledge base concerning the impacts of land use on travel behaviour is also being investigated and developed from the transportation perspective (Mokhtarian & Cao, 2008; Litman, 2008).

Road development is a primary mechanism responsible for habitat, ecosystem, and overall biophysical fragmentation, replacing or modifying pre-existing land cover such as wetlands, creating edge habitat and altering landscape structure and function (Saunders et al., 2002). While conserving the remaining natural environment as well as restoring environmentally impacted areas is vital for natural sustainability, transportation corridor development is required by society and results in our modern transportation infrastructure and travel patterns.

Previous lessons learned show that environmental issues should be considered early the transportation planning process in order to balance economic, engineering and natural sustainability perspectives (Amekudzi & Meyer, 2006). A highway design that meets the transportation corridor needs, while minimizing environmental impacts, requires cooperation and compromise among different parties. It is a pressing challenge for researchers and practitioners to develop and validate novel methods for transportation planning that deliver streamlined planning approaches and improved environmental benefits beyond those possible through traditional approaches (Spellerberg, 1998; Stefanakis & Kavouras, 2002; Mongkut & Saengkhao, 2003; Huang et al., 2003; Gregory et al., 2005). The integration of transportation demand, current and long term development plans, and economic and ecological impacts in time-series scenarios by using land cover and land use analysis is a good way to provide promising results (Saunders et al. 2002; Forman & Alexander, 1998). The use of Multi-Criteria Decision Making (MCDM) as a decision-making framework for transportation infrastructure planning, which can accommodate, model, and combine varying stakeholder values and help to resolve conflicting opinions, is an area that has only been recently explored. Initial results offer significant promise to streamline the National Environmental Policy Act (NEPA) process (Nobrega et al., 2009).

MCDM can facilitate the integration of different planning scenarios as well as the combination of different approaches for environmental sustainability in transportation planning. In modern transportation projects, considerations of both landscape analyses and natural-economic sustainability are mandatory under programs such as NEPA and similar state and local-level laws (Corrales et al., 2000). In 2003, Burnett and Blaschke demonstrated that advances in informatics and geographic information tools have made it possible to segment the complex environments supported by the ecological theory into factors that may be considered in a landscape analysis approach. Current reviews about geospatial landscape analysis in ecology reflect the relatively recent trend towards the use of remote sensing through object-based image analysis (Blaschke et al., 2001; Burnett & Blaschke, 2003; Aplin, 2005). Geographic Object-Based Image Analysis (GEOBIA) employs polygons as bounding areas which delimit the landscape and enable data and image analyses that transcend traditional per-pixel approaches such as spectral-based analysis (Nobrega, 2007; Hay & Castilla, 2008). The use of object-based segments for landscape analysis enable the generation of a large number of parameters based not only on intrinsic values extracted from the polygons, but also extrinsic values computed from the geometry, texture, and context of the objects. This information can be used to form a classification decision hierarchy and provide results that may be combined with existing GIS information to offer significant and innovative results to benefit transportation planning and management and streamline the Environmental Analysis processes (Nobrega, 2007).

2. Background

2.1. Watersheds as natural biophysical landscape segments

Hydrological watersheds are natural subdivisions of the landscape and exercise influence on other natural and man-made features. Wetlands are among the most sensitive of natural features and are vital components of the habitat requiring protection from adverse impacts that may be caused by human development and infrastructure projects. Indeed, NEPA requires transportation planners to consider possible impacts on the hydrological system including stream crossings, flood plains, land cover, and wetlands as part of maintaining the ecological and biophysical balance within the local watersheds (Amekudzi & Meyer, 2006).

This research describes the use of a collaborative, interactive, and iterative multi-scale approach to assess and rank hydrologically segmented features and wetlands to deliver enhanced understanding of how these biophysical systems are affected by transportation infrastructure projects. This chapter addresses a two-level object-based landscape analysis computed from hydrological sub-watersheds from Hydrologic Unit Code 12-level (HUC-12), wetlands, and a subsegment of the proposed Interstate 269 (I-269, a proposed bypass around Memphis, Tennessee, in the southern United States of America) as major objects of interest. Firstly, parameters are extracted per watershed from percentage of wetlands, zoning, existing and current developments, and density of perennial and intermittent streams. Watersheds are ranked according the potential for risk on the natural environment, as described below. The watersheds are considered as primary objects in this hierarchical landscape analysis. After ranking these objects, the next step in the hierarchical analysis process is identifying and ranking wetlands based on potential for adverse impact. For each watershed, topographic analysis (computed from LiDAR elevation data) and computer-assisted image interpretations are performed to enhance the delineation of the wetlands. Wetlands are analyzed according their distance from planned developments, planned roads and the I-269 corridor.

It should be recognized that there are limitations inherent to geospatial data and their analysis within any research framework, and the practical implementation of innovative contributions for geospatial analysis depends upon properly designing and structuring approaches that may be implemented in a practical and feasible framework available in readily available GIS software. In this paper, a top-down GIS framework for landscape analysis is proposed using hydrological watersheds as reference objects for segmentation of the landscape. This segmentation facilitates the geographical analysis of biophysical subdivisions of the landscape based on a watershed approach to conduct contextual, geometrical, and hierarchical analysis. The overall idea is quite similar to standard approaches in object-oriented landscape analysis; however, the use of watersheds as a segmentation layer enables the analysis to consider biophysical subdivision as parts of transportation corridor planning and enables the use of output results in cumulative cost surfaces that may be employed to refine land use and corridor plans and improve agency coordination during the NEPA process.

2.2. Landscape analysis

Landscapes are shaped by the interaction of social and ecological systems (Brunckhorst, 2005). Current and future use of land, productivity and patterns of sustainability are continually modified by humans within the landscape in spatial scales across time in different magnitudes (Ono et al., 2005). For environmentally-focused transportation planning, eco-regions and hydrological watersheds are keys concepts that must be considered in landscape analysis. Understanding landscape and watershed characteristics, the geographic context of sensitive environmental resources, and the services provides by natural systems, is vital to providing balanced solutions for sustainable development amidst natural resources that face economic and social issues (Figure 1). Despite the similarity in some points of view between creating subdivisions of eco-regions and watersheds, a common misunderstanding of each of these landscape subdivision frameworks has resulted in inconsistency in their use and, ultimately, to ineffective application in addressing landscape analysis (Omernik & Bailey, 1997).

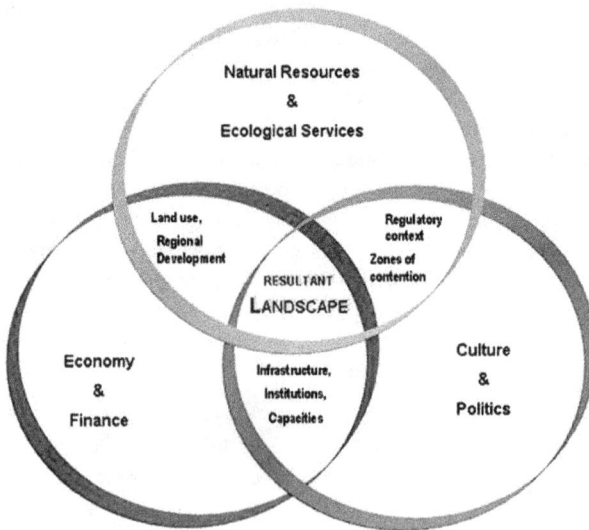

Figure 1. Complex spheres of interaction reflecting human values, identity, and activities affecting landscape change (Brunckhorst, 2005).

2.3. Geographic object-based analysis

The traditional methods of classifying remote sensing data are based upon statistical and cluster-based classification of single pixels in a digital image (Lillesand & Kiefer, 2004). Recent research indicates that pixel based classification methods may be less than optimal in producing high-accuracy land use / land cover maps since they do not consider the spatial relationships of landscape features (Schiewe et al., 2001). For example, a significant proportion of the reflectance recorded for a single pixel is derived from the land area immediately surrounding the pixel (Townshend et al., 2000). Analyzing at the polygon object scale enables imagery classification to move beyond this traditional problem.

Contemporary object-based landscape analysis uses parameters derived from hierarchy, context and geometry of the image objects rather than pixel values. Despite a successful history with remote sensing, the accuracy of pixel-based image analysis can be compromised when applied to high resolution images (Nobrega, 2007).

The development and practice of object-based classification has grown as have the variety of methods and approaches of incorporating spatial context into the classification process. Most object-based approaches compliment the axiom of landscape ecology; that it is preferable to work with a meaningful object representing the true spatial pattern rather than a single pixel (Blaschke et al., 2001). Furthermore, the development or use of objects (at one or multiple scales) is always an initial primary phase of the analysis which emphasizes capturing, extracting, or refining the size, shape, and distribution of features of interest.

Object-based classification can be functionally decomposed into two major steps: segmentation and classification. In the segmentation step, relatively homogeneous image objects (polygons) are derived from both spectral and spatial information (Benz et al., 2004). In the classification phase, image objects are labeled as to their class membership by using established classification algorithms, knowledge-based approaches, fuzzy classification membership degrees or a combination of classification methods (Civco et al., 2002).

The commercial software package, Trimble eCognition Developer (formerly *Definiens Developer*), has been well received as a tool for performing object-based classifications of land cover (an example list of scientific papers using eCognition for various land cover mapping tasks is available at *http://www.ecognition.com/learn/resource-center/show-more?type=Scientific%20Paper*). For automated generation of segmentation objects, the application uses a region growing multi-scale segmentation algorithm for the delineation of image objects. The application also enables pre-existing spatial features to be used as objects within which segmentation may be constrained. eCognition provides two different classification methods that may be used separately or combined: a sample-based nearest neighbor classifier with fuzzy logic capabilities and a classifier that enables the development of hierarchic class-membership through a set of rule-based fuzzy logic membership functions.

This chapter presents an implementation of constrained segmentation in which naturally occurring objects provide the initial basis for identifying relevant features on the landscape within which classification and analysis that implement GEOBIA theory are explored. No segmentation objects are computed, since the objects of interest (watersheds and wetlands) already exist, and segmentation statistics are generated for these areas and used in subsequent phases of analysis. The method combines intrinsic and extrinsic information extracted from the objects and the analyses are organized hierarchically.

2.4. Spatial MCDC-AHP in transportation planning

Driven by the need to find a balanced solution among conflicting scenarios and because of the vast and growing availability of geospatial data, decision making theory has been explored by the environmental assessment community, including transportation planners.

Multi-Criteria Decision Making is a systematic methodology to generate, rank, compare, and make a selection from multiple conflicting alternatives using disparate data sources and attributes (Gal et al., 1999; Nobrega et al., 2009). The applicability of MCDM is being extended to many different fields including GIS, which is capable of handling massive amounts of geospatial data. Analytical Hierarchy Process is a decision making approach introduced by (Saaty, 1994) based on pair-wise comparisons among criteria and factors in different hierarchical levels. AHP is presented as an effective technique for combining heuristic inputs from stakeholders to achieve a consensus-based decision. The technique allows competing agency expert views as well as stakeholder opinions to be considered quantitatively in a decision making approach (MacFarlane et al., 2008). In keeping with the spirit of NEPA, AHP does not pre-select any specific alternative; it exposes all potential alternatives to the analysis and selection process.

AHP is robust and easily implemented in GIS for geospatial analysis. Results demonstrated in (Sadasivuni et al., 2009) and (Nobrega et al., 2009) showed that AHP can provide significant benefits in facilitating multi-criteria decision-making for planning. AHP is a tool useful for planning and can lead to stakeholder buy-in on planning approaches that consider resource allocation, benefit/cost analysis, the resolution of critical conflicts, and design and optimization. This chapter explores a practical application of spatial MCDM-AHP for transportation planning. The solution presents a semi-automated approach based on an adaptation of Dr. Saaty's theory.

3. The study area: Initial processing

The Interstate 69 is a proposed 1,600-mile long corridor that connects Canada to Mexico. The entire corridor is divided into 32 Segments of Independent Utility (SIU) for transportation planning and construction purposes. SIU-9 ranges from Millington, TN down to Hernando, MS, crossing the metropolitan area of Memphis, TN and reusing some existing roads such as I-55. However, a new I-269 bypassing the metropolitan Memphis, TN area to the east has been approved through an Environment Impact Statement (EIS) process and is entering the construction phase (Figure 2). The I-269 bypass is the test-bed for a series of research projects sponsored by the National Consortium for Remote Sensing in Transportation - Streamlined Environmental and Planning Process- (NCRST-SEPP). This work is concentrated in Desoto County-MS, which is traversed by the designed I-269.

The NCRST-SEPP project (*http://www.ncrste.msstate.edu/*) applied remote sensing technology and geospatial analysis to streamlining the EIS process for a specific on-the-ground transportation project. NCRST-SEPP research was designed to demonstrate the innovative application of commercial remote sensing and spatial information technologies in specific environmental and planning tasks and activities, validating the use of those technologies by conducting rigorous comparison to traditional methods (Dumas et al., 2009).

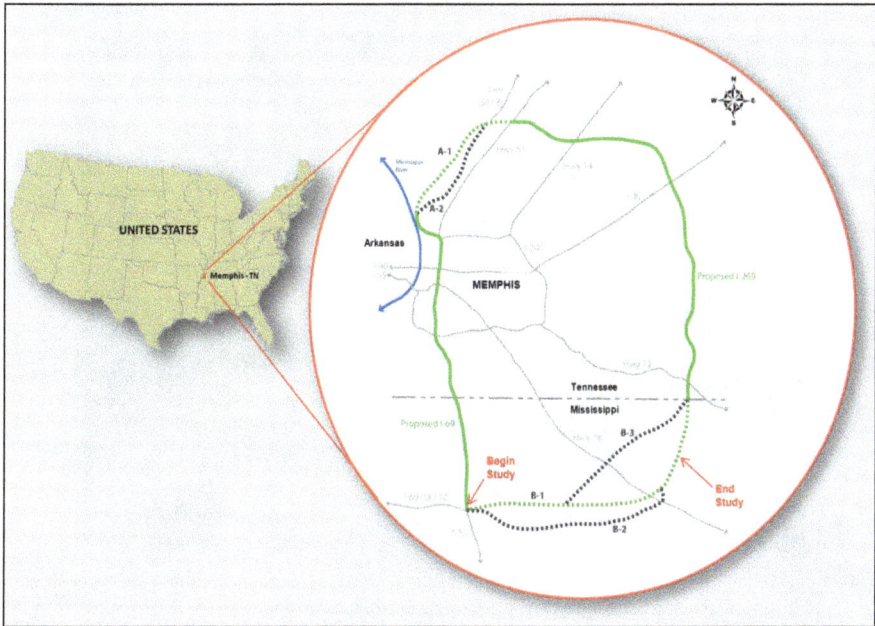

Figure 2. The route of the I-269 bypass including alternatives considered during the EIS process. The study extends along the I-269, in Desoto County, Mississippi, near Memphis-Tennessee.

To make the proposed top-down watershed-wetlands framework analysis useful, this work utilized local geodata provided by Desoto County, MS, such as the transportation network, hydrographical data, LiDAR elevation data, zoning and the county comprehensive plan. A large collection of three-inch resolution aerial images provided support to enhance evaluation of wetland locations. Additionally, wetlands and hydric soil information extracted from satellite radar imagery were used to cover the lack of National Wetlands Inventory federal wetlands data for this specific area (Brooks et al., 2009).

3.1. Overcoming the lack of NWI information in North Mississippi

In our investigation of efficient methods to provide early assessment to wetlands potentially impacted by transportation corridors, we adopted existing findings of woody wetlands in North Mississippi. According to (Brooks et al., 2009), the motivation in improved methods of mapping forested (or "woody") wetlands areas was two-fold: National Wetlands Inventory (NWI) digital mapping information of wetlands location is unavailable for approximately ¼ of the lower 48 U.S. States, including northwest Mississippi, based on the U.S. Fish and Wildlife Service NWI "*Wetlands Online Mapper*"; and forested wetlands are very poorly mapped using traditional mapping methods including optical remote sensing (Sader et al., 1995; Bourgeau-Chavez et al., 2001).

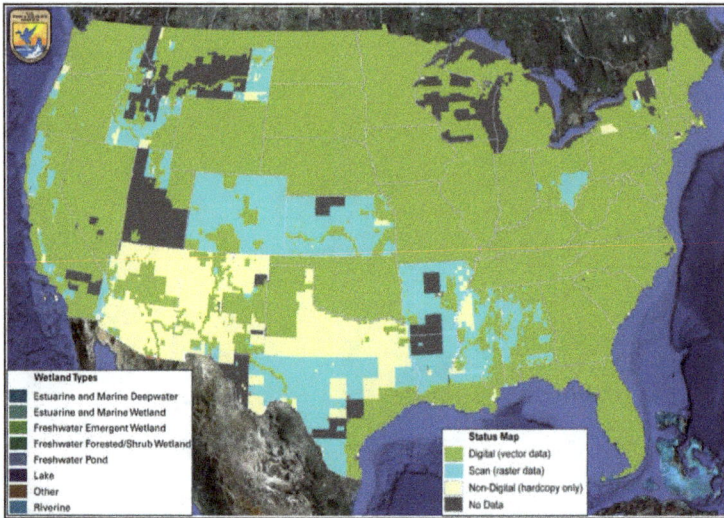

Figure 3. Availability of national wetland data (Modified from U.S. Fish and Wildlife Service, National Wetland Inventory, background image source: Google Earth.

Given this data gap and problems with available traditional sources, we adopted the results described by (Brooks et al., 2009) that used a combination of radar remote sensing data with object-based techniques to compute potential woody wetlands and create a soil moisture index map for the NCRST-SEPP project (Figure 4).

4. The top-down watershed-based landscape analysis

The partition of the landscape into hydrological watersheds was a logical ecologically-focused way to explore the context interactions between the natural and the man-made features. The methodology employed concepts of object-based geographical analysis to evaluate the level of landscape impact of the proposed transportation corridor scenarios. The focus on hydrological watersheds as principal objects made the main difference in comparison with the traditional object-oriented landscape analysis. Two levels of hierarchy were addressed in this work:

1. Watersheds were identified and ranked according certain criteria as a significant percentage of unfavourable zoning, density of streams, wetlands and future man-made constructions.
2. Wetlands identified and ranked for each watershed. This used topographical LiDAR data, image interpretation and the wetlands impacted by the designed I-269 corridor.

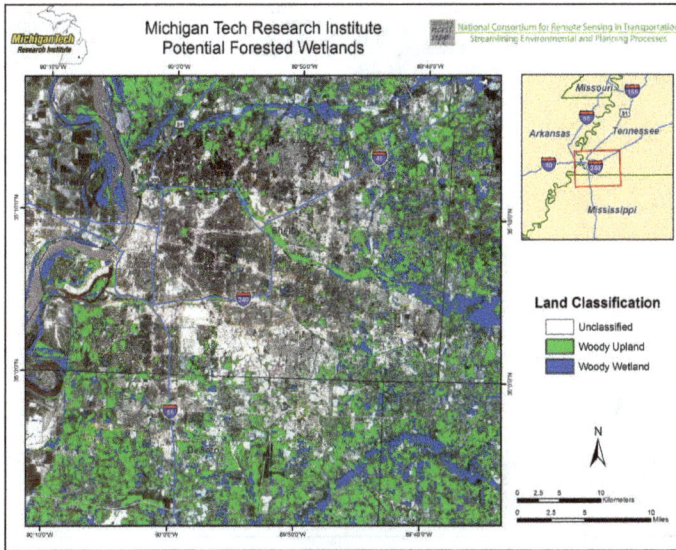

Figure 4. Potential woody wetlands product for the NCRST-SEPP I-269 study area.

4.1. Defining objects in a hierarchical landscape analysis in GIS

An I-269 area GIS was developed to improve the capabilities of geographical analysis by providing ways to access, process, store and disseminate large amounts of information in comparison with human tasks. The traditional GIS features (points, lines and polygons) enable a series of spatial operations as union, overlapping, intersection, etc. Some of these operations were used when integrating the watershed polygons and the landscape layers of information on the first level and when assessing and refining wetlands on the second level (Figure 5).

4.2. Level I: Identification of watersheds

The first step was the identification of the HUC-12 watersheds intersected by the part of the I-269 bypass that included alternative routes in the southeast part of the area located in northwest Mississippi. A simple spatial intersection operation highlighted ten watersheds as shown in Figure 6.

Selecting the watersheds intersected by the I-269 corridor options area caused a significant reduction in the field study area and, consequently, the optimization of the data to be processed. Thus, the next step assessed the numerical criteria that enabled ranking the selected watersheds. At this point, the polygons of the 10 selected watersheds were intersected with other layers of information such as 100 year floodplain, hydrograph, existing roads and urban features, planed roads and developments and zoning in order to extract features to quantify the system. Figure 7 illustrates the layers used in this intersection.

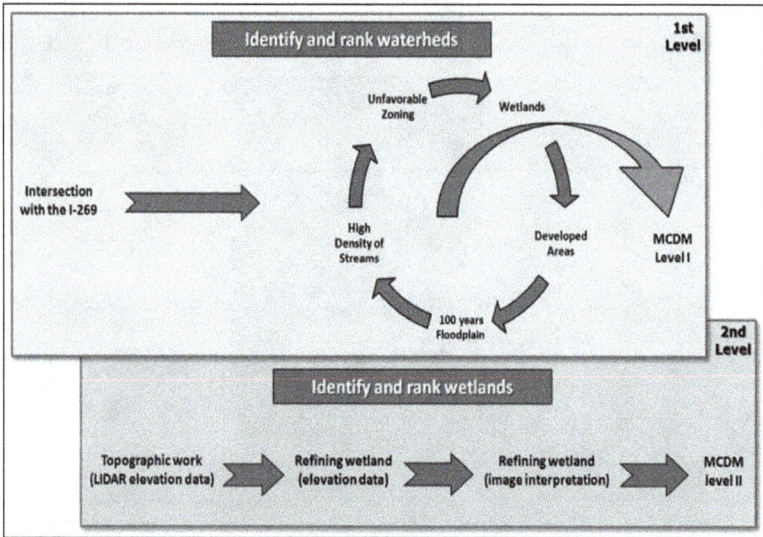

Figure 5. The basic workflow for the two levels of the hierarchical process.

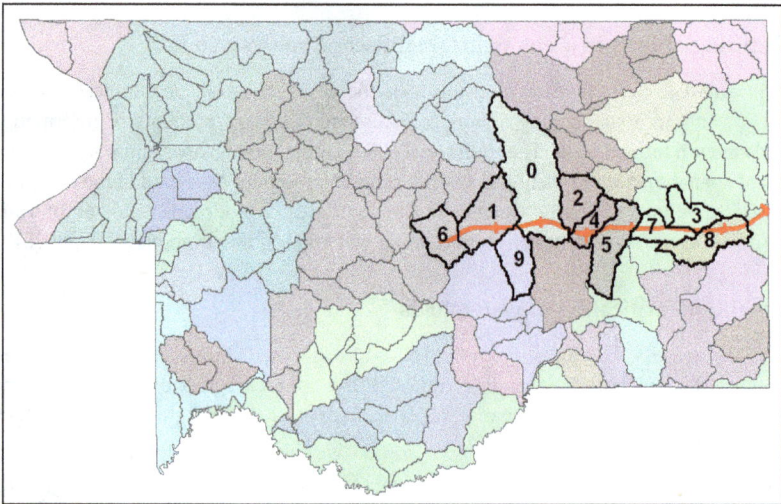

Figure 6. The HUC-12 watersheds intersected by the part I-269 analyzed in Desoto County-MS, which was the section with corridor options assessed as part of the EIS.

In order to make straightforward the process, the zoning map in particular was previously reorganized into 5 classes: agriculture (green), residential (yellow), agriculture-residential (light green), commercial (red) and industrial (orange). Similarly, the maps of existing roads and existing buildings provided by Desoto County GIS Department were combined to produce a density map that reflects the urbanized areas. These steps were necessary since no map of this kind was found to be available in existing GIS databases.

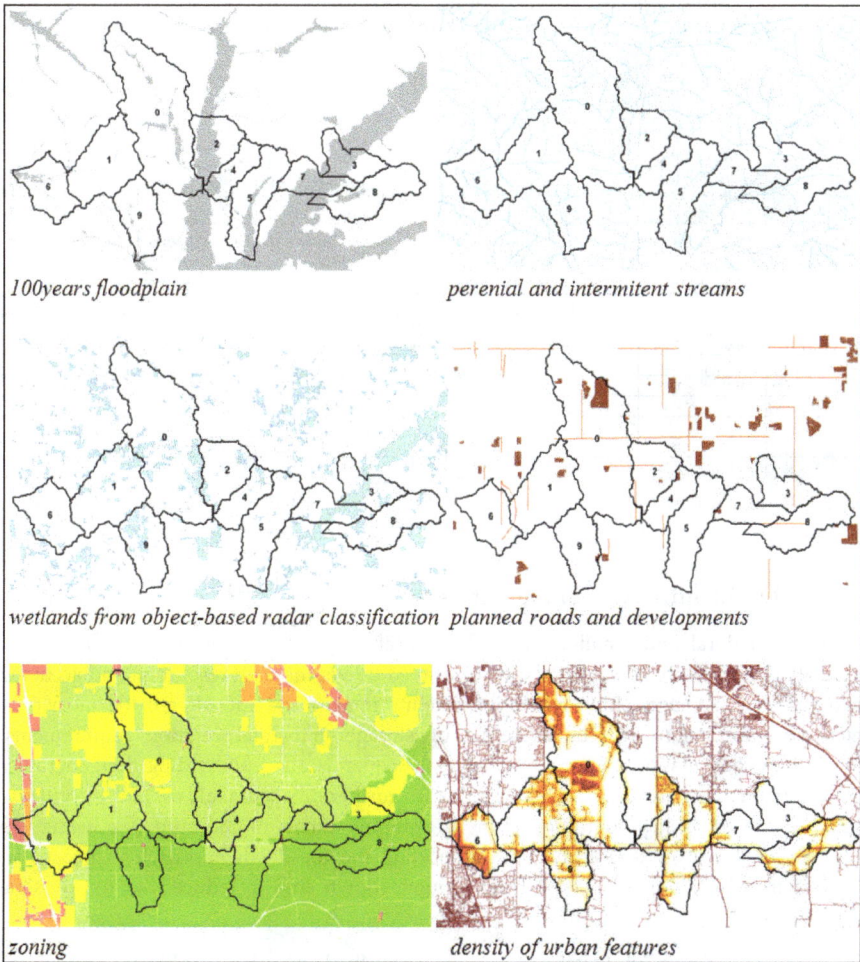

Figure 7. Spatial intersections between the selected wetlands and the layers of interest to assess watershed characteristics to be used in the MCDM process.

Aiming to simplify the decision making process, quite a few different impact factors were assigned to the layers of interest, ranging from 1 (low impact) to 9 (high impact). These values were hypothetical, but reflected the importance of the features due to the potential environmental impact upon existing wetlands. The percentage of covered areas was computed per watershed for the following GIS layers: *watersheds, 100 year floodplain, dense urban, future developments, residential, agriculture, agriculture-residential, commercial and industrial.* Similarly, the density of linear features (km per Km²) was computed per watershed for the layers *perennial streams, intermittent streams* and *planned roads.* Table 1 presents the relative values extracted per layer from the selected watersheds. These numbers were used to compute the ranking through the weighted average, as show in the Equation 1.

$$\text{Rank} = \left[(A+C+E+F+H) + 3*(J) + 5*(B+I) + 7*(K+L) + 9*(D+G) \right] / 50 \quad (1)$$

	Layer A	Layer B	Layer C	Layer D	Layer E	Layer F	Layer G	Layer H	Layer I	Layer J	Layer K	Layer L
Impact factor	1	5	1	9	1	1	9	1	5	3	7	7
watershed	Floodplain %	Dense Urban %	wetlands %	Future Develop. %	Streams km/km2	km/km2	planned roads k/km2	Agriculture %	Residencial %	Agri-Resid %	Comercial %	Industrial %
0	10.18	35.42	10.28	5.56	0.99	7.96	0.30	5.24	24.08	66.51	0.57	0.07
1	1.58	10.91	10.06	1.75	0.95	8.27	0.25	15.00	10.31	116.31	0.00	0.00
2	65.57	20.24	9.14	0.44	1.79	5.21	0.52	0.00	0.00	128.18	0.00	0.00
3	62.66	3.96	21.09	0.00	0.99	4.64	0.27	33.41	24.21	42.02	0.00	0.00
4	17.08	11.16	5.46	1.50	0.67	6.84	0.16	0.56	0.00	96.59	0.19	0.00
5	18.11	17.03	3.36	0.04	0.93	7.10	0.00	35.53	0.00	62.21	0.13	0.00
6	9.80	43.17	3.00	0.00	0.99	7.26	0.37	0.03	29.40	111.16	3.89	4.30
7	50.85	5.47	23.47	1.39	0.68	3.33	0.14	60.16	0.00	38.57	0.43	0.00
8	11.66	20.24	6.09	0.11	1.01	7.38	0.21	96.51	0.00	0.00	0.04	0.00
9	3.63	28.06	15.05	0.59	0.85	8.79	0.00	95.54	0.00	1.99	0.42	0.00

Table 1. Relative values computed per layer per watershed (percentage of area and density of linear features)

4.3. Level II: Identification of the wetlands

Unlike the federal and small-scale geodata, local (large-scale or ground-level) geodata normally demand substantial time to be computed due to the high resolution and accuracy involved. Aerial images, high resolution satellite images and LiDAR are the most data intensive information in GIS in terms of storage and interpretation requirements. Minimizing computational efforts by analyzing the landscape, subdividing the geography into semi-homogeneous units, selecting units for further detailed analysis, and prioritizing areas of interest is key to reducing the geographic extent of the study, reducing the computational cost of the study, and supporting the top-down approach in geospatial analysis in which the analysis funnels options down into a reduced set of possible alternatives.

Given the completion of Level I processing described, a series of GIS analyses using information extracted from the topographic surface, such as topographic depressions and flat areas, as well as image analysis such as land cover, provided enhanced inputs to refine wetlands feature geometry as well as classifications based on the radar-based wetland mapping results.

The following processes were developed for the watershed #0, which is second in the ranking as shown in the results section (Figure 8); it serves as a representative example for more detailed examination in this chapter. The reason is that the top-ranked watershed covers a small area and is mostly composed by developed areas and does not present a large variety of landscape features (including wetlands) to illustrate the exercise proposed in this work.

The geospatial analysis was performed using map algebra, so features in vector formats were converted to raster format. Due to landscape analysis considerations and the potential implications of I-269 and planned roads, the layers of information selected in this level (Level II) emphasize the hydrographic and physical aspects, which are basis for engineering construction perspectives.

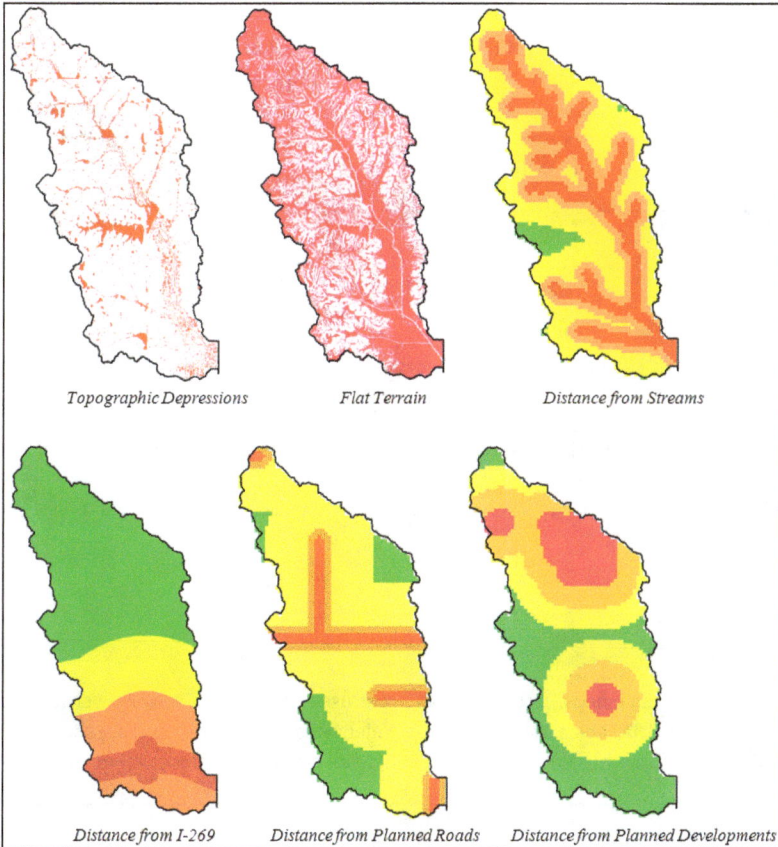

| Topographic Depressions | Flat Terrain | Distance from Streams |
| Distance from I-269 | Distance from Planned Roads | Distance from Planned Developments |

Figure 8. The layers employed to refine and rank the wetlands per watershed. For display purposes, the impact factor ranges from green (low) to red (high).

Table 2 presents the layers used to refine the wetlands features, their respective criteria for classification and the weight to be used on MCDM. Distance criteria and weights are hypothetical; however, they reflect the goal of the paper on assessing potential impacted wetlands.

LAYER	CRITERIA	WEIGHT
Topographic depressions		5
Flat terrain	Slope equal or less than 5%	3
Distance from perennial streams	0-100m, 100-300m, 300-1000m, > 1000m	5
Distance from I-269	0-300m, 300-1500m, 1500-3000m, > 3000m	9
Distance from planned roads	0-300m, 300-1500m, 1500-3000m, > 3000m	7
Distance from planned develop.	0-300m, 300-1000m, 1000-3000m, > 3000m	7

Table 2. Layers, criteria and weights used on level II analysis

The weights are included in the multi-criteria decision tool as input rankings. The tool was developed as part of the SEPP-NCRST project and implemented based on Saaty's AHP method (Figure 9). The normalized weights are used as factors in the map algebra equation that is responsible to produce the cumulative cost surface, where high "cost" would represent higher environmental impact.

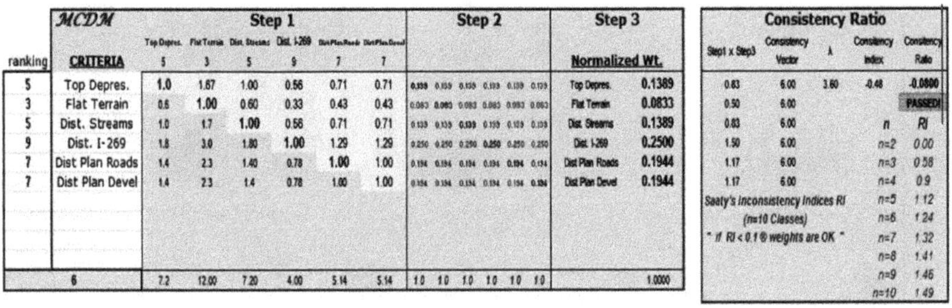

Figure 9. Multi-criteria decision making tool developed to compute normalized weights for the map algebra.

5. Results

5.1. Step 1

For the selected watersheds, impact factors were used to calculate a first-level ranking for watershed and wetland areas impact ranking. Watersheds were ranked and are shown in table 3 in relative order from highest of 15.6 (left-most) to lowest of 4.5 (right-most) on the table.

Watershed ID	6	0	2	1	4	3	5	7	9	8
Ranking	15.6	11.8	11.5	10.2	7.9	7.8	6.8	6	5.6	4.5

Table 3. Computed ranking of potentially impacted wetlands intersected by I-269 in the analysis area

5.2. Step 2

For a selected watershed, shown in figure 10 as the watershed #0, the cumulative cost surface shows that impacts are greatest in the lower part of the watershed (Figure 11). In this area, the amount of upland drained is highest; the floodplain is broader and the wetlands are more frequent as are areas of likely ponded water (surface depressions). Over and above the actual number of stream crossings or acres of wetland impacted by a proposed transportation system, this analysis step illustrates that the landscape and hydrologic context of the ecologic and hydrologic features impacted can be shown to play a significant role in assessing the overall impacts of a transportation project on the hydrologic and biophysical systems traversed.

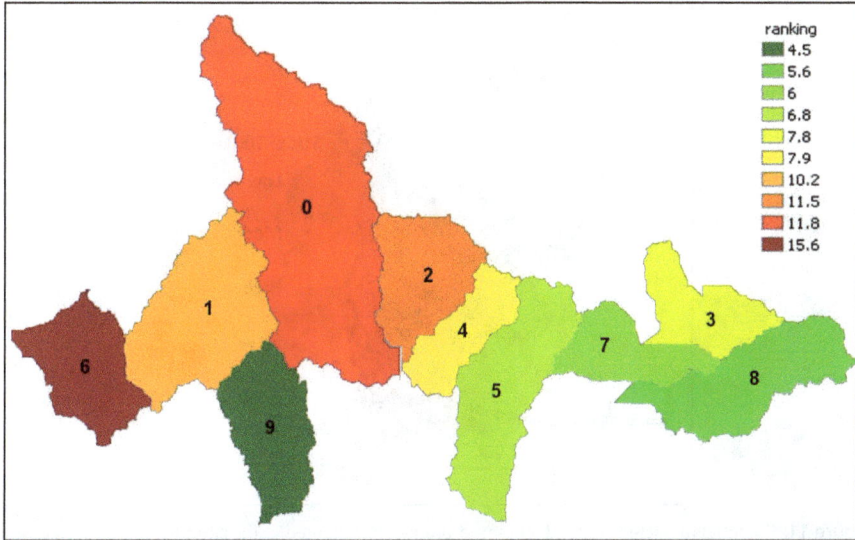

Figure 10. Potentially impacted watersheds intersected by the I-269 – Level I of proposed methodology.

6. Discussions and contributions

In addition to the landscape analysis and transportation planning issues, the results of this investigation showed that a top-down analytical framework based on GIS and MCDM offers value to the early assessment of potentially impacted areas affected by future transportation networks. The work was developed using a set of geospatial data ranging from federal to county, and were intentionally selected to be included in a multi-scale geographic object-based analysis. The hierarchical decision making framework supported the top-down approach through a simple customization of AHP in a GIS environment. The methods and rankings were hypothetically selected (and intentionally made simple) in order to encapsulate the idea and test the MCDM methodology for analysing the impacts of the I-269 study area.

Indeed, in an actual implementation, the relative weights assigned to factors and the rankings associated with factor properties would be subject to alternative assignments of values which would produce results that could significantly depart from those presented and enable a rich evaluation of potential alternatives, including ones based on agency and public inputs. The results presented show a single scenario to illustrate the process rather than an exhaustive exploration of possible scenarios which might arise from collecting a cross-section of objective and subjective values from stakeholders.

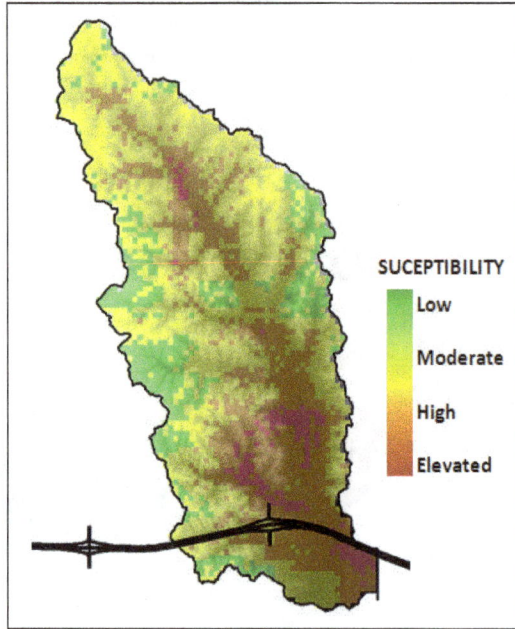

Figure 11. Potential impacted wetland areas (red-orange) highlighted in the cumulative cost surface.

The results demonstrate that the proposed top-down approach is a practical screening process valid to the early assessment and ranking of the impacted wetlands.

Desoto County (MS) is an example of many areas that are not fully mapped or adequately covered by Federal mapping efforts such as the detailed county-based soil surveying program and state-based wetland inventories. Therefore, the methodology demonstrates that the complementary use of wetlands computed from radar-based remote sensing can be used to overcome gaps of in the National Wetlands Inventory (NWI) (Bourgeau-Chavez et al., 2009). Indeed, a specific benefit of using remote sensing data is that they can enable the identification of features of interest where ground-based observations or surface-mapped results are limited or absent. For this reason, it is important to highlight that NWI and other relevant data, such as soil survey GIS layers, are not available nation-wide in high detail. Thus, the methodology presented in this paper can be reproduced from environmental and landscape applications, in particular for areas where other map-based products are not available.

The methods presented in this paper were intentionally simplified to highlight a set of framework approaches to help demonstrate the collection of technologies implemented, especially MCDM. Some concepts of geographic object-based analysis such as neighbourhood relationships, contextual analysis, and others could be explored in more depth; however, such an effort would overshadow the desired explanation and synthesis of the more innovative characteristics highlighted in the methodology that focus on MCDM and AHP and their application to context-sensitive landscape analysis in the transportation planning and NEPA processes. Furthermore, it should be noted that the methods presented are both flexible and extensible. They may be adapted to other purposes, transferred to other geographic areas and transportation corridors, and extended to include additional data, steps, and analysis procedures. Follow-on studies are suggested to further explore the application and extension of the methods presented.

7. Conclusion

This chapter presents novel methods that leverage spatial implementation of MCDM-AHP in the integrated application of geospatial data to assist transportation decision making throughout the NEPA process. A significant finding is that advanced technologies in geographic object-based analysis can be used to partition the landscape into hydrological watersheds as a basis for context- and object-based analysis. The methodology employed object-based approaches to analyze the landscape and considered a plurality of data layers to derive ranking and weights for understanding the impacts of transportation infrastructure relative to the watershed as a whole as well as to the landscape position of possible transportation alignments within the watershed. The focus on hydrological watersheds as principal objects highlights an important difference in this new and innovative approach as compared to traditional environmental impact analysis.

Watersheds provide subdivisions which are biophysically and ecologically focused, enabling the application of spatial analysis methods which explore the context-sensitive interactions between natural and man-made features in a landscape. The results indicate that the object-based analysis of landscape context and position can provide understanding and insight for assessing transportation corridor impacts on the environment and ecosystems that extend beyond traditional approaches which simply quantify the number of stream crossings and the areas of wetlands impacted. The results show that example hypothetical but reasonable values can be assigned to various landscape features and that these values may be considered in the context of spatially enabled MCDM-AHP. The hierarchical decision making framework implemented through top-down GIS-based analysis enabled the adoption of the segments of the landscape by hydrologic areas. The combination of data, methods and values per level delivers results significant to making decisions, assessing impacts, and designing mitigation strategies that are contextually aligned and indicate an environmentally responsible attitude and sustainable focus for anthropogenic impacts on the environment

Author details

Rodrigo Nobrega, Charles O'Hara and Bethany Stich
Geosystems Research Institute, Mississippi State University, USA

Colin Brooks
Environmental Science Laboratory, Michigan Tech Research Institute, USA

Acknowledgement

The authors would like to thank the US Department of Transportation Research and Innovative Technology Administration (USDOT-RITA) for funding the research, as well as the Desoto County GIS Department for providing much of the GIS data.

8. References

Amekudzi, A. and M.M. Meyer. 2006. Considering the environment in transportation planning: review of emerging paradigms and practice in the United States. *Journal of Urban Planning and Development*. Vol. 132 (1), 42-52.

Aplin, P. 2005. Remote sensing: ecology. *Progress in Physical Geography*. Vol. 29 (2), 104-113.

Benz, U., Hofmann, P., Willhauck, G., Lingenfelder, I. and Heynen, M. 2004. Multiresolution, object-oriented fuzzy analysis of remote sensing data for GIS-ready information. *ISPRS Journal of Photogrammetry and Remote Sensing*. Vol 58, 239-258.

Blaschke, T., Conradi, M. and Lang, S. 2001. Multi-scale image analysis for ecological monitoring of heterogeneous, small structured landscapes. *Proceedings of SPIE*, Toulouse, 35-44.

Blaschke, T., Lang, S., Lorup, E., Strobl, J. and Zeil, P. 2000. Object-oriented image processing in an integrated GIS/remote sensing environment and perspectives for environmental applications. In: Cremers, A. and Greve, K. (eds.): *Environmental Information for Planning, Politics and the Public*. Metropolis, Marburg. Vol 2, 555-570.

Boarnet, M. G. and Chalermpong, S. 2001. New highways, house prices and urban developments: a case study of Toll Roads in Orange County, CA. *Housing Policy Debate*. Vol 12 (3), 575-605.

Bouman, B.A.M., Jansen, H. G. P., Schipper, R.A., Nieuwenhuyse, A., Hengsdijk, H. and Bouna, J. 1999. A framework for integrated biophysical and economic land use analysis at different scales. *Agriculture, Ecosystems and Environment*. Vol 75, 55-73.

Bourgeau-Chavez, L.L, K. Riordan, R.B. Powell, N. Miller, and M. Nowels. 2009. Improving wetland characterization with multi-sensor, multi-temporal SAR and optical/infrared data fusion. *Advances in Geoscience and Remote Sensing*, InTech, G. Jedlovec, ed. pp. 679-708.

Bourgeau-Chavez, L.L., E.S. Kasischke, S.M. Brunzell, J.P. Mudd, K.B. Smith and A.L. Frick. 2001. Analysis of space-borne SAR data for wetland mapping in Virginia riparian ecosystems. *International Journal of Remote Sensing*, 22(18): 3665-3687.

Brinckhorst, D. J. 2005. Integration research for shaping sustainable regional landscapes. *Journal of Research Practice*, v. 1, n. 2

Brooks, C., Dajos, T., Bourgeau-Chavez, L. 2009. MTRI forested wetlands and soil moisture: inputs to the multi-criteria decision model. In: NCRST - SEPP: May 2009 Memphis

Workshop, 2009, Memphis, TN - USA. *National Consortium for Remote Sensing and Transportation Annual Meeting,* 2009 http://www.ncrste.msstate.edu/sepp_workshops/memphis_workshop_2009/presentatio ns/mtri_memphis_yr2_rev2.pdf (accessed in July 10, 2010)

Burnett, C. and Blashke, T. 2003. A multi-scale segmentation/object relationship modelling methodology for landscape analysis. *Ecological Modelling,* V. 168, pp. 233-249.

Corrales, M.; Grant, M. and Chan, E. 2000. *Indicators of the environmental impacts of transportation: highway, rail, aviation and marine transport.* Diane Publishing, 244p.

Dumas, J., O'Hara, C and Stich, B. 2009. Context sensitive solutions and transportation planning: the detection of land-use, historical and cultural resources, community features and their integration into the planning process. *Proceedings of Transportation Research Board,* Washington DC, January 2009.

Fletcher Jr, R. J., Hutto and Richard L. 2008. Portioning the multi-scale effects of the human activity on the occurrence of riparian forest birds. *Landscape Ecology.* Vol 23, 727-739.

Forman, R.T.T. and L.E. Alexander. 2002. Roads and their major ecological effects. *Annual Review of Ecology and Systematics.* Vol. 29, 207-233.

Forman, R.T.T., Sperling, D., Bissonette, J.A., Clevenger, A.P., Cutshall, C.D., Dale, V.H., Fahrig, L., France, R., Goldman, C.R., Heanue, K., Jones, J.A., Swanson, F.J., Turrentine, T. and Winter, T.C. 2003. *Road Ecology Science and Solutions.* Island Press, Washington, DC.

Gal, T., T.J. Stewart, T. Hanne (eds.). 1999. *Multicriteria decision making: advances in MCDM models, algorithms, theory, and applications.* Kluwer Academic Publishers, Norwell, MA. 560 pp.

Gregory A. Kiker, Todd S. Bridges, Arun Varghese, Thomas P. Seager, and Igor Linkov. 2005. *Application of Multicriteria Decision Analysis in Environmental Decision Making, Integrated Environmental Assessment and Management.* Vol 2, 95–108.

Hay, G.J. and G. Castilla. 2008. Geographic Object-Based Image Analysis (GEOBIA): A new name for a new discipline. *Lecture Notes in Geoinformation and Cartography.* Section 1, 75-89.

Hilty, J.A., C.Brooks, E.Heaton, and A.M. Merenlender. 2006. Forecasting the effect of land-use change on native and non-native mammalian predator distributions. *Biodiversity and Conservation.* Vol 15 (9), 2853-2871.

Huang, B., Cheu, R.L., and Y.S. Liew. 2003. GIS-AHP Model for HAZMAT Routing with Security Considerations. *Proceedings of IEEE 6th International Conference on Intelligent Transportation Systems,* October 2003, Shanghai, China.

Keller, I.; Largiardèr and C. R. 2003. *Recent habitat fragmentation caused by major roads leds to reduction of gene flow and loss of genetic variability in ground beetles.* The Royal Society, v. 270, pp 417-423.

Kuitunen, M., Rossi, E. and Stenroos, A. 1998. Do highways influence density of land birds? *Environmental Management,* Vol 22 (2), 297-302.

Lillesand, T.M, and Kiefer, R. W. 2004. *Remote Sensing and Image Interpretation* (4th ed.), N.Y., John Wiley & Sons.

Litman, T. A. 2008. *Land Use Impacts on Transport: How Land Use Factors Affect Travel Behavior.* [online]. Victoria Transportation Policy Institute, 58 p. Available from: http://www.vtpi.org/landtravel.pdf [Accessed 15 May 2009]

McGarigal, K., and B. J. Marks. 1995. FRAGSTATS: spatial pattern analysis program for quantifying landscape structure. [online]. Forest Science Department, Oregon State University, Corvallis-OR. Available from:
http://www.umass.edu/landeco/pubs/Fragstats.pdf [Accessed 15 May 2009]

Mokhtarian, P. L. and Cao, X. 2008. Examining the impacts of residential self-selection on travel behavior: A focus on methodologies. *Transportation Research B,* v.42, 204-228.

Nobrega, R. A. A. 2007. Deteccao da malha viaria na periferia urbana de Sao Paulo utilizando imagens de alta resolucao espacial e classificação orientada a objetos. *Tese de Dourotado,* Escola Politecnica da Universidade de Sao Paulo, Sao Paulo. 165p.

Omernik, J. M. and Bailey, R. G. 1997. Distinguishing between watershed and ecoregions. *Journal of the American Water Association.* Vol 33 (5), 935-944.

Ono, S., Barros, M. T.L. and Conrado, G. 2005. A Utilizacao de SIG no planejamento e Gestao de Bacias Urbanas. *Proceedings of AbrhSIG.* Sao Paulo, 2005.

Saaty, T. L. 1995. Transport planning with multiple criteria: the analytic hierarchy process applications and progress review. *Journal of Advanced Transportation.* Vol 29, 81–126.

Sadasivuni, R., O'Hara, C.G., Nobrega, R. and Dumas, J. 2009. A transportation corridor case study for multi-criteria decision analysis. *Proceedings of American Society of Photogrammetry and Remote Sensing Annual Conference,* March 11-14, 2009, Baltimore-MD.

Sader, S., D. and Ahl, W. 1995. Accuracy of Landsat-TM and GIS rule-based methods for forest wetland classification in Maine. *Remote Sensing of Environment.* Vol 53 (3), 133-144.

Saunders, S. C., Mislivets, M, R., Chen, J. and Cleland, D. T. 2002. Effects of roads on landscape structure within nested ecological units on the Northern Great Lakes Region, USA. *Biological Conservation.* Vol 103, 209-225.

Schiewe, J., Tufte, L. and Ehlers, M. 2001. Potential and problems of multi-scale segmentation methods in remote sensing. GIS – *Zeitschrift für Geoinformationssysteme.* Vol 6, 34-39.

Spellerberg, I. F. 1998. Ecological effects of roads and traffic: a literature review. *Global Ecology and Biogeography Letters.* Vol 7, 317–333.

TownshendWalker, J. R.., Huang, G., Kalluri, DeFries, N. V. and Liang, S. 2000. Beware of per-pixel characterisation of land cover. *International Journal of Remote Sensin*g. Vol 21, 839– 843

Wheeler, A. P., Angermeier, P., Rosenberguer, A. 2005. Impacts of new highways and subsequent landscape urbanization on stream habitat and biota. *Reviews in Fisheries and Science.* Vol 13, 141-164.

Chapter 2

Raster and Vector Integration for Fuzzy Vector Information Representation Within GIS

Enguerran Grandchamp

Additional information is available at the end of the chapter

1. Introduction

This chapter deals with a vector model for fuzzy data sets within a GIS and with the raster and vector combination to refine vector objects. The model will be useful to represent information such as natural phenomenon (forest, desert, geology, separation between mountain and valley, rain, inundation, wind, etc.), social phenomenon (population density changes, poverty, etc.) or physical phenomenon (hurricane, etc.), which are diffused in space (Guesgen, 2000). The use of raster information will be useful to split vector units when the vector division reaches its limits.

The main sources of raster data are raw images (airborne and satellite images) and results of treatments (geostatistical, pixel classification, etc.). Vector information is often obtained by manual measurements (using GPS receptor for example), or is the result of the vectorisation of a raster treatment (such as classification, etc.).

The main goal is to manipulate vector information instead of raster because their manipulation within GIS engenders many problems. Indeed, the raster data are not adapted to GIS treatment (Benz, 2004) because of a lack of contextual information, the size of the data and the time consuming algorithm to produce information. The raster information is not split into identified objects and the vector representation is more flexible and gives the possibility to be easily combined with other information layers. For these reasons, we aim to split a wide image into several small units and convert the raw image information into a vector of features characterizing the unit in a raster view.

Another observation is that the vector data structures are not adapted to model fuzzy information in GIS (Shneider, 1999). More than 10 years later, the only way to use fuzzy representation in GIS is to build a raster map (Bjorke, 2004), computed with different raster or vector sources (Kimfung, 2009), (Ruiz, 2007). The only approach dealing with fuzzy vector representation uses a series of regular buffers inside and outside a polygon boundary to represent different belts of membership function values (Lewinski, 2004). The main drawback of this approach is that for most of the natural phenomena that engender fuzzy data (such as forest, population movement, physical phenomena like hurricane, etc.) the membership function doesn't have the same behaviour in each spatial direction.

We propose in this chapter a vector model adapted to fuzzy information without the previous limitations. The chapter is organized in XX sections including this introduction. Section 2 details the fuzzy vector representation by successively introducing *(i)* a state of the art for fuzzy information representation within GIS *(ii)* the proposed fuzzy model *(iii)* an illustration of the model. Section 3 deals with the use of raster information to split vector units within certain hypothesis. Then a conclusion is given in section 4.

2. Fuzzy vector representation

2.1. State of the art

In 1997, in the conclusions of the special issue of *Spatial Data Types for Database Systems in Lecture Notes in Computer Science* the authors underline the importance of specific data structures for GIS and the lack of adaptation of the existing one to several data including fuzzy data. As if many propositions have been made, more than 10 years later the advance in this field are not relevant. The fuzzy modeling is only approached with strict sets.

Historically the first approaches to solve the problem were to arbitrary decide of a strict border between fuzzy sets and to model them with classical data structures. But the raise of more complex problematics integrating many parameters reached the limits of this model which have guided the conception of the datastructures within the GIS.

The first studies about 2D fuzzy sets have been made by Peter Burrough in 1986 (Burrough, 1986) 106. After this, many studies with the definition of 2D fuzzy operators or fuzzy spatial relation (Bjorke, 2004), (Kimfung, 2009) have been made. But all this studies are using a raster representation of the data (Zhu, 2001), (Mukhopadhyay, 2002), (Sunila, 2004), (Bjorke, 2004), (Guo, 2004), (Ruiz, 2007), (Sawatzky, 2008), (Sunila, 2009), (Gary, 2010), (Wolfgang, 2011) and it was still the same in 2010. In all this studies it was s necessary to rasterise the data in order to apply the fuzzy operators. This implies a loss of precision when choosing a scale of analysis, a time consuming process and the lack of fuzzy representation of sources data. This last point has been underline by GIS community as a main drawback (Altman, 1994), (Shneider, 1999), (Fisher, 2000), (Cross, 2001), (Yanar, 2004), (Kainz, 2011).

The actual challenge is then to directly deal with the vector data in order to improve precision, reduce computation time and to ensure a better abstraction of the data. Nowadays, only marginal studies include a vectorial approach to fuzzy spatial problems (Benz, 2004) (karimi, 2008). These approaches consist on a series of regular buffers around a strict polygon to represent different level of the membership function values. The main drawback of these approaches is that the regular evolution of the fuzzy membership function in each direction doesn't translate the reality of most of the observed phenomenon.

2.2. Objective and data

In the study case, used to illustrate this chapter, we built a fuzzy representation of different forest in Guadeloupe Island. A forest is typically the kind of data adapted to the modelling with fuzzy sets. Indeed, the transition between two kinds of forest is mostly a gradient than a strict transition. Moreover, the gradient depends on many parameters and could be relatively short if environmental conditions change quickly or long if there is a smoother change. In some particular conditions there is a strict border for a forest, for example at the interface with agriculture or if a road, river, crest etc. interact with the forest. In any case, the transition gradient is locally defined and not uniform in every direction.

The classes are semantically and numerically defined using floristic information collected over 47 areas (about 250 m² each, *Fig. 1*-a). This step is based on a Principal Component Analysis (PCA, *Fig. 1*-b) and an ascending hierarchical clustering method (AHC, *Fig. 1*-c). These first steps allow regrouping the floristic information into significant clusters by sorting the numerous parameters describing the 47 areas. The AHC is used to select the number of classes and also serves as base for the semantic definition of the classes given through an ontology (*Fig. 1*-d) (Jones, 2002), (Kavouras, 2005), (Fonseca, 2002, 2006,2008), (Baglioni, 2008), (Gutierrez, 2006). The ontology is useful for a high level use of the fuzzy representation of the forest and particularly when integrating it to a shared conceptual layer (Eigenhofer, 1991), (Cruz, 2005), (Bloch, 2006), (Grandchamp, 2011).

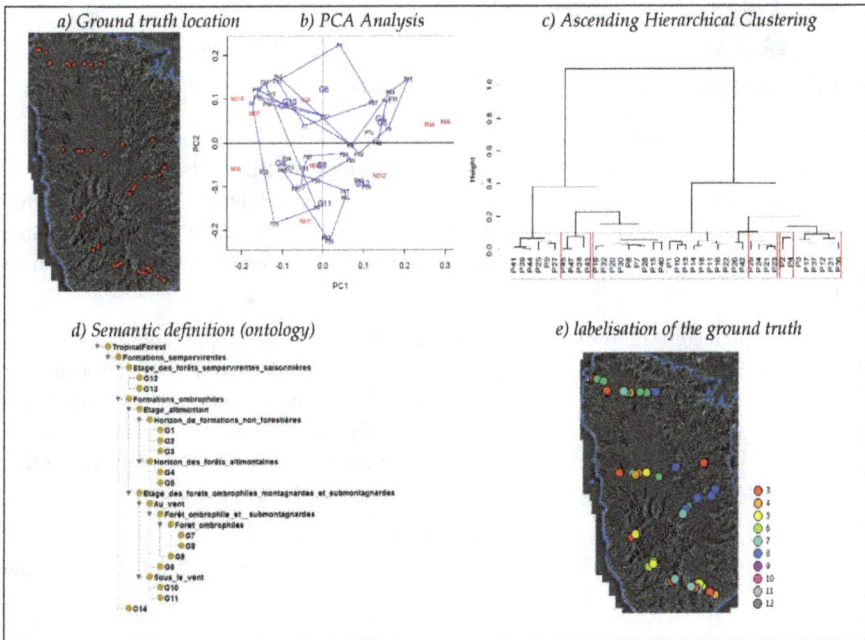

Figure 1. Clustering on floristic data and class definition

This step allows defining 14 kinds of forest over the main part of Guadeloupe Island (*Fig. 1-*d and e).

2.3. Description of the treatments

The previous step allows labelling each of the 47 areas according to the 14 classes (*Fig. 2-a*). But the fuzzy representation of the whole territory is not possible using floristic information because we didn't have this information for any other area. Under the hypothesis that the environment directly influences the formation of the forests we project the 47 areas in a topographic and environmental space. This space includes information such as general ground occupation, elevation, exposition, slope, humidity or latitude. Each of this data is stored within a vector information layer and their fusion leads to the division of the territory into elementary areas called Vector Units (VU, *Fig. 2-b*). Each of the 47 ground truth areas is contained in a VU.

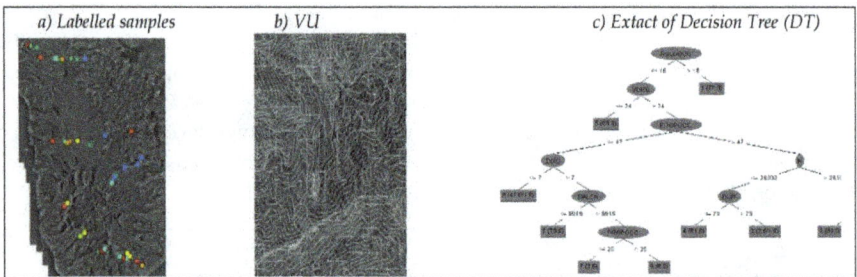

Figure 2. Learning on topographic and image data based on decision tree

Each VU represents a uniform area regarding each of the fused layers and we add to these features information extracted from a raster view (Raster Unit RU) of the area (texture and colour characterization using co-occurrence matrices, Law filters, Gabor filters, Hue moments, fractal dimension, etc.). We use satellite images (IKONOS, Spot5 and Quickbird) and airborne images. A total of about 20 features including topographic and image features are used.

This step of characterisation of the VU is the first Raster-Vector cooperation. It includes a partition of the image according to heterogeneous vector data and not to spectral or structural properties of the image. This approach is more simple and quicker than image segmentation and the returned VU have a semantic signification. Moreover, the adjunction of raster and vector information allows combining a theoretical uniformity of an area and a raster view of the reality.

With these 47 labelled VU we are now able to classify the whole territory in a fuzzy way. Indeed, we apply a supervised classification based on a decision tree (*Fig. 2-b*) to obtain the different kind of forest.

The decision tree is built after a learning step based on both topographic and image features. We use different kind of decision trees such as functional decision trees (FT) (Gama, 2005)

and C4.5 decision tree (Quinla, 1993). The best results were statistically obtained with FT, so we keep this approach to illustrate the method in the rest of the chapter.

The fuzzy model requires the computation of a membership function for each elementary unit (VU). This membership function is derived from the reliable coefficient returned by the decision tree. Indeed, each VU is analysed using the FT and a reliable coefficient is returned for each class. Commonly a strict classification using DT will choose to label the VU with the label of the class having the highest value of reliable coefficient.

By retaining only the highest value, we totally ignore the gradient nature of the transitions. By keeping all values we are able to build a fuzzy representation of each class and different representations of the resulting map by defining rules for the transitions. Moreover, in case of wide transition area, this approach could be used to reveal full transition classes.

Fig. 3 shows the membership function of each VU to different classes among the 14 identified classes of Guadeloupean forest. The lower the membership function value is, the darker is the colour. We remark that some classes are clearly localised, such as class 9 which represents high mountain forest around the sulphur mine, or class number 4 which is a typical forest kind oriented to the west and where a dry hot wind is blowing or else class 1. But other classes are more dispersed over the whole territory. This reveals wide transitions between classes. We remind that these maps are not raster data but each coloured element is a VU.

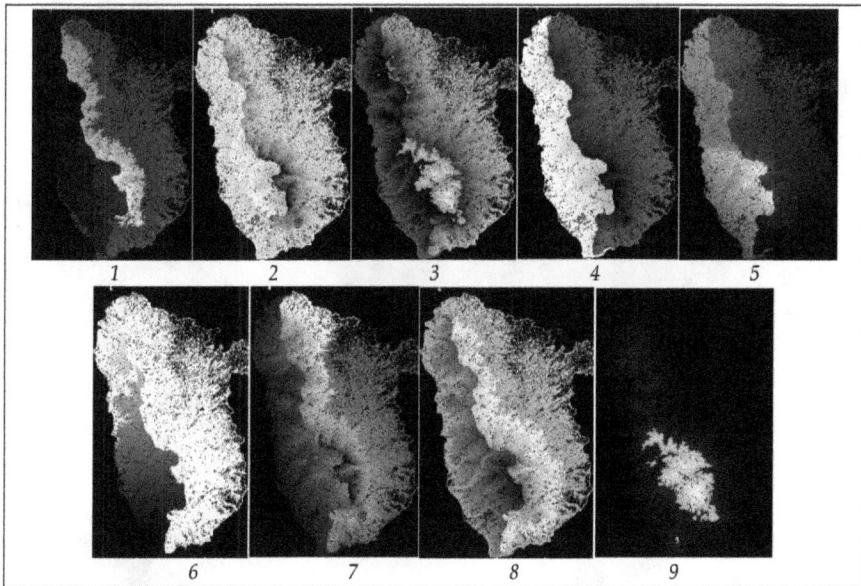

Figure 3. Membership function

2.4. The fuzzy model

Now we have all necessary information to build the fuzzy model of the forest. The simplest model is to store the vector of the membership degree values in each VU. This is also the more precise model because we don't lose any data. But this model is not easily useful. In order to simplify the representation of the different transition gradients we decide to build different belts of membership degree values for each class. The spatial and topologic information used in the fuzzy classification process ensure the spatial coherence and compactness of the membership degree values. The models will differ from the number of belts and also the value of the threshold between the belts. These values will influence the treatments in two ways: *(i)* the more belts there is, the more precise are the results and the more time consuming are the treatments, *(ii)* the values of the threshold allow focusing on some parts of the transition.

So we will now see different ways to set the belts. *Fig. 4* and *Fig. 5* show different fuzzy representations of the class number 9. The differences are linked to the number of belts (5 for *Fig. 4* and 10 for *Fig. 5*) and their positions: uniformly distributed (left), centred on most represented values (middle), centered on highest membership degree values (right).

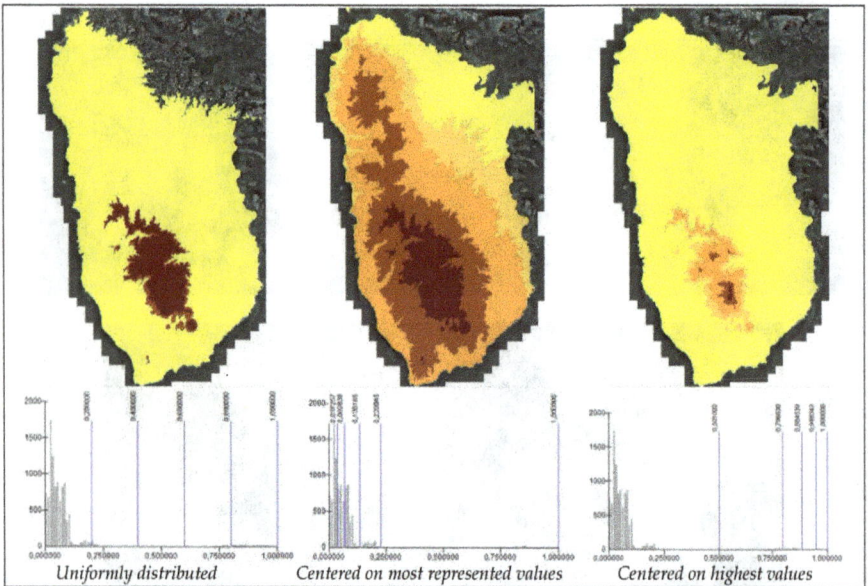

Figure 4. Different fuzzy vectorial respresentation of a class 9 with 5 buffers

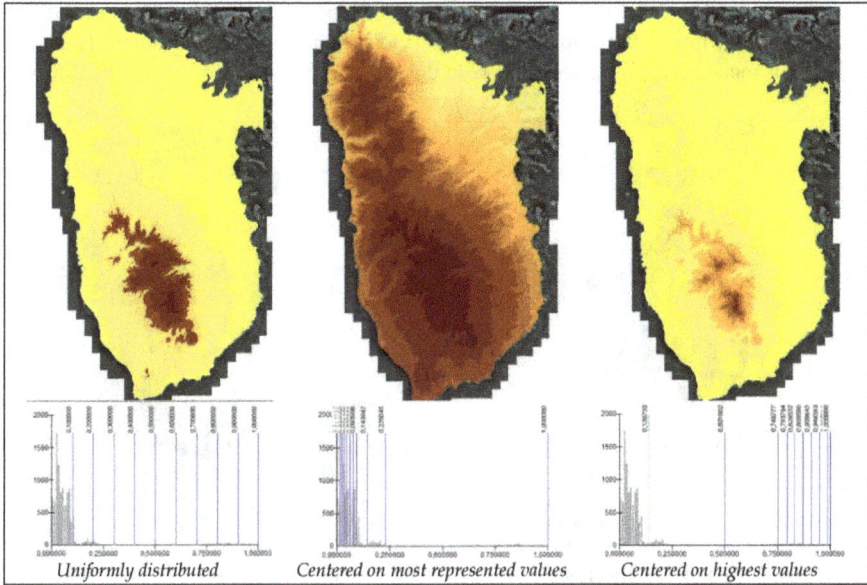

| *Uniformly distributed* | *Centered on most represented values* | *Centered on highest values* |

Figure 5. Different fuzzy vectorial respresentation of class 9 with 10 buffers

The choice of the number and positions of the belts is made by the user depending on the application. These values could be set manually or automatically computed using rules such as: natural breaks, equal intervals, geometrical interval, standard deviation, etc. This fuzzy data structure allows a fine and faithful representation of the heterogeneous evolution of the class in each direction. Indeed this model translate this property because the heterogeneity of the different information layers used (resulting on VU with totally different shapes) leads to a non uniform evolution of the membership degrees in each direction.

Moreover, the choice of the thresholds influences the topology of the model and some resulting belts could be a unique connected polygon or on contrary a set of disjointed polygons.

2.5. Validation and strict view of the model

The only way to validate the model is to compare it to an existing model. However there isn't any fuzzy model of the studied classes. So in order to validate the model we are going to compare a strict view of the model with an existing strict classification of the concerned forests. The referred classification is an ecological map obtained manually by biologists in 1996 (Rousteau, 1996).

Fig. 6 shows in left the reference map and in right the map obtained with a conversion of the fuzzy map into a strict map. The conversion has been made in the simplest way by attributing to each VU the class label of the class having the highest membership degree value. Areas in white in the map are not taken into account in the classification. The main confusion is made between classes 9 and 12 in the east part of the Island.

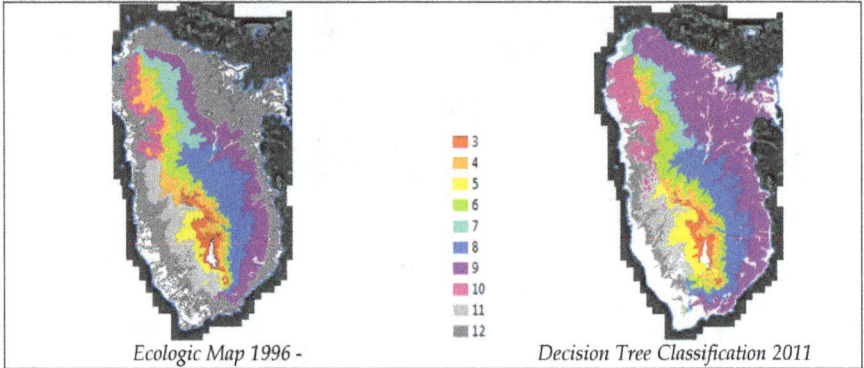

Figure 6. Strict classification comparison

Fig. 7 shows a zoom of the map on a particularly complex area: the National parc. We observe the similarity of the two maps with a localisation of the forest at the attended places. Fig. 7 -c) shows the differences (inblack) between the ecologic map and the strict classification. The differences are localized at the limits between each kind of forest and this is exactly what is criticable in a strict classification of diffuse data. Taking into account that the map on the left (Fig. 7 -a) has been made manually with arbitrary decision concerning the limits of the forest and that the map in the middle (Fig. 7 -b) is the result of a complex and complete modelling, learning and classification process we can estimate that the results are of good quality.

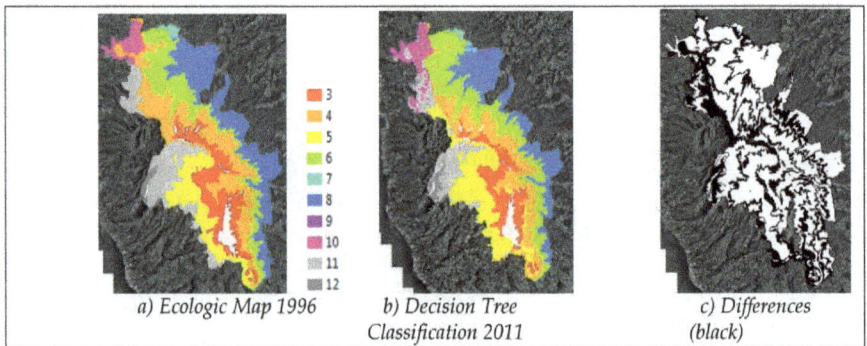

Figure 7. Resulting strict classification zoomed on Natural parc of Guadeloupe

Now we are interested on the transition between classes and want to focus on wide transition area in order to eventually decide to create transition classes. *Fig. 8* shows different transitions between classes. We display in black all areas were the membership degree has a value under a fixed threshold. So in the first map, only few areas have a reliable coefficient under 0.5, and in the third map more areas but relatively few are under 0.9 or 0.95. These results tend to show that this uncertainty is relative low concerning the classes and that the transition gradient is high, that is to say that we quickly go from one forest to another.

We also observe that the width of the transition is not the same between every forest (for example compare transition between blue and green classes and between red and orange classes). This illustrates the good and attended property of the model which should translate the reality of the transition which is heterogeneous.

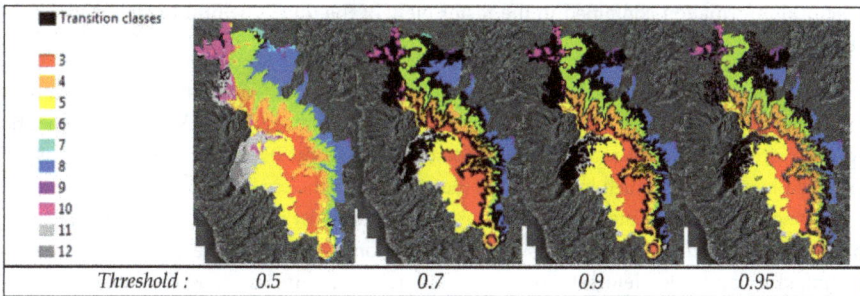

Figure 8. Full transition classes

This first part of the chapter shows a new fuzzy model to represent data having diffuse and uncertain limits. The advantage of this model is that *(i)* it doesn't require a rasterisation of the data which implies a loss of precision *(ii)* it doesn't treat every transition in the same way and allows different transition width *(iii)* the model is enough flexible to be adapted for different applications. The building of the model uses a raster-vector cooperation to define the features during the classification process but the model is based on a vector reperesentation.

It has been successfuly apply to forest modelling and classification and gives results close to a reference manual classification made by biologists.

3. Vector unit refinement using raster information

3.1. Principe

This step aims to refine the Vector Units (VU) by using the Raster information computed on each Raster Unit (RU) separately (Grandchamp, 2009), (Grandchamp, 2010).

At this step, we consider that the VU couldn't be refined using external vector information. This could be the case if all vector layers have been used to produce the VU or if the adjunction of any other vector information haven't any sense (for semantic reasons or else) (Gahegan, 2008). For example if we split environmental VU with administrative borders. We also remind that a VU is a uniform area according to some specific vector information and that we expect that the VU will also have a uniform visual aspect. If not, or if we are looking for some specific area within a VU we have to split it into sub-VU.

So the only way to split the VU into sub-VU is to use the Raster information (satellite images, airborne images, etc.). Let us now consider two cases leading to the same process. Firstly we are looking for some specific objects within a VU and we hope their spectral or structural information is enough discriminating to detect them. Secondly we detect a non homogenous VU (regarding criterions detailed later) and we hope the homogenous sub-areas are semantically significant. In both cases, we make the hypothesis that the spectral or structural information contained in the raster view of the unit is helpful to split it into sub-VU.

The process is based on an homogenous criterion computed on some specific features describing the spectral or structural information of the image. As a VU has a semantic signification and a small size it's often not divisible using its raster view if there is no external phenomenon which interact with the unit. Nevertheless, if the unit has been altered it's often composed of a main sub-VU representing the most important part of the VU, one (or at most two) sub-VU having a semantic definition (in the case of ecological units, they are linked to landslide, deforestation, etc.) and a reject sub-unit composed of non identified or useless area (such as shadow).

The localisation and identification of the sub-VU are done in two main ways. Firstly in a supervised way, where both localisation and identification are done at the same time by a semantic and numeric definition of each sub-VU. We use in this case a classifier where each class is a sub-VU. Secondly in a semi-supervised or unsupervised way by applying a clustering method where at most the number of clusters is set. In this way, the identification must be done in a second step.

3.2. Process

We will now present the details of this Raster-Vector cooperation. Starting from a raw image (*Fig. 8*-a), and a vector splitting of the space (*Fig. 8*-b), we fuse both information (*Fig. 8*-c). The image is an extract of a very high resolution IKONOS satellite image (1m resolution) allowing an accurate classification (Gougeon, 2001) of vegetation (Junying, 2005), (Yu, 2006), (Johansen, 2007).

We compute an homogeneity coefficient on each Raster Unit (RU) (*Fig. 8*-d) in order to localise and later split the non uniform ones. The homogeneity coefficient is based on the measure of the compactness of the feature vectors computed on each pixel of the RU. This vector is composed of mean and standard deviation of each Color band (Angwine, 1998) and textures features (over each color band too): co-occurrence matrices (Gotlieb, 1991), Gabor filters (Manjunath, 2010), Laws filters (Laws, 1980), Hue moments (Hu, 1962) and fractal dimension (Mandelbrot, 1977). There is a total of 25 normalized features including geometrical, statistical, frequential and fractal description of the RU (Abadi, 2008), (Scarpa, 2006).

Then the decision to split or not the VU depends on an empirical threshold of the homogeneity coefficient. In this case, the threshold has been set to 0.6 leading to one non homogenous unit (*Fig. 8*-d) with a coefficient of 0.37.

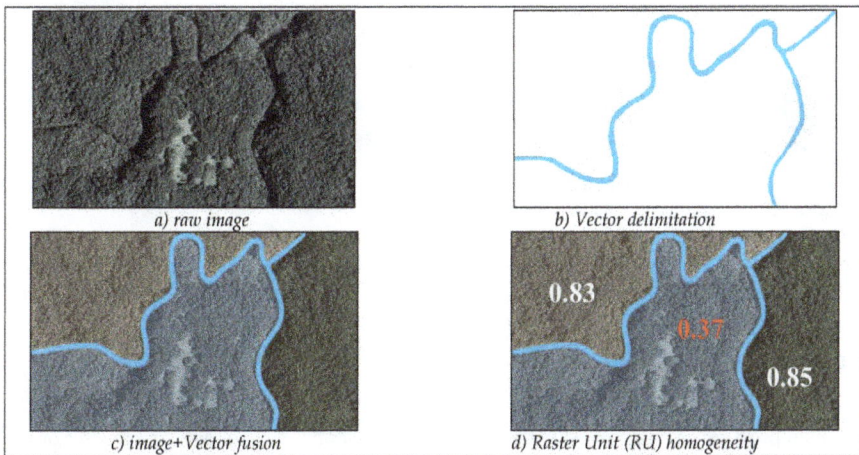

Figure 9. Non uniform VU localisation

Now we will localize and extract the sub-VU by analyzing the spectral and structural information contained in the RU. In this case, a semi-supervised classification is applied on the non homogenous RU extraction (*Fig. 9* a). We use a K-Nearest-Neighbour (KNN) algorithm and set the number of classes to 3, with the aim to obtain classes corresponding to shadow, main ecosystem and secondary ecosystem. The *Fig. 9*-b) shows the classification results after a filtering post-treatment to eliminate isolated pixels. The class repartition is as follow: class 1 in blue (81%), class 2 in red (13,8%) and class 3 in yellow (5,2%).

The interesting sub-RU is identified has being the class 3 and corresponding to the a deforestation (*Fig. 9*-c). The identification has been made manually taking into account expert knowledge and environmental information. For example, this sub-VU could be a landslide or a deforestation, the choice has been made according to the slope of the Unit. Then a sub-VU is produced (*Fig. 9*-d).

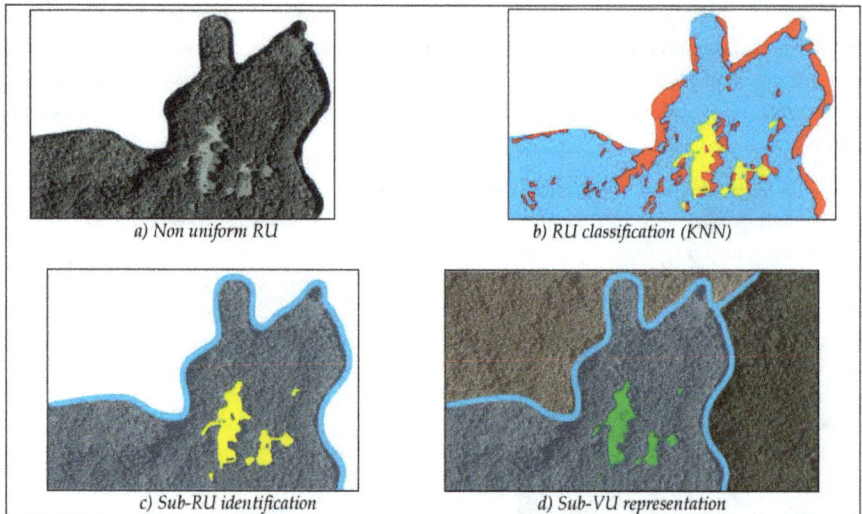

a) Non uniform RU

b) RU classification (KNN)

c) Sub-RU identification

d) Sub-VU representation

Figure 10. Vector Unit splitting using raster information

The use of raster information to both evaluate the homogenity of the VU and to split the VU into sub-VU is an efficient way to produce new vector information. The predivision of the image with the VU combined with the homogeneity evaluation considerably reduce the complexity of the classification problem by replacing the classification of the whole image by a selection of reduced classifications applied to non homogeneous RU.

The main problem of this process is the threshold the user has to fixe to decide to split or not a RU. A common way to use this tool and which avoids fixing the threshold is to sort the units according to their homogenity value and to let the operator take the decision. Indeed, as a post identification of the sub-VU must be done by an operator, the best way is to let him appreciate the relevance of a sub division.

4. Conclusion

This chapter presents in a first part a vector model for fuzzy data sets within a GIS and in a second part a raster and vector cooperation to divide the space into elementary units represented in a vector way and called Vector Units (VU). The fuzzy vector model offers a new way to represent data which have diffuse borders without the limitations of classical approach which are based on a rasterisation of the information or on a regular evolution of the transitions in each direction. This model tries to combine both the flexibility of the vector representation and the accuracy of the raster representation. This fuzzy model allows revealing full transition classes that a strict model doesn't allow. The raster-vector cooperation is linked to the mixing of both raster features (colour or textures) and vector features (humidity, elevation, etc.).

In a second part, this chapter deals with the use of raster features to split VU into sub-VU. The raster features are used to evaluate the homogeneity of the VU and in case of insufficient homogeneity to localise the region of interest in order to produce new sub-VU.

These two raster-vector cooperations are ways to produce flexible and useful information. They have been successfully used to classify the forest of a Caribbean Island named Guadeloupe and to localise landslide within its Natural parc.

Some improvement could be done concerning the model and particularly concerning the thresholds to define the different fuzzy belts. But the possibility to store the membership degree vectors in each VU allows producing as much fuzzy models as required.

Concerning the use of raster information to subdivide VU, a comparison (or a combination) with object oriented classification could be done (Maillot, 2008), (Forestier, 2008), (Hudelot, 2008). Object oriented classifications directly start with the raw images divided into elementary raster units (RU) according to a segmentation process (such as Watershed). The main drawback of this approach is the lack of semantic of the RU and the difficulty to retrieve semantically defined objects.

Author details

Enguerran Grandchamp
University of Antilles and Guyana, France, Guadeloupe

5. References

Abadi, M., Phd Thesis, 2008, Couleur et texture pour la représentation et la classification d'images satellite multi-résolutions,

Altman, D., 1994, Fuzzy set theoretic approaches for handling imprecision in spatial analysis, Internat. J. Geographical Inform. Systems 8 (3) (1994) 271–289.

Angwine, S. J., and Horne, R. E. N, 1998,. The colour Image Processing Handbook (oelectronics, Imaging and Sensing). Springer-Verlag New York, Inc., Secaucus, NJ, USA

Baglioni M., et al., 2008, "Ontology-supported Querying of Geographical Databases", In Transactions in GIS, vol. 12, issue s1, pp. 31–44,

Benz, U., and al., 2004, Multi-resolution, object-oriented fuzzy analysis of remote sensing data for GIS-ready information, *ISPRS Journal of Photogrammetry & Remote Sensing, Vol.* 58, pp 239– 258

Bjorke, J. T., 2004,Topological relations between fuzzy regions: derivation of verbal terms, Fuzzy Sets and Systems 141 449–467

Bloch I. et al., 2006, "Modeling the spatial relation *between* for objects with very different spatial extensions", In Proceedings of Reconnaissance des Formes et Intelligence Artificielle (RFIA'06), Tours, France,

V.V., 2001, Cross, Fuzzy extensions for relationships in a generalized object model, International Journal on Intelligent Systems 16 843–861.

Cruz, I., Sunna, W., and Ayloo, K., 2005, "Concept-level matching of geospatial ontologies", In Proceedings of GISPlanet, Lisbon, Portugal,

Egenhofer, M., and Franzosa, R. D., 1991, "Point-Set Topological Spatial Relations", In International Journal of Geographical Information Systems, vol. 5(2), pp. 161-174,

Fisher, P., 2000, Sorites paradox and vague geographies, Fuzzy Sets and Systems 113,7–18.

Fonseca, F., Egenhofer, and al., 2002, "Using ontologies for integrated geographic information systems", In Transactions in GIS, vol. 6, pp. 231–57,

Fonseca, F., Camara, G., and Monteiro, M., 2006, "A framework for measuring the interoperability of geo-ontologies", In Spatial Cognition and Computation, vol. 6, pp. 309–31,

Fonseca, F., Rodríguez, M. A., and Levashkin, S., 2007, "Building geospatial ontologies from geographical databases", In Proceedings of the International Conference on GeoSpatial Semantics (GeoS 2007), Springer Lecture Notes in Computer Science, vol. 4853, pp. 195–209, Berlin

Forestier, G., Derivaux, S., Wemmert C. and Gançarski, P., 2008, "An Evolutionary Approach for Ontology Driven Image Interpretation", In Proceedings of the International Conference on Applications of Evolutionary Computation (Evo'08), Springer, Lecture Notes in Computer Sciences, vol. 4974, pp 295--304, Italy,

Gahegan, M., et al., 2008, "A Platform for Visualizing and Experimenting with Measures of Semantic Similarity", In Ontologies and Concept Maps, Transactions in GIS, vol. 12 issue 6, pp. 713-732,

Gama, J, 2005, Functional Trees. Niels Landwehr, Mark Hall, Eibe Frank. Logistic Model Trees.

Gary L., 2010, New fuzzy logic tools in ArcGIS 10, Raines and al., ESRI Communication,

Gotlieb, C. C., and Kreyszig, H. E, 1991 Texture descriptors based on cooccurrence matrices. Computer Vision Graphic Image Process 51, 1, 70–86.

Gougeon, A., and al, 2001, Individual Tree Crown Image Analysis – A Step Towards Precision Forestry. Natural Resources Canada,

Grandchamp, E., 2009, GIS information layer selection directed by remote sensing for ecological unit delineation, *IGARSS*

Grandchamp,E., 2010, Raster-vector cooperation algorithm for GIS, *GeoProcessing*

Grandchamp,E., 2011, Specification for a Shared Conceptual Layer in GIS, *GeoProcessing*

Guesgen, H. W., and Albrecht, J, 2000, "Imprecise reasoning in geographic information systems", In Fuzzy Sets and Systems, vol. 113, pp. 121-131,

Guo, D., Guo, R., Thiart, C., 2004, Integrating GIS with Fuzzy Logic and Geostatistics: Predicting Air Pollutant PM10 for California, Using Fuzzy Kriging

Gutierrez, L., 2006, "WP 9: Case study eGovernment D9.9 GIS ontology, "http://dip.semanticweb.org/documents/D9-3-improved-eGovernment.pdf   ",   last access: 17/01/2011,

Hu, M.-K, 1962, Visual pattern recognition by moment invariants. Information Theory, IEEE Transactions on 8, 2, 179–187.

Hudelot, C., Atif, J., and Bloch, I., 2008, "Ontologie de relations spatiales floues pour le raisonnement spatial dans les images", In proceedings of Représentation et Raisonnement sur le Temps et l'Image (RTE),

Johansen, K., and al., 2007, Application of high spatial resolution satellite imagery for riparian and forest ecosystem classification, Remote Sensing of Environment, Vol. 110, No. 1., pp. 29-44

Jones C. B. et al., 2002, "Spatial Information Retrieval and Geographical Ontologies: An Overview of the SPIRIT project", In Special Interest Group on Information Retrieval (SIGIR), ACM Press, pp.387 – 388, Finland,

Junying, C., Qingjiu, T., 2005, Vegetation classification model based on high-resolution satellite imagery, Proceedings of SPIE, Remote sensing of the environment, 19-23 August, Guiyang City, China

Kainz, Introduction to Fuzzy Logic and Applications in GIS – Example,

Karimi, M., and al., 2008, Preparing Mineral Potential Map Using Fuzzy Logic In GIS Environment, *The International Archives of the Photogrammetry, Remote Sensing and Spatial Information Sciences. Vol. XXXVII. Part B8. Beijing 2008*

Kavouras, M., Kokla, M., and Tomai, E., 2005, "Comparing categories among geographic ontologies", In Computers and Geosciences, vol 31, pp 145–54,

Kimfung L., Wenzhong S., 2009, Quantitative fuzzy topological relations of spatial objects by induced fuzzy topology, *International Journal of Applied Earth Observation and Geoinformation*, vol. 11 38–45

Laws, K. I., 1980, Rapid texture identification. In SPIE Image Processing for Missile Guidance 238, 376–380.

Lewinski, S., Zaremski, K., 2004, Examples of Object Oriented Classification Performed On High Resolution Satellite Images. *Miscellanea Geographica*. Volume 11.

Maillot, N. E., and Thonnat, M., 2008, "Ontology based complex object recognition", In Image and Vision Computing, vol 26, pp 102–113,

Mandelbrot, B. B., 1977 Fractals: Form, Chance and Dimension. W. H. Freeman and Co.

Manjunath, B. S., and Ma, W. Y, 1996. Texture features for browsing and retrieval of image data. IEEE Trans. Pattern Anal. Mach. Intell. 18, 8, 837–842.

Mukhopadhyay, B., 2002, Integrating exploration dataset in GIS using fuzzy inference modeling, GISdevelopment

Quinlan, R., 1993, C4.5: Programs for Machine Learning. Morgan Kaufmann Publishers, San Mateo, CA.

Rodriguez, M. A., and Egenhofer, M., 2003, "Determining semantic similarity among entity classes from different ontologies", In IEEE Transactions on Knowledge and Data Engineering, vol. 15, pp. 442– 56,

Rousteau, A., 1996, Carte écologique de la Guadeloupe. 3 feuilles au 1/75.000ème et notice (36 p.). *Conseil Général de la Guadeloupe, Office National des Forêts et Parc National de la Guadeloupe.*

Ruiz, C., and al, 2007, The Development of a New Methodology Based on GIS and Fuzzy Logic to Locate Sustainable Industrial Areas, *10th AGILE International Conference on Geographic Information Science*

Sawatzky, D., and al., 2008, Spatial Data Modeller, Technical Report,

Scarpa, G. Haindl, M., and Zerubia, J., 2006, Hierarchical finite state modeling for texture segmentation with application to forest classification. Research Report 6066, INRIA, France,

Schneider, M., 1999, Uncertainty management for spatial data in databases: fuzzy spatial data types, in: *Advances in Spatial Databases, Lecture Notes in Computer Science,* vol. 1651, Springer, Berlin, pp. 330–351.

Schuurman, N., et al., 2006, "Ontology-Based Metadata", In Transactions in GIS, vol. 10, pp. 709–726,

Smith, B., 2001, "Ontology and information systems", http://www.loa-cnr.it/Papers/FOIS98.pdf, last access : 17/01/2011,

Sunila, R., 2004, Fuzzy modelling and Kriging for modelling imprecise soil polygon boundaries. Proceedings 12th International Conference on Geoinformatics-Geospatial Information Research: Bridging the Pacific and Atlantic, Gävle, Sweden, pp. 489 – 495

Sunila, R., and Horttanainen, P., 2009, Fuzzy Model of Soil Polygons for Managing the Imprecision. Interfacing GeoStatistics and GIS

Viegas, R., and Soares, V., 2006, "Querying a Geographic Database using an Ontology-Based Methodology", In Brazilian Symposium on GeoInformatics (GEOINFO 2006), pp. 165-170, Brazil

Wiegand, N., et al., 2007, "A Task-Based Ontology Approach to Automate Geospatial Data Retrieval", In Transactions in GIS, vol. 11, issue 3, pp. 355–376,

Wolfgang Kainz, Fuzzy Logic and GIS, book chapter 1. Web document, last access 01/09/2011

Yanar, T. A., and Akyürek, Z., 2004, The Enhancement of ArcGIS with Fuzzy Set Theory ESRI International User Conference

Yu, Q., and al., 2006, Object-based Detailed Vegetation Classification with Airborne High Spatial Resolution Remote Sensing Imagery, Photogrammetric Engineering & Remote Sensing Vol. 72, No. 7, July 2006, pp. 799–81

Zandbergen, P.A., 2008, "Positional Accuracy of Spatial Data: Non-Normal Distributions and a Critique of the National standard for Spatial Data Accuracy", In Transactions in GIS, vol. 12, issue 1, pp. 103–130,

Zhu, A. X., and al., 2001, Soil Mapping Using GIS, Expert Knowledge, and Fuzzy Logic, , Soil Sci. Soc. Am. J. 65:1463–1472

Do Geographic Information Systems (GIS) Move High School Geography Education Forward in Turkey? A Teacher's Perspective

Süleyman İncekara

Additional information is available at the end of the chapter

1. Introduction

Geographic Information Systems (GIS) can be defined as a comprehensive mapping system designed for capturing, storing, analyzing, synthesizing, querying, editing, retrieving, manipulating and displaying spatial data obtained from earth's surface in the form of charts, tables, 3D images and maps based on the richness of the information entered into the GIS database.

Numerous studies done on GIS in education in the international context indicated that in addition to increasing student and teacher motivation, GIS are very effective tools for incorporating project-based teaching and learning, and promoting students' geographic skills such as thinking geographically, analyzing and synthesizing spatial data, map reading and interpreting (Tinker, 1992; Geography Education Standards Project [GESP], 1994; Palladino, 1994; Audet & Abegg, 1996; Lemberg & Stoltman, 1999; Pottle, 2001; Shin, 2006). Tinker (1992), Palladino (1994), and Audit & Abegg (1996), who were the leading academicians, conducted the first research on education with GIS and underlined the positive relationships between education with GIS and the development of spatial skills of the students.

Wanner & Kerski (1999) found that GIS have the potential to accelerate the geographical inquiry skills of the students as well as analyzing and displaying geographic data. According to Pottle (2001), a GIS is a beneficial tool for learning and motivating students in the process of gaining important skills and knowledge from geography curricula. The outcomes of a study by Bednarz & Van der Schee (2006) show that there are three main benefits to geography teachers of integrating GIS into their geography courses, including the support of GIS on teaching and learning, the help investigating geographical problems at different levels, and their widespread use in business in the existing century. GESP (1994) suggested that the presence of GIS in geography education is enhancing many geographic skills of the students including acquiring, organizing, and analyzing geographic information, and asking and answering geographic questions.

However, there are some other studies questioning whether education with GIS is certainly beneficial in every case or enhancing the spatial skills of the students. According to Bednarz (2004), there is no sufficient proof that a GIS enhances the spatial skills. The optimum conditions are needed to further expand its applications and we need to know whether there are any easier and better methods to reach the same objectives. In a study conducted on the effectiveness of GIS in courses, Kerski (2003) concluded that in spite of reinforcing the standard-based skills and spatial analysis, the inquiry-oriented lessons with GIS did not consistently increase geographic skills. Shin (2006) also argued that it is not possible to insist that using GIS was the best or the only medium to teach geography subjects; rather, education with GIS has a potential student learning that most other tools do not have. Some other research pointed out the importance of methods which will be applied during the process of education with GIS, and in the case of choosing wrong educational methods, GIS in education may result in failure of student learning (King, 1991; Walsh, 1992; Wanner & Kerski, 1999; Johansson, 2003; Baker & White, 2003; Bednarz, 2004).

2. Background

2.1. Adoption of GIS by education and other sectors in Turkey

Almost 50 years later then its first initiation in Canada (Yomralioglu, 2000), today GIS has been an extensively used technology in various sectors which use spatial data, including environmental protection, urbanization, planning, engineering, environmental protection, remote sensing, municipal works, transportation, forestry, and all levels of education. As for the education, the adaptation of GIS into education was even slower compared to other sectors. The utilization of GIS in education was first initiated at university level at US and Canadian universities in 1980s (Zhou et al., 1999). With the beginning of the 1990s, GIS found more place in secondary school curricula in North America and Europe (Johansson & Pellikka, 2006). Bednarz & Ludwig (1997) stated that most geography and social studies teachers were in the "pre-awareness" category in terms of adaptation of GIS in course environments. However, the studies of Kerski (2003, 2007) revealed that the teachers reached the "understanding" stage (Table 1).

Phase	Key question
Phase 1: Pre-awareness	
Phase 2: Awareness	What is GIS?
Phase 3: Understanding	How can I teach geography with GIS
Phase 4: Guided practice	How do I do GIS?
Phase 5: Implementation	

Table 1. Five phases in adaptation of GIS. Adapted from Bednarz & Ludwig, 1997.

However, in spite of providing an important potential to enrich geography education, the use of GIS has not been an integral part of geography education and its diffusion remained slow at secondary schools. The statistics showed that slightly less than 2% of American high schools and only 20% of the teachers out of 1,520 who have GIS software and knowledge use it in more than one lesson in more than one class (Kerski, 2003). The level of utilization of GIS in secondary education remained almost the same in the UK (Office of Standards in Education [OFSTED], 2004). In a study conducted on the Netherlands' geography teachers, 12% of the respondents stated that they were using GIS in their courses and 81% of them supported GIS having a greater role in teaching geography (Korevaar & Schee, 2004). The figure is almost the same in Singapore schools (Yap et al., 2008). In another study conducted in Turkey on attitudes of geography teachers to using GIS, Demirci concluded that almost all teachers that filled out the questionnaire form expressed that they haven't attended an education program about GIS, and don't know how to use GIS software, despite the fact that all of them they believe the in necessity of using GIS in geography lessons (Demirci, 2008).

Some researchers sought the factors of why GIS technology is being integrated very slowly into the K-12 curriculum. We can classify these factors including:

- Limited time: Learning and implementation of GIS takes a long time (Kerski, 2003). Allocation of limited time and balancing other demands from the school system (Shin, 2006). Insufficient time in the curriculum to better incorporate GIS into education (Meyer et al., 1999).
- Curriculum problems: Limited extent of GIS in the curriculum; lack of necessary digital data, lesson plans, learning objectives, and instruction problems (Yap et al., 2008; Demirci, 2009).
- Teacher problems: Lack of necessary GIS skills of teachers. Teachers' attitudes, perceptions of technology and lack of consciousness about the benefits of GIS or the pre-awareness status of teachers in incorporating GIS into their courses (Bednarz & Ludwig, 1997). Some teachers have negative conceptions of geography in that they view geography as memorizing names and features, so they do not believe that GIS do any good for spatial analysis (Patterson et al., 2003; Bednarz & Ludwig, 1997).
- In-service training: Unavailability of GIS training and exposure, insufficient peer support and inadequate lesson demonstrations by experienced GIS teachers (Yap et al., 2008).
- Issues on physical conditions: Lack of access to appropriate hardware and software, GIS-based resource packages, etc. (Meyer et al., 1999).

Unfortunately, the integration of GIS into public or private sectors of Turkey was almost 20 years later than North American countries in the 1980s. The integration of GIS into the Turkish education sector was even slower (Table 2). The adoption of GIS into universities was initiated in the 1990s by a graduate thesis aimed at exploring the GIS potential in Turkey. The first departments that dealt with GIS were Geodesy and Photogrammetry Departments. The number of departments employing GIS courses then increased sharply between 1990 and 2004 (Olgen, 2004, Yomralioglu, 2002). The first undergraduate level GIS course and fully equipped GIS education laboratory in a geography department was established in 1998. In the following years, the number of geography departments providing

GIS courses and GIS laboratories increased slowly. In a study conducted in 2009, Demirci reported that there were only 6 geography departments in Turkey, out of 36 either in Art and Science or Education Faculties, which had a GIS laboratory (Demirci, 2009).

Cornerstones of GIS	Year
The first use of GIS by a private company	1981
The first use of GIS by General Command of Mapping	1986
The first use of GIS in the public sector, such as General Directorate of Land Registry and Cadastre, Turkish Statistical Institute, and State Meteorological Service	1990
The first use of GIS in higher education	1991
The first national conference on GIS	1994
The first use of GIS by municipalities (Metropolitan Municipality of Bursa)	1996
The first GIS course in a geography department	1998
The first GIS education laboratory in a geography department	1998
The first national GIS conference organized by a geography department	2001
The first GIS for teachers workshop	2004
The first national geography curriculum including GIS-related activities	2005
The first international conference on GIS organized by a geography department	2008
The first GIS course materials (books, CDs, and course activities) for secondary schools	2008
The first GIS-based civil involvement project concerning secondary school students aiming at integrating GIS-based activities into geography courses	2009
The first GIS education certificate program for geography teachers	2011

Table 2. Cornerstones of GIS in Turkey. Adapted from İncekara, 2010; İncekara & Karakuyu, 2010; Demirci, 2009, Yomralioglu, 2002.

2.2. GIS in Turkish secondary school curriculum

The 2005 geography high school curriculum has been playing an important role in terms of adoption of GIS into geography courses in that it is the first geography curriculum of Turkey in which the concept of GIS was included. By 2005, The Turkish Ministry of National Education initiated an overall reform for the national curriculum, including geography. The new geography course curriculum suggested maximum adoption of technology into geography courses and particularly underlined the importance and necessity of the extensive integration of GIS into geography education. By the initiation of the 2005 curricula, many efforts were ongoing in order to make GIS a widely used tool and method in geography education through a number of activities, including conferences, panels, in-service GIS training, certificate programs, student projects, course materials, papers, workshops, etc.

In the content of the 2005 high school geography curriculum, GIS applications are recommended for some attainments. The new geography program has a constructivist base and spiral structure. Attainment is examined consecutively. Content foreseen by the attainment is provided (Yasar & Seremet, 2009). Geography teachers may develop GIS activities, projects, panels, etc., or analyze existing ones (Milli Eğitim Bakanlığı [MEB], 2005). GIS is an important part of the "Geographic Skills and Applications" learning module of the new curriculum along with map skills, use of information technologies skills, critical thinking skills, and field trip skills of the students. By investigating the 2005 geography curriculum content, there are 21 GIS activities. These GIS-related course activities are placed in different grades and learning modules (Tables 3 and 4). The general characteristics of these activities are:

- Irregular distribution from ninth to twelfth grade.
- With 9 activities, the tenth grade includes the highest number of GIS-related activities.
- Each GIS-related activity refers to two or more skills such as observation, map, critical thinking, field trip, perception of time, organizing the geographical data skills, etc.
- The two modules which have the highest number of GIS activities are "A spatial synthesis: Turkey "and "Human systems" modules. "Environment and society" is the only learning module in which there are no GIS activities suggested.

Activities, attainments, and learning modules	Grades				Total
	9th Grade	10th Grade	11th Grade	12th Grade	
Number of GIS-related activities suggested	5	9	3	4	21
Number of related attainments	6	10	5	6	27
Number of GIS activities in learning modules					
Natural systems	3	1	-	-	4
Human systems	-	5	2	-	7
A spatial synthesis: Turkey	2	3	-	2	7
Global environment: regions and countries	-	-	1	2	3
Environment and society	-	-	-	-	-

Table 3. Distribution of suggested GIS-related activities, attainments, and learning modules by grades in the 2005 secondary school geography curriculum. İncekara, 2010; MEB, 2005.

When we look at the subject-based distribution of GIS activities, we see that they are also irregular to each grade. Subjects including political and cultural alliances, relationships and patterns, maps, economy, population, international economics, natural and human systems in Turkey, and population characteristics have more than one GIS-related activity while there are no activities recommended for soil, water, vegetation cover, and tourism units (Table 4).

Subjects	Grades				Total
	9th Grade	10th Grade	11th Grade	12th Grade	
Maps and map components	3	-	-	-	3
Natural and human systems in Turkey	2	3	-	1	6
Landforms	-	1	-	-	1
Population	-	4	-	-	4
Economy	-	1	1	-	2
Settlements	-	-	1	-	1
International economic, politic, and cultural alliances, relationships, and patterns	-	-	1	1	2
Global and regional connections of Turkey	-	-	-	1	1
Regions and countries	-	-	-	1	1
Total	5	9	3	4	21

Table 4. Subject-based distribution of GIS-related activities in the new geography curriculum. İncekara, 2010; MEB, 2005.

2.3. Use of technology and GIS in geography courses in Turkey

There are a number of studies in Turkey on the availability of educational technologies in geography, the most used technologies in geography education, the attitudes of geography teachers towards using educational technologies in their lessons, and to what extent the technology use is included by geography teaching programs and textbooks. One of these studies, Özel (2007), concluded that geography teachers don't use educational technologies sufficiently and they are partly competent to use these technologies in their courses. Moreover, the study revealed that computer, VCD players, and LCD projectors were the most used technologies by geography teachers. Sonmez et al. (2009) and Demirci (2009) stated that along with the negative effect of the lack of necessary technologies in public schools, many technologies are unknown to many geography teachers, and that this may only be overcome by more in-service training opportunities provided by the ministry of education.

In another study by Tas at al. (2007), it was found that almost all of the geography teachers seemed to have enough consciousness that the technology is beneficial and necessary to enhance student learning and motivation; however, the diversity of the up-to-date technologies couldn't be reflected to the classroom environment due to a number of factors, including physical infrastructure of schools, competency, etc. The studies by Demirci (2008, 2009) supported the results of previous studies that LCD projectors and Power Point Presentations were among the most used technologies by geography teachers. Moreover, he found that 53% of the geography teachers expressed that they didn't have a computer in their classrooms, 49% of them didn't have an LCD projector, and 63% of the teachers didn't have an internet connection in their classrooms.

There are two up-to-date studies done by Incekara (2011a, 2011b) specifying the use of technology in geography teaching and the learning process in Turkey. In one of them, Incekara (2011a) investigated the technology use of geography teachers in their courses and the effect of the 2005 program on their attitudes to using technology in their courses. The results indicated that a large majority of geography teachers (%84) agreed that the geography teaching program of 2005 supports the greater use of technology in geography teaching and learning. Additionally, they expressed that they started utilizing the technology more than they used it before the initiation of the 2005 program with a big majority of 80.5%. It is quite encouraging that almost 90% of respondent geography teachers recognize the significance of educational technologies in ideal geography teaching. However, almost half of them stated that the limited physical infrastructures and facilities of their schools prevented the efficient use of technology in their courses. Almost one-third of the respondents stated that they didn't have enough knowledge about how to integrate technology into their courses (Table 5).

	Statements		Level of agreement			
			Strongly disagree/ Disagree 1/2*	Neutral 3*	Agree/ Strongly agree 4/5*	Total
1	The 2005 geography teaching program supports the greater use of technology	n	12	16	152	180
		%	6.7	8.9	84.4	100
2	I utilize the technology more in my courses with the 2005 geography teaching program	n	17	18	145	180
		%	9.5	10	80.5	100
3	Technology must be used for ideal geography teaching	n	12	8	160	180
		%	6.7	4.4	88.9	100
4	The limited facilities of the school prevent me from using technology sufficiently	n	68	22	89	179
		%	37.9	12.3	49.7	100
5	I have enough knowledge about how to incorporate technology into my courses	n	29	29	120	178
		%	16.3	16.3	67.4	100
*1: Strongly disagree 2: Disagree 3: Neutral 4: Agree 5: Strongly agree						

Table 5. Opinions of geography teachers about the use of technology in geography education. Incekara 2011a.

The research results also indicated that public school teachers tend to support the notion that the 2005 teaching program supports more use of technology more than private school and private course (these are the private educational institutions preparing students for different exams, including university and high school entrance exams) teachers. Moreover, public school teachers are more positive than private course teachers about the idea that "technology must be used" for ideal geography teaching. Additionally, public and private course teachers suffer the most from the lack of infrastructure in using technology, compared with private school teachers. According to the further analyses, the belief of geography teachers in the importance of technology use is increasing as the teachers' English level increases (Incekara, 2011a).

Incekara (2011b) also studied the most and least used technologies in geography education and to what extent (frequency) these technologies are being used. Research results indicated that the educational technologies used in geography teaching have been diversified and have changed in the last few decades. Maps, atlases, and globes are among the most popular technologies used in geography education. The moderate use of Google Earth, various models, Internet, and computer animations shows that geography teachers make an effort to visualize their courses. However, spatial technologies including satellite images, air photos, global positioning systems (GPS), and GIS are among the most rarely used technologies in geography courses.

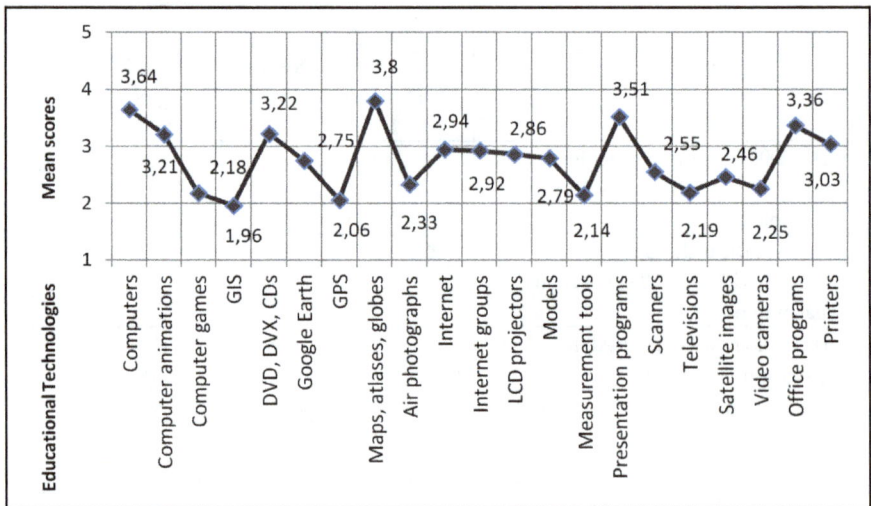

Figure 1. The mean scores of geography teachers in using educational technologies. Incekara, 2011b.

Research results also revealed that the measuring tools which are subjected to use both in laboratories and field trips and computer games are among the most rarely used technologies in geography education. This shows that field trips, laboratories, and computer games haven't become and integral part of geography education. Statistical test results suggested that younger geography teachers tend to use computer games more than older ones, and public and private school teachers are using educational technologies more than

Do Geographic Information Systems (GIS) Move High School Geography
Education Forward in Turkey? A Teacher's Perspective

45

private course teachers. However, public school teachers use educational technology in their courses as much as private course teachers as a result of the fast expansion of computers, Internet and Internet-based technologies, and Internet groups based on geography profession (Incekara, 2011b).

Another important result that the statistical results proved was that as the diploma level (undergraduate, graduate) and foreign language competencies of teachers increase, the success of integrating the technologies into education also increases. As a final assessment of technology use in geography courses we can say that the mean score of geography teachers regarding the use of technology is 2.73 out of 5 which corresponds to "sometimes" on a 5-point Likert scale, which means the geography teachers are using educational technologies sometimes in their courses. This frequency is not enough if the status of geography and the geography education level on global scale is considered (Incekara 2011b). The use of GIS technologies in Turkish secondary schools can't be isolated from the general state of technology use in geography courses which is very low if we consider that geography is a suitable course to integrate all kinds of educational technologies. Even worse, GIS is being used more rarely than any kind of educational technologies in spite of the existence of a big consensus among the geography teachers that GIS is a beneficial tool for geography courses and must have more place in teaching and in-service training programs.

For example Demirci (2008) reported that almost all geography teachers that attended his questionnaire study replied that they haven't attended any GIS education program, and they didn't know how to use GIS software or how to use GIS in their courses. 93% stated that they haven't had a computer laboratory in their schools; however, all of them supported the use of GIS in geography education. The situation is better in his 2009 study, where 66% of teachers surveyed didn't know what GIS was, 82% of them didn't know how to use it in their courses, and 80% of them had never attended a training course about GIS. However, 84% of the geography teachers stated that they hadn't used GIS software before while only 16% they used it on a basic level (Demirci, 2009). Artvinli (2009) underlined advantages and disadvantages of using GIS from a teacher's perspective and concluded that while most of the teachers appreciated the advantages of using GIS, they mostly complained about technical insufficiencies, time limitations, lack of in-service training, and over class-sizes as the most expressed limitations in using GIS in their courses.

3. Methodology

3.1. Content analysis of related studies

To provide a comprehensive understanding about GIS in Turkey's geography education, we provided an extensive review of literature in the previous parts of the chapter including the definition of GIS and benefits of using it in education, the advent of GIS and its diffusion in Turkey's education and other sectors, GIS in the secondary school curriculum, and use of technology and GIS in geography courses in Turkey. In the first part, the definition of GIS and the benefits of using GIS in geography learning and teaching processes was explored from an international perspective. In addition to its benefits, some controversial issues were underlined regarding whether education with GIS is certainly beneficial or best in every case or enhancing the spatial skills of students. In the second part, a detailed process of GIS adoption to education sectors along with other private and public sectors in Turkey and a table entitled "milestones of GIS in Turkey" was given a place. In this chapter, the difficulties and limitations in front of an effective integration of GIS into education sector were discussed item by item. Then, the GIS in the Turkish secondary school curriculum were underlined in order to reveal the relationships between the use of GIS in Turkish secondary education level and curriculum priorities. In this part, the distribution of GIS activities by subjects was also given to understand the irregularity of distribution.

There is a strong relationship between the overall use of educational technologies and GIS use in geography education. In other words, as the teachers' tendency to use technology increases, their tendency to use GIS also increases. So, we think that it is appropriate to look at the attitudes of geography teachers in Turkey towards technology use in geography education under the light of previous studies realized on this subject. In the following part, the literature on GIS in geography education in Turkey was reviewed to give the chance to compare the previous studies and this study in terms of teachers' attitudes towards using GIS technology in their schools.

3.2. Research questions and objectives

The chapter will seek the insights into what changes were made by the GIS applications which were integrated into new high school geography curricula from geography teachers' perspective and to what extent the GIS have been an integral part of geography education in Turkish high schools. The study aims at determining whether high schools have the required physical infrastructure for sufficient adoption of GIS into geography courses, whether the geography teachers have enough knowledge and skills to incorporate GIS in their courses, how often the geography teachers use GIS in their teaching process, and their beliefs and attitudes towards using GIS in geography education.

3.3. Statistical analysis

The "Teachers' attitudes towards using GIS" questionnaire constitutes the main method of the study. A 29-item questionnaire was prepared and distributed to the 183 geography teachers who voluntarily accepted to fill out the form from almost 50 different provinces in Turkey. The questionnaire was prepared as 4 parts, including demographic questions which were prepared to investigate gender, age, professional experience, education level, English level, the school type, etc. The second part consists of questions to determine the infrastructure of the schools in which the teachers are employed in terms of effective use of GIS in geography courses. The third part is developed an understanding of whether the geography teachers have enough skills to apply GIS applications in their courses. The fourth part aims at measuring the tendency of teachers to use GIS in their courses. The last part was allocated to determine the attitudes of geography teachers towards using GIS in their courses including the benefits of using GIS, fundamental factors preventing them from

Do Geographic Information Systems (GIS) Move High School Geography
Education Forward in Turkey? A Teacher's Perspective

47

using GIS, etc. For the descriptive analysis, frequencies and crosstabs are used throughout the study.

4. Findings

4.1. Demographic features of respondents

Demographic features of 183 geography teachers who responded to the questionnaire from almost 50 different provinces of Turkey (nearly half of them were employed in Istanbul and Ankara) revealed that most of the teachers were male (n= 128). The number of female teachers was 55. As the statistics revealed, almost half of the teachers were between the ages of 33 and 40, more than half of the respondents were employed in public schools, and 123 teachers out of 183 had more than 10 years of professional experience. It was found that the big majority of respondents (n= 130) knows a beginner level of English, while 46 of them announced that they had intermediate level of English and just 7 of them expressed that they know English at advanced level. With regard to the educational level of teachers, we can say that more than two thirds (n= 134) of them had an undergraduate diploma while 47 of them had a graduate diploma including master and doctoral education. 138 teachers out of 183 reported that class-size that the respondents teach changes between 16 and 30.

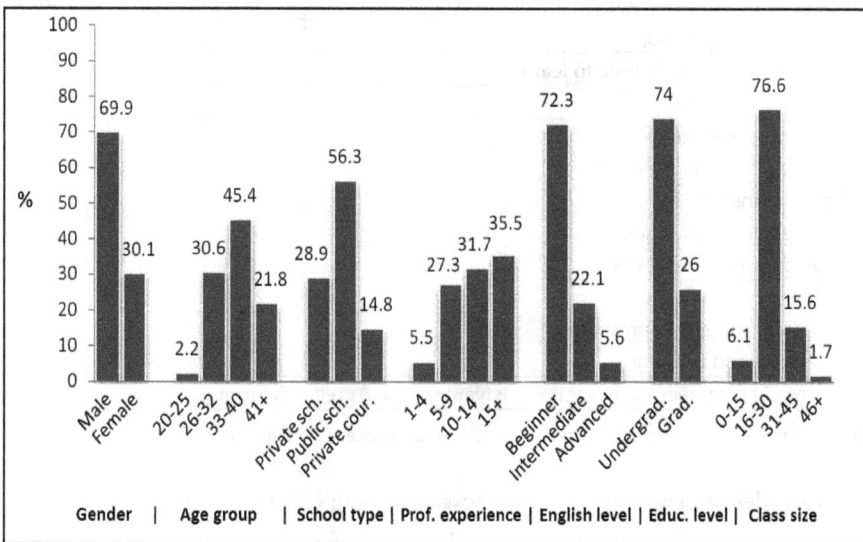

Figure 2. Demographic features of respondents.

4.2. The opinions of geography teachers regarding the importance and limitations of GIS

By assessing the composition of Table 5, the following inferences may be drawn:

- The big majority of teachers seem to appreciate the importance and necessity of using GIS in geography education, as almost 82% of them agreed or strongly agreed that GIS is an important tool in geography education and slightly more than 60% of them thought that a geography course without using GIS is missing.
- Almost half of the teachers think that limited time, insufficient infrastructure of schools and weak background of teachers are significant restraints in terms of an efficient use of GIS in their courses.
- More than two thirds of respondents complained about insufficient support from the ministry of education in terms of GIS programs and GIS training (Table 6).

Statements			Level of agreement			
			Strongly disagree/ Disagree 1/2*	Neutral 3*	Agree/ Strongly agree 4/5*	Total
1	GIS is an important tool for geography courses	n	9	24	148	183
		%	4.9	13.3	81.8	100
2	A geography course without using GIS is missing	n	29	43	109	181
		%	16.1	23.8	60.1	100
3	I don't have enough time to learn and teach with GIS	n	68	37	76	181
		%	37.6	20.4	42	100
4	It is impossible to use GIS in geography courses with the present insufficient infrastructure of schools and background of teachers	n	44	37	100	181
		%	23.3	20.4	55.3	100
5	Ministry doesn't support the teachers enough in providing software and training	n	24	28	129	181
		%	13.3	15.5	71.2	100
*1: Strongly disagree 2: Disagree 3: Neutral 4: Agree 5: Strongly agree						

Table 6. Opinions of geography teachers about the importance and limitations of GIS.

4.3. Attitudes of geography teachers towards using GIS in their courses:

The answers of the teachers to the related questions and their agreement level to the given statements provide us with very little reasons to be optimistic regarding the use of GIS in geography education in Turkey. This is because 75% of teachers or more stated that:

- They didn't have GIS software,
- Their schools didn't have GIS software,
- They didn't know how to use GIS software,

- They didn't have required GIS documents in order to develop a course activity,
- They couldn't develop a GIS course activity,
- They didn't use GIS in their courses,
- Their students didn't use GIS in their projects and assignments,
- They didn't attend a GIS training program, and 51.4% of them reported that they didn't have a computer laboratory to use in the courses developed by the help of GIS (Table 7).

	Questions and statements		Answers		
			Yes	No	Total
1	Do you have a computer laboratory which can be used in geography courses?	n	88	93	181
		%	48.6	51.4	100
2	Do you have a GIS software?	n	21	159	180
		%	11.7	88.3	100
3	Does your school have a GIS software?	n	6	173	179
		%	3.4	96.6	100
4	I heard the term "GIS" first time?	n	17	164	181
		%	9.4	90.6	100
5	I know what the GIS is and its area of usage.	n	135	46	181
		%	74.6	25.4	100
6	I know how to use a GIS software.	n	38	143	181
		%	21	79	100
7	I have required documents such as data, map, and images, etc to develop GIS activities in my courses.	n	33	149	182
		%	18.1	81.9	100
8	I can develop a GIS course activity by using GIS.	n	39	143	182
		%	21.4	78.6	100
9	I use GIS in my courses.	n	24	158	182
		%	13.2	86.8	100
10	My students use GIS in their assignments and projects.	n	12	170	182
		%	6.6	93.4	100
11	I know where to get a GIS training.	n	89	92	181
		%	49.2	50.8	100
12	I have attended a GIS training program so far including in-service training, course, panel, conference, workshop, etc.	n	40	141	181
		%	22.1	77.9	100
13	I want to attend any GIS in-service training program that will be organized by ministry of education	n	152	28	182
		%	83.5	15.4	100

Table 7. Attitudes of geography teachers towards using GIS.

However, more than half of the geography teachers stated that they know where to get a GIS education and more than 83% of them expressed that they are willing to attend any GIS training program by the ministry. This shows that they want to develop themselves in GIS and are planning to use it if the negative circumstances are changed by the authorities.

5. Conclusion

The research results revealed that before assessing the use of GIS in geography education, we have to focus on the technology adaptation into geography education. In spite of a common consensus among the educators that technology has innumerable benefits for teachers and students, there are some issues to be taken into account by all stakeholders of education: almost half of the teachers in Turkey suffer from the insufficiency of a technological infrastructure in their schools and at least one third of them have difficulties with how to integrate the technology in their courses, although the 2005 geography program of Turkey suggests more integration of technology in education (Incekara, 2011a, Table 7). It is not realistic to expect GIS to be a widely used technology, as the schools and geography teachers experience many discouraging problems on technological infrastructure and competencies for incorporating technology into education.

With regards to GIS in geography education, the integration of GIS into the geography curriculum of Turkey goes back to the secondary school curriculum reform realized by the ministry of education in 2005. Turkey's geography curriculum of 2005 suggested maximum use of technology and particularly underlined the extensive incorporation of 21 GIS activities into geography courses. Nevertheless, almost 6 years since the ministry of education launched the new program, it is quite clear that GIS has failed to become an integral part of geography education in Turkey: research results indicated that only 13% of geography teachers use GIS in their courses and just 6.6% of students benefit from GIS for their projects and courses. These results unfortunately correspond to the previous studies' outcomes that the usage of GIS in geography education is very limited on international and national levels despite fact that the support for GIS among geography teachers is very high (Kerski, 2003; Korevaar & Schee, 2004; Yap et al., 2008; Demirci, 2008; Artvinli, 2009; Demirci; 2009; Incekara, 2011b).

The research outcomes showed that 74.6% of the respondents stated that they knew what GIS is and its area of usage and almost 79% of them expressed that they didn't know how to develop a GIS activity for their courses. This means that most of them know what GIS is, but they don't have enough knowledge of how to teach geography with GIS. These outcomes exposed the fact that Turkey is placed in between the "awareness" and "understanding" stages based on Bednarz & Ludwig's four stages in the adoption of educational innovation (Bednarz & Ludwig, 1997, Table 1). The research results also revealed that geography teachers will walk alone in the way of using GIS in their courses because they believe that (71.2%) the ministry of education doesn't support them enough in providing software and training. According to geography teachers, the limited time to teach and learn GIS, the insufficient technological infrastructure of schools, and lack of necessary knowledge of GIS and how to incorporate it into education are among the most critical restraints that prevent an effective use of GIS in courses to be mitigated.

As closing remarks, when we seek the answer to the question "Do Geographic Information Systems (GIS) move high school geography education forward in Turkey?" from a teacher's perspective, we found that the answer is quite clearly "not yet", at least "not in the short-run". However, the research results give us enough evidence to be optimistic in the long-run because:

- More than 80% of the teachers agreed or strongly agreed that GIS is an important tool for geography education,
- More than 60% of them think that a geography course without using GIS is missing, and
- More than 83% of geography teachers expressed that they want to attend any GIS training program if organized by the ministry of education.

Author details

Süleyman İncekara
Fatih University, Department of Geography, Turkey

6. References

Artvinli, E. (2009). Approaches of geography teachers to geographical information systems (GIS). *Balikesir University Journal of Social Sciences Institute*, Vol. 12, No. 22, pp. 40-57.

Audet, R.H. & Abegg, G.L. (1996). Geographic information systems: Implications for problem solving. *Journal of Research in Science Teaching*. Vol. 33, No. 1, pp. 21-45.

Baker, T. & White, S. (2003). The effects of GIS on students' attitudes, self-efficacy, and achievement in middle school science classrooms. *Journal of Geography, Vol.* 102, No. 6, pp. 243-254.

Bednarz, S.W. & Ludwig, G. (1997). Ten things higher education needs to know about GIS in primary and secondary education. *Transactions in GIS*, Vol. 2, No. 2, 123-133.

Bednarz, S.W. (2004). Geographic information systems: a tool to support geography and environmental education? *GeoJournal*, Vol. 60, No. 2, pp. 191-199.

Demirci, A. (2008). Evaluating the implementation and effectiveness of GIS-based applications in secondary school geography lessons. *American Journal of Applied Sciences*, Vol. 5, No. 3, pp. 169-178.

Demirci, A. (2009). How do teachers approach new technologies? Geography teachers' attitudes towards geographic information systems (GIS). *European Journal of Educational Studies*, Vol. 1, No. 1, pp. 57-67.

GESP (Geography Education Standards Project). (1994). *Geography for Life: National Geography Standards*. Washington, DC: National Geographic Research and Exploration.

Incekara, S. (2010). The place of geographic information systems (GIS) in the new geography curriculum of Turkey and relevant textbooks: is GIS contributing to the geography education in secondary schools? *Scientific Research and Essays*, Vol. 5, No. 6, pp. 551-559.

Incekara, S. & Karakuyu, M. (2010). How are geographic information systems (GIS) conferences at Fatih University contributing to education in Turkey? *Scientific Research and Essays*, Vol. 5, No. 19, pp. 2975-2982.

Incekara, S. (2011a). The Turkish geography teaching program (2005) and technology use in geography courses: an overview of high school teachers' approach. *Educational Research and Reviews*, Vol. 6, No. 2, pp. 235-242.

Incekara, S. (2011b). Technology use in secondary geography courses from teachers' perspective. *Buca Faculty of Education Journal* (Under review).

Johansson, T. (2003). GIS in teacher education facilitating GIS applications in secondary school geography. ScanGIS'2003 On-line Papers, pp. 285-293, 13.09.2011, Available from <http://www.scangis.org/scangis2003/papers/20.pdf>.

Johansson, T. & Pellikka, P. (2006). Geographical information systems applications for schools. *9th AGILE International Conference on Geographic Information Science. Shaping the future of Geographic Information Science in Europe*, Visegrád, Hungary, April, 2006.

Kerski, J. (2003). The implementation and effectiveness of geographic information systems technology and methods in secondary education. *Journal of Geography*, Vol. 102, No. 3, pp. 128-137.

Kerski, J. (2007). Geographi information systems in education. In: *Handbook of Geographic Information Science*, John P. Wilson, A. Steward Fotheringham, pp. 540-557, Blackwell Publishers, Singapore.

King, G.Q. (1991). Geography and GIS technology. *Journal of Geography*, Vol. 90, No. 1, pp. 66-72.

Korevaar, W. & Schee, J.V. (2004). Modern aardrijkskundeonderwijs met GIS op de kaart gezet. *Geografie*, Vol. 13, No. 9, pp. 44-46.

Lemberg, D. & Stoltman, J.P. (1999). Geography teaching and the new technologies: opportunities and challenges. *Journal of Education*, Vol. 181, No. 3, pp. 63-76.

MEB (Milli Eğitim Bakanlığı). (2005). *Coğrafya Dersi Öğretim Programı (9., 10., 11. ve 12. Sınıflar)*. Talim ve Terbiye Kurulu Başkanlığı, Ankara.

Meyer, J.W., Butterick, J., Olkin, M. & Zack, G. (1999). GIS In the K-12 curriculum: a cautionary note. *Professional Geographer*, Vol. 51, No. 4, pp. 571-578.

OFSTED (Office of Standards in Education). (2004). *ICT in schools, the impact of government initiatives: secondary geography.* May 2004, Manchester, UK.

Olgen, M.K. (2005). Türkiye'de CBS eğitimi. *Ege Coğrafi Bilgi Sistemleri Sempozyumu Bildiriler Kitabı,* İzmir, April 2005.

Ozel, A. (2007). How social science and geography teachers perceive educational technologies that have been integrated in educational program. *Journal of Applied Sciences,* Vol. 7, No. 21, pp. 3226-3233.

Palladino, S. (1994). A role of geographic information systems in the secondary schools: an assessment of the current status and future possibilities. 13.09.2011, Available from <http://www.ncgia.ucsb.edu/~spalladi/thesis/title.html>

Patterson, M.W., Reeve, K. & Page, D. (2003). Integrating geographic information systems into the secondary curricula. *Journal of Geography,* Vol. 102, No. 6, pp. 275-281.

Pottle, T. (2001). *Geography and GIS: GIS Activities for Students.* Irwin Publishing Ltd., Toronto.

Shin, E. (2006). Using geographic information system (GIS) to improve fourth graders' geographic content knowledge and map skills. *Journal of Geography,* Vol. 105, No. 3, pp. 109-120.

Sonmez, O.M., Cavus, H. & Merey, Z. (2009). The usage levels of educational technologies and materials of geography teachers. *Journal of Social Science Research,* Vol. 4, No. 2, pp. 213-228.

Tas, H.I., Ozel, A. & Demirci, A. (2007). Geography teachers's perspectives on technology and the level of utilization of technology. *Dumlupınar University Journal of Social Science,* Vol. 2007, No. 19, pp. 31-51.

Tinker, R.F. (1992). Mapware: educational applications of geographic information systems. *Journal of Science Educational and Technology,* Vol. 1, No. 1, pp. 35-48.

Van der Schee, J. (2006). Geography and new technologies, In: *Geographical Education in a Changing World: Past Experience, Current Trends and Future Challenges,* John Lidstone, Michael Williams, pp. 185-193, Springer, Netherlands.

Walsh, S.J. (1992). Spatial education and integrated hands-on training: essentials foundations of GIS instruction. *Journal of Geography,* Vol. 91, No. 2, pp. 54-61.

Wanner, S. & Kerski, J. (1999). The effectiveness of GIS in high school education. *Proceedings of the 1999 ESRI User Conference,* Colorado, USA, 1999.

Yap, L.Y., Tan, G.C.I., Zhu, X. & Wettasinghe, M.C. (2008). An assessment of the use of geographical information systems (GIS) in teaching geography in Singapore schools. *Journal of Geography,* Vol. 107, No. 2, pp. 52-60.

Yasar, O. & Seremet, M. (2009). An evaluation of changes to the secondary school geography curriculum in Turkey in 2005. *International Research in Geographical and Environmental Education,* Vol. 18, No. 3, pp. 171-184.

Yomralioglu, T. (2000). *Coğrafi Bilgi Sistemleri: Temel Kavramlar ve Uygulamalar.* Seçil Ofset, Istanbul.

Yomralioglu, T. (2002). GIS activities in Turkey. *Proceedings of International Symposium on GIS*, Istanbul, Turkey, September 2002.

Zhou, Y., Smith, B.W. & Spinelli, G. (1999). Impacts of increased student career orientation on American college geography programmes. *Journal of Geography in Higher Education*, Vol. 23, No. 2, pp. 157-165.

Chapter 4

Map Updates in a Dynamic Voronoi Data Structure

Darka Mioc, François Anton, Christopher M. Gold and Bernard Moulin

Additional information is available at the end of the chapter

1. Introduction

Within a traditional geographic information system, it is currently difficult to ask questions about data that possess positions in both space and time. Information about objects that have spatial position, spatial relationships with nearby objects, and the time of their existence needs to be stored within a computer system, making them available for queries concerning spatial locations, dates, and attributes.

In existing GIS software spatio-temporal queries cannot be answered easily; the absence of a temporal component implies that analysis of past events and future trends is difficult or impossible [37].

In traditional geographic information systems, the temporality of spatial data has been treated separately from their spatial dimensions. The lack of mechanisms for recording incremental changes poses a serious problem for the integration of temporal data inside a GIS. In current GISs, the spatial changes affect the whole map. Therefore, the "snapshot model[1]" with global changes of map states is considered to be a basis of any spatio-temporal model [6].

Nowadays, there is a growing demand from the user community for a new type of GIS that will be able to support temporal data and spatio-temporal facilities such as: spatio-temporal queries [41] and interactive spatial updates [45].

Spatio-temporal facilities would be useful in many GIS applications such as harvesting and forest planning, cadastre, urban and regional planning as well as emergency planning.

In all these fields, there is a need for a GIS technology which can manage the history of spatial objects and their evolution, and which can efficiently answer spatio-temporal queries. In forest management, there is a need for GIS technology which can manage the history of spatial objects, map versioning and spatio-temporal queries. An example is a forest inventory map that is being frequently updated with the latest information about forest roads, cut areas, fire, etc. It would be desirable to preserve all the previous states, and to be able to track the evolution of the spatial objects.

---

[1] The snapshot model is a map at a single moment in time.

While in the past, most of the approaches concentrated on the extension of existent GIS models with temporal data, recent research [8] shows that dealing with time as calendar time and mapping it onto an integer domain is feasible, but does not capture the semantics of time and leaves out most of its important properties. According to Frank [8], the understanding of how time and temporal reasoning processes are conceptually structured is a prerequisite to building support for temporal reasoning into current GISs. Frank [8] further emphasized the need for formal models to determine the representation of temporal information and temporal and spatio-temporal reasoning methods.

2. Previous research

Several authors [25], [40], [5] proposed spatio-temporal data models based on extending existing GIS models, in order to include temporal information. The problem they faced is that most commercial GISs are closed systems which cannot be extended nor modified in order to include temporal information. Therefore, the solutions they propose are based on a "dual architecture" [45], which is composed of two subsystems: a commercial GIS and a relational DBMS. Van Oosterom [46] emphasized the problem of the complexity of the maintenance of the proposed spatio-temporal data models.

There are several spatio-temporal functionalities such as retroactive map updates [20] and the incorporation of temporal data without exact temporal information [7] that cannot be handled with such hybrid models. Another problem with existing GISs, is that the semantics of map topology construction [46] are lost. This problem is better known under the term of "long transactions in GIS" [38]. This is due to complex models of spatial topology which need to be processed globally after being updated. When a map is processed in batch mode, its topology is built, and it is not possible to go back to previous states, nor to reuse the past map states, because the information about spatio-temporal objects and the operations that have been executed upon them is lost. Batch processing of spatial topology is managed through "long transactions" which can be composed of nested transactions and could last for hours or days, and system or user generated aborts will not be permitted during that time [29].

Newell and Batty [38] stated that current GISs differ from standard DBMS in that only "long transactions" (and not short ones) are possible. This restricts access to intermediate map states, which is a limitation when the concern is with local updates and temporal queries. They also state that short transactions are required for problems such as emergency planning, vehicle tracking, fault logging, and other real-time system problems.

The problem of long transactions is so acute (time consumption, database inconsistency problems, etc.), that the underlying problem is not well defined. The problem lies in the fact that "dynamic topology" [14] is not used: geometric algorithms cannot be executed interactively upon spatial objects while maintaining topological relationships. Recently, several GIS researchers (Van Oosterom [45], [46], Chrisman [6], Newel and Batty [38], Kraak [23], [24]) tackled the same problem from several different perspectives.

2.1. A fundamental problem in GIS

Recently, the research of Newell and Batty [38], Gold [14], and independently Chrisman [6] shows that "batch processing" of spatial data is a fundamental problem in commercial GISs. The "batch processing" of static spatial topology used by line intersection based spatial systems does not support mechanisms for recording incremental change and poses a serious problem for the integration of temporal data inside GISs.

The "batch processing" of spatial data cannot support incremental (local) addition and deletion of spatial objects, and they cannot support the temporal evolution of spatial data.

Briefly, we will explain now what led to the batch processing of spatial data that imposed the snapshot or time-slicing approach in GIS.

The snapshot model of map time [6] was always emphasized by cartographers, resulting in current commercial GIS models. Current GISs can represent a space evolving in time only through a series of map snapshots. The snapshot approach of traditional geographic information systems, where independent coverages are generated for each time step, cannot easily maintain the incremental changes of cartographic data evolving in space and time. The limitations of this approach include high data redundancy, due to the inability of conventional GIS models to support incremental changes, causing difficulties in the maintenance of long series of cartographic snapshots [23].

The snapshot model misses the key nature of change [6], which can be seen as a composition of events. The "time slicing" idea leading to the snapshot approach collapses many events, each of which occurred separately [6], causing difficulties in determination of spatio-temporal processes. Map snapshots tend to be created independently at specific intervals, rather than incrementally, and thus there is no preservation of topological relationships between map elements in different time slices, and no effective way of determining the continuity of existence of map elements and their neighbours between snapshots. Although a snapshot method is an important capability, this historical model represents only a single point in time, and does not reveal the sequence of events, or history of the area.

For change detection, the snapshot method uses two distinct maps [6] which are measured with some error. When the map overlay is used to compute differences between these two maps [6], the errors are confounded with the actual changes, and it is difficult to distinguish error from change when using snapshots.

2.2. Overview of recent research on spatio-temporal models

Van Oosterom proposed a new approach for handling spatio-temporal information [45], [46]. He proposed a data model in which both time and topology are consistently maintained in the database updates. His topological data structure is based on the "CHAIN" method [46], similar to the winged edge data structure[2]. He also addresses the problems of long transactions in GIS, the loss of transactional semantics and possible database inconsistency problems [45], [46]. In his approach, the history of complex map objects can be obtained

---

[2] The winged edge data structure or a polygon data structure is a way of representing a geometric graph in the plane. The winged edge data structure was proposed by Baumgart. It stores a record for each vertex, face and edge, with the edge record being the most important. The record for a vertex v stores v's coordinate position and a pointer to one arbitrary edge that is incident upon v. Likewise the record for a face f stores the name of f and a pointer to one arbitrary edge that lies on the boundary of f. The record for an edge e stores many fields including:
Pointers to the two vertex endpoints v_1 and v_2 of e.
Pointers to the two faces f_1 and f_2 on either side of e.
Pointers to the predecessor and the successor edges of e on the boundaries of f_1 and f_2 respectively. They are the so called "wing edges".
Thus, the total number of pointers needed is $|V| + |F| + 8|E|$.
Most importantly, just by following the appropriate pointers, one can traverse the boundary of a face in either clockwise or counterclockwise order in time proportional to the number of edges on the boundary. Likewise one can list all the edges incident upon a vertex v in clockwise or counterclockwise order in time proportional to their number.
Source: http://www.ugrad.cs.ca/spider/cs414/winged-edge.html

only by using spatial overlap queries, with respect to the given objects over time. Indeed, the model he proposed can be queried for historic versions only for simple object changes: it does not work for splits, joins nor more complicated spatial editing, unless spatial overlay is performed. Within his data structure, retroactive map updates are not implemented, due to possible consistency problems.

Frank [7] provides a theoretical framework for spatio-temporal reasoning based on the relative order of events[3] and he proposed the integration of ordered event structures in a GIS. In his paper titled "Different Types of 'times' in GIS" [8], he made two very interesting points that will be further explored in this section. The first point is related to the exploration of the relations[4] between temporal order relations and the order relation in the lattice of partitions in geometric space [8], which he considered a challenging research question.

The second point is related to the different spatio-temporal models that he described, which will now be mentioned briefly. Some of the models he describes are not explored (and implemented) yet in current GISs. An example of such models is the movement along a path[5], that is an interaction between temporal and spatial reasoning. Another example of spatio-temporal model that cannot be implemented within current GISs is spatial data without precise temporal information, which has been described by Frank [8]. This example occurs often in history and geology related mapping, where some of the spatio-temporal sequences of the data are missing and need to be interpolated. Further, he stated a "need for more realistic GISs" that will unify different models presently used for spatio-temporal reasoning. The "realistic GIS" [8] should be able to deal with error correction and other improvements of existing data. In other words, there is a need for retroactive updates in spatio-temporal models, and they are difficult to achieve. Gold [16] proposed another spatio-temporal model based on event ordering, which responds instantly to map construction commands given by the user, changing the state of map objects and their spatial topology, and storing all the changes in topological states. In the past few years, his research efforts showed that the Voronoi diagram offers "a more intelligent, more realistic, and semantically richer" model for spatial representation. In his model, based on the Voronoi diagram for sets of points and line segments, map topology states can be reconstructed at any time because all the operations on his data structure are reversible. The Voronoi diagram for sets of points and line segments is the generalization of the ordinary Voronoi diagram (for sets of points), where the set of sites may contain not only points, but points and line segments as well (see Figure 1).

3. The dynamic spatial Voronoi data structure

Cartographic objects on maps are composed of points, curves and surfaces. The ordinary point Voronoi diagram does not allow us to accurately model linear or areal objects (curves and surfaces) needed in many GIS applications.

---

[3] Orderings of events in which it is not known for every event if it is before or after another one are called partial orderings [8]. Completely ordered sets of events are known as totally ordered [8] with a single viewpoint of all events.

[4] *"A set with a partial order relation is called a **partially ordered set** or **poset**. Subsets of a poset may be totally ordered; this is, for example, the case for each sequence of events that apply to a single parcel (and all its predecessors). The elements in a poset can be linearly ordered, but there is more than one possible solution (topological sorting). A special case of "a poset" is a **lattice**. The intersection of the successors of two points have a single earlier point. The same concept can be applied to a family of partitions of the plane, where the partitions of different levels of subdivisions form a lattice (ordered by a "refinement" relation)."* from [8]

[5] Movement along a path is one of the facilities needed in GIS, for applications related to emergency planing, robotics, and marine GIS.

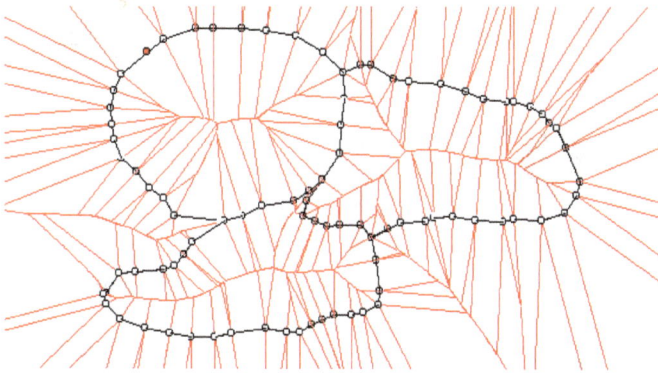

Figure 1. A line Voronoi diagram, from [33]

Therefore, in order to represent the different kinds of cartographic objects, we need to use the Voronoi diagram for sets of points and line segments instead of just the ordinary point Voronoi diagram.

Gold [11], [18] developed a dynamic spatio-temporal data structure based on the Voronoi diagram for a set of points and line segments (also known as a line Voronoi diagram, see Figure 1). His data structure is dynamic: objects may be added to and removed from the data structure. This data structure is also kinematic: objects are created by splitting the nearest point into two and then moving the newly created object to its desired location, and deleted by moving it to the nearest point location and merging it with the nearest point. In Section 3.1, we present the properties of this data structure: the dynamic spatio-temporal Voronoi data structure [33]. In Section 3.2, we develop a formalism for specifying the operations upon this data structure and their changes in topology [33], [36],[35].

The Voronoi diagram for a set of map objects (points and line segments) is the tessellation of space where each map object is assigned an influence zone (or Voronoi region), that is the set of points closer to that object than to any other object (see [39] and Figure 1).

The algorithm used to construct the Voronoi vertices has been described in [4]. The boundaries between the regions of this tessellation form a net (the Voronoi diagram), whose dual graph (the Delaunay quasi-triangulation or Delaunay graph) stores the spatial adjacency (topology) relationships among objects. Within such a dynamic Voronoi spatial data structure, as developed by Gold [12], map objects (points and/or line segments) are stored as nodes of the dual spatial adjacency (topology) graph: the Delaunay triangulation. The underlying data structure used is the Quad-Edge data structure [19].

The Quad-Edge data structure was used for computing the line Voronoi diagram [17], which is the basis of the dynamic Voronoi data structure for points and line segments. The Quad-Edge data structure was introduced by Guibas and Stolfi [19] as a primitive topological structure for the representation of any subdivision on a two-dimensional manifold. The Quad-Edge data structure is the implementation of an edge algebra [19], which is the mathematical structure that defines the topology of any pair of dual subdivisions on a two-dimensional manifold. In the context of the application of the Quad-Edge data structure to the computation of Voronoi diagrams, both a primal planar graph (the Voronoi diagram) and its dual graph (the Delaunay triangulation) are stored in the Quad-Edge data structure - see [19].

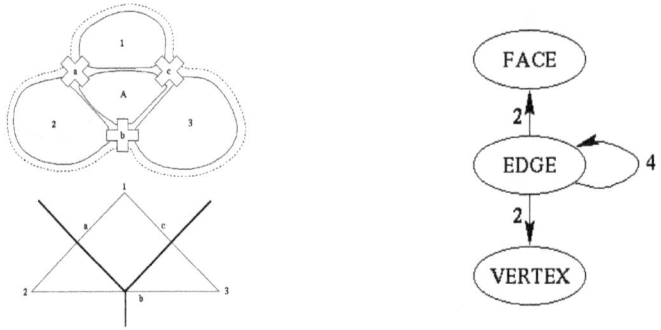

Figure 2. Left: a simple Voronoi diagram and its corresponding Quad-Edge; right: the PAN graph of the Quad-Edge data structure on the left

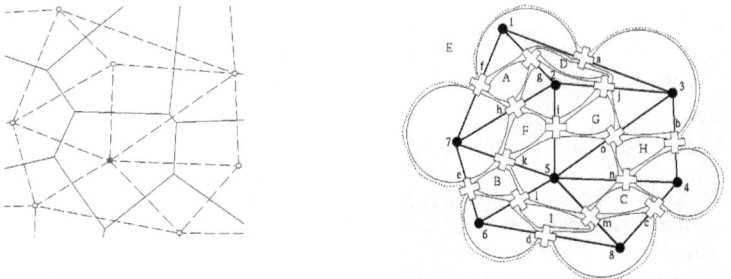

Figure 3. Left: a Voronoi diagram; right: the corresponding Quad-Edge data structure

The PAN graph [10] of the Quad-Edge data structure gives a representation more suitable within a GIS context (see Figure 2). The Quad-Edge data structure (see Figure 3) represents a graph and its geometric dual. In the context of the application of the Quad-Edge data structure to the computation of Voronoi diagrams, both a primal planar graph (the Voronoi diagram) and its dual graph (the Delaunay quasi-triangulation) are stored in the Quad-Edge data structure.

The definition of spatial adjacency relationships within the Voronoi model is the adjacency of the Voronoi regions of two objects. In such a case, this spatial adjacency relationship is stored in the dual representation of the Voronoi diagram: the Delaunay quasi-triangulation.

The main characteristic of the topology[6] within the Voronoi spatial data model which distinguishes it from other models such as the vector model, is that it does not need any other computation than the incremental construction of the Voronoi data structure. Topology in the Voronoi spatial data model is given by the fact that two objects have adjacent Voronoi regions, which is stored in the dual representation of the data structure.

The dynamic Voronoi spatial data model is based on an event-condition-action paradigm [16], which seems to provide many advantages over traditional GIS data models. The main

---

[6] Topology is the branch of mathematics, that deals with the properties of points of some space that are invariant under some continuous transformations. Topological relationships are spatial adjacency relationships.

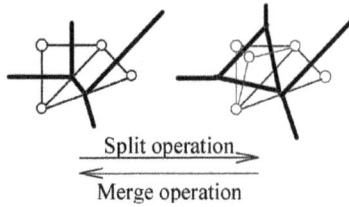

Split operation

Merge operation

Figure 4. The topological changes due to the Split and Merge operations

advantage of the dynamic Voronoi data structure is its dynamic, incremental and explicit topology, which allows one to automatically keep track of each event and change of map state [16].

The changes in this data structure are therefore the changes in the spatial adjacency relationships, that is to say the changes in the Delaunay triangulation [39]. Within this data structure, the user's commands are changing the map incrementally and locally, and the map objects and their spatial adjacency relationships are all visualized at any point in time [2].

Furthermore, this approach allows real-time dynamic maintenance of the spatial data structure, as well as dynamic sequential processing of events [16].

3.1. The atomic actions on the dynamic Voronoi data structure

These map state changes are produced by map commands [12], that are composed of atomic actions. Each atomic action in the map command executes the geometric algorithm for addition, deletion or change of map objects and corresponding Voronoi cells.

The *atomic actions* are:

- the *Split* action inserts a new point into the structure by splitting the nearest point from the pointed location into two points (see Figure 4);
- the *Merge* action deletes the selected point by merging it with its nearest neighbour (see Figure 4);
- the *Switch* action is performed when a point moves and a topological event occurs (i.e. the moving point enters or exits a circle circumscribed to a Delaunay triangle, see Figure 7), switching[7] the common boundary of two adjacent triangles (see Figure 5). On Figure 7 we can see the topological event caused by the "Switch" atomic operation.
- the *Link* action adds a line segment[8] between the points obtained after a Split action (see Figure 6). A Link action must occur after a Split action, and adds a line segment between the point selected for splitting and the newly created point.

---

[7] The *Switch* action will be used in the construction of the *Move* action. The *Move* (topological event) action moves the selected point from its current position to a new position or until the next topological event.

[8] A line segment is composed of two half-line segments, whose Voronoi regions are on each side of the line segment, having the line segment as a common boundary (see Figure 6).

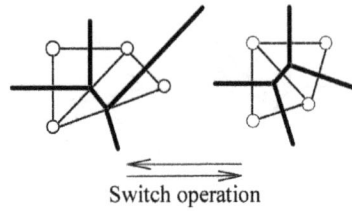

Switch operation

Figure 5. The topological changes due to the Switch operation

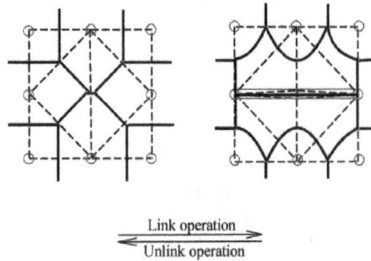

Link operation
Unlink operation

Figure 6. The topological changes due to the Link and Unlink operations

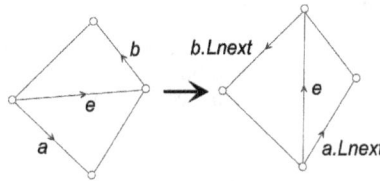

Figure 7. The topological event caused by a "swap" atomic operation

- the *Unlink* action removes the selected line segment. An Unlink action must occur before a Merge action, and removes the line segment between the selected point and its nearest object.

Figure 6 shows the succession of these atomic actions. These actions compose the set of atomic actions of the dynamic spatial Voronoi data structure [32].

3.2. Topological changes in the Voronoi data structure induced by the atomic actions

The map changes produced by the atomic actions on this data structure are the changes in the spatial adjacency relationships among spatial objects. The spatial adjacency relationships are defined as follows: two objects are Voronoi neighbours if, and only if, their Voronoi regions have one portion of the Voronoi diagram (a Voronoi edge) in common (see Figure

Atomic Operation	Symbol	Newly created Voronoi regions	Inactivated Voronoi regions
Split	S	1	0
Merge	M	0	1
Switch	N	0	0
Link	L	2	0
Unlink	U	0	2

Table 1. The atomic actions and their associated changes in topology expressed in the numbers of newly created and inactivated Voronoi regions

3). Therefore, the only map state changes of the dynamic Voronoi data structure produced by events are the changes in the Delaunay triangulation/Voronoi diagram preserved in the Quad-Edge structure. These events are ruled by the Delaunay triangulation empty circumcircle criterion (see Figure 7).

When a point comes inside a circumcircle or exits from a circumcircle - a "topological event" occurs - the boundary between the two triangles inscribed in the circumcircle "switches" [44].

Within the dynamic Voronoi spatio-temporal data model, all the operations are local and "kinematic": the addition of a new point is performed by splitting the nearest point into two and moving the newly created point to its destination; and the deletion of an existing point is performed by moving it to its nearest point and merging them. It is easy to see that the two actions described previously are mutually reversible: the reverse of a split being a merge and the reverse of a merge being a split (see Gold [12]). A Split action takes the Voronoi cell of a "parent" point and splits it into two, generating a "child" point that may then be moved to the desired destination. A Merge action reverses this process, combining two adjacent cells into one.

Each *atomic action* produces different changes in spatial topology. The possible changes are:

- the triangle switches (topological events) changing the corresponding Voronoi edges (see Figure 7),
- the creation of a new map object (point or line) and the corresponding appearance of its Voronoi region,
- the inactivation of a map object and the corresponding disappearance of its Voronoi region. Objects and spatial adjacency relationships are not removed, but inactivated, in order to be able to record all the history information.

The atomic actions of the dynamic spatio-temporal Voronoi data structure, their reverse actions, and their corresponding changes in topology are described in Tables 1 and 2. We also introduce in Tables 1 and 2 the symbols for each atomic action (N, S, M, L and U) that will be used latter for specifications of complex map operations. The topological changes for each atomic action (see Table 2) in the map are represented by the numbers of newly created and inactivated spatial adjacency links (i.e. Voronoi edges or Quad-Edges). Each atomic action is uniquely characterized by the numbers of new Quad-Edge (or Voronoi) edges and inactivated Quad-Edge (or Voronoi) edges (see Table 2). This means that from changes in topology we can determine which atomic action was applied, and vice versa. In other words, the actions on the data structure have a deterministic behaviour. More precisely, we can say that the set

Atomic Operation	Symbol	New Edges	Inactivated Edges
Split	S	6	3
Merge	M	3	6
Switch	N	1	1
Link	L	11	5
Unlink	U	5	11

Table 2. The atomic actions and their associated changes in topology

of atomic actions is naturally isomorphic to the set of the number of new edges as well as to the set of the number of inactivated edges.

3.3. The map construction commands

The *atomic actions* are the basis upon which *map commands* have been built. All the map construction commands [12] of this dynamic Voronoi data structure are complex operations composed of atomic actions (illustrated in Figures 5, 4 and 6). The composition of atomic actions into map commands is provided by syntactic rules. The meaning of the word "syntax" is based on the theory of formal languages and grammars [22]. In the theory of formal languages, the semantics of the basic operations that can be applied on the set of objects is described by a grammar. A grammar provides a set of rules, known as production rules, specifying how the sequence of atomic actions will be applied to the elemental map object (currently a point or a line segment).

Rewriting is a useful technique for defining complex objects by successively replacing parts of elemental map objects using a set of rewrite rules or production rules [42]. Given a set of productions we can generate an infinite number of map objects [1]. In the Voronoi spatial data system there is more than one rule we can apply, and the user is given the freedom of selecting the production rule appropriate to the map update needed. Rewriting context is extended from topological context to include geometrical (spatial) context[9] (position). Therefore rewriting is done sequentially at the specific locations selected by the user or given by coordinates. Thus, the update of the Voronoi data structure given by map commands can be interpreted as the execution of production rules which constitute a map grammar [42]. Graph-grammars may be used as a natural and well-established syntax-definition formalism [43] for languages of spatial relationships graphs. The map grammar shows the hierarchical presentation of the production rules and spatial objects. For example, the move command is a part of any other production rule described as a map command.

The *map construction commands* are illustrated in Figure 8. On the left side of Figure 8, we can see the map objects on which the map command will be applied, and on the right side, the map objects that have been rewritten. In the graphical illustration of map commands, the topological part of the model is left out for better understanding of the general principle, and

---

[9] In context sensitive systems the selection of the production rule is based on the context of the predecessor. A context sensitive system is needed to model information exchange between neighboring elements [47].

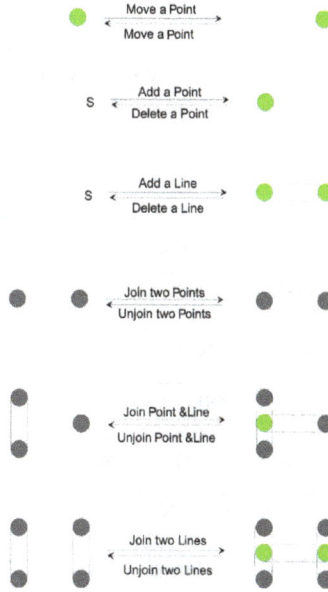

Figure 8. The map commands (S is the starting symbol in the production rule

also the line-line collisions and their effects are not shown. The detailed description together with the graphical illustrations will be presented in Section 4.

The *map commands* (see Figure 8) are composed of atomic operations, and the exact decomposition of map commands into sequences of atomic actions is given in Table 3. The atomic operations are denoted by the symbols (N, S, M, L and U) from Table 2. For example, the map command *"Move a Point"* corresponds to the sequence of movements of the point from its initial position to its destination through all the intersections of its trajectory with circumcircles, and the corresponding triangle switches ("N") in the Voronoi data structure.

The map command *"Move a Point"* is possible in this Voronoi data structure because the Voronoi data structure is kinematic: one point may move at a time, and this point is called the *"moving point"* [14]. In fact, all the operations on this kinematic Voronoi data structure use this concept of the moving point. For example, when a point is to be created at some location, the nearest point from that location is split into two (S term in the decomposition of *"Add a Point"* operation SN^t), and then the newly created point is moved to its final destination (N^t term in the decomposition of *"Add a Point"* operation SN^t). In fact, the triangle switch operation incorporates the movement of the moving point to the intersection of the trajectory of the moving point with the circumcircle that induced the triangle switch. In Table 3, the exponents denote how many times the operation is executed repeatedly, e.g. N^t denotes N executed t times, where t denotes the number of topological events. Whenever more than one connected sequence of topological events is executed in a map command, such as in the *"Add a Line"* command $(SN^{t_1} \ SLN^{t_2} \ (SLN^{t_{2i+1}} \ MSLN^{t_{2i+2}}))$, the total number of topological events is broken down into the number of topological events in the first connected sequence (N^{t_1}), the number

Map construction command	Decomposition (the terms in parentheses appear at each line-line collision, i = collision index, $i \in \{1,...,c\}$, c = number of collisions; t, t_x denote numbers of topological events
Move a Point	N^t
Add a Point	SN^t
Delete a Point	$N^t M$
Add a Line	$SN^{t_1} SLN^{t_2} \left(SLN^{t_{2i+1}} MSLN^{t_{2i+2}} \right)$
Delete a Line	$\left(N^{t_{2i+2}} UMSN^{t_{2i+1}} UM \right) N^{t_2} UMN^{t_1} M$
Join 2 Points	$SLN^{t_1} \left(SLN^{t_{2i}} MSLN^{t_{2i+1}} \right) M$
Unjoin 2 Points	$\left(N^{t_{2i+1}} UMSN^{t_{2i}} UM \right) N^{t_1} UM$
Join Pt & Line	$SLN^{t_1} \left(SLN^{t_{2i+1}} MSLN^{t_{2i+2}} \right) SLN^{t_2} M$
Unjoin Pt & Line	$SN^{t_2} UM \left(N^{t_{2i+2}} UMSN^{t_{2i+1}} UM \right) N^{t_1} UM$
Join 2 Lines	$SLN^{t_1} SLN^{t_2} \left(SLN^{t_{2i+2}} MSLN^{t_{2i+3}} \right) SLN^{t_3} M$
Unjoin 2 Lines	$SN^{t_3} UM \left(N^{t_{2i+3}} UMSN^{t_{2i+2}} UM \right) N^{t_2} UMN^{t_1} UM$

Table 3. The map commands and their decomposition into atomic actions

of topological events in the second connected sequence (N^{t_2}), and so on. The parameter i denotes the number of times the line segment being added has already intersected existing line segments. This type of intersection with an existing line segment is called a collision, and i is called the collision index. The terms in parentheses are repeated for each intersection with an existing line (i.e. each collision).

We will now briefly explain the decomposition of each map command. We have already seen the description of *"Move a Point"* and *"Add a Point"* map commands. Map command *"Delete a Point"* is exactly the reverse of *"Add a Point"* map command: the point to be deleted is moved to the location of the nearest point (N^t), and then they are merged with this nearest point (M).

The remaining map commands involve the addition or removal of one or more new line segments. For all these map commands, the decomposition includes a fixed sequence of atomic actions that is executed only once (the sequence outside the parenthesis), and a sequence that is executed at each collision (replicating sequence).

In the case of *"Add a Line"* and all the join map commands, the replicating sequence has always the same pattern in terms of atomic operations $((SLN^{t_{2i+1}} MSLN^{t_{2i+2}})$, although the actual indices may vary). This corresponds to the splitting of the existing line $(SLN^{t_{2i+1}})$, the merging of the newly created point (by the S atomic action in this last sequence) with the extremity of the line segment being added (M), and the continuation of the new line segment after collision ($SLN^{t_{2i+2}}$).

In the case of *"Delete a Line"* and all the unjoin map commands, the replicating sequence has always the same pattern in terms of atomic operations (($N^{t_{2i+2}}UMSN^{t_{2i+1}}UM$), although the actual indices may vary). This is exactly the reverse of the previous replicating sequence.

Now, we will explain the fixed sequence for all these map commands. In order to *"Add a Line"*, the nearest point to the starting extremity location has to be split into two (S), then it has to be moved to the starting extremity location (N^{t_1}). Then, the ending extremity has to be created by splitting the starting extremity into two (S). At this point the two extremities must be linked (L) in order to form a line segment. Finally, the ending extremity has to be moved (N^{t_2}) to its expected location. The sequence for *"Delete a Line"* is exactly the reverse of the preceding sequence.

In order to join two points with the *"Join two points"* map command, the first point must be split into two (S) in order to create the ending extremity of the line segment that starts at the first point. Then, these two points must be linked (L) in order to form a line segment. Then, the ending extremity must be moved (N^{t_1}) to the location of the second point (including eventually the replicating sequence in case of collisions). Finally, the ending extremity must be merged with the second point (M).

The sequence for the *"Unjoin two points"* map command is exactly the reverse of the sequence for *"Join two points"* map command. The sequences of the remaining map commands follow immediately from the sequence of the *"Join two points"* map command. Indeed, the other join map commands fixed sequence involve several sequences corresponding to the same atomic actions as the SLN^{t_1} sequence already encountered in the fixed sequence of *"Join two points"* map command. The unjoin map commands are the exact reverse of their join counterpart.

3.4. Reversibility of the map commands in the dynamic spatio-temporal Voronoi data structure

For each map command, the reverse map command is composed of reverse atomic actions in exactly the reverse order [34]. Due to the local scope of its spatio-temporal topology, all the atomic actions of the dynamic Voronoi spatio-temporal model are reversible. Indeed, each atomic action has its reverse atomic action shown in the Table 4. The consequence of the property of reversibility of the atomic actions inside the Voronoi dynamic data structure is that a sequence of atomic actions applied in a map construction command can be reconstructed from the predecessor and successor map states. This proves in another way that the atomic actions are reversible: the input can be deduced from the output; or, in other words, computation happens without any loss of information [9].

The resulting complex operations (map commands) are reversible (see Figure 8 and Table 5), as long as their decomposition into atomic actions is exactly known (including the numbers of topological events and the number of line-line collisions).

The reversibility of the addition and deletion of intersecting line segments has been studied in [3]. This strictly showed that in order to perform backwards visualization through reverse execution, we need to access the sequences of atomic actions stored in a log file. This can become cumbersome due to the potentially large number of line-line collisions and of their

Atomic action	Reverse atomic action
Split	Merge
Switch	Switch is self-reversible
Link	Unlink

Table 4. The reversibility of the atomic actions

Map construction command	Reverse map construction command
Move a Point	Self-reversible
Add a Point	Delete a Point
Add a Line	Delete a Line
Join 2 Points	Unjoin 2 Points
Join Pt & Line	Unjoin Pt & Line
Join 2 Lines	Unjoin 2 Lines

Table 5. The reversibility of the map commands

associated replicating sequence (see terms in parenthesis in Table 3 and [32]). Moreover, if we want to perform spatio-temporal queries or spatio-temporal analysis, we need to access the previous map states. However, the undoing and redoing of atomic actions consume large amounts of time and storage and are not spatially localized. This is the consequence of the total ordering of the map construction events, that does not allow spatially local redoing or undoing map commands. Therefore, in order to avoid these problems, we keep a spatio-temporal structure which captures the execution traces of map commands in the dynamic Voronoi data structure [32]. This spatio-temporal structure, along with the reversibility property of the atomic actions, guarantees exact reverse execution of map construction.

3.5. Logging the transactions

The incremental updates (given by the user) are recorded in the log file. Transactions are logged at two levels: the log file that stores map command transactions, and the history of the map (the hierarchical Voronoi data structure) that is logging the changes in topology as newly created, deleted and modified Voronoi regions. The operations recorded in a log file have a direct translation into the spatio-temporal topology as the hierarchical Voronoi data structure. Temporal (parent-child, or modification) links between two update levels correspond to the applied (executed) map commands (see Figure 9).

In the next subsection we will briefly introduce a theoretical approach needed for the execution of map commands or transaction processing.

3.6. Trace systems

Execution traces [28] give us a formalism to represent map history at a more abstract level. The theory of traces established by Mazurkiewicz [28] deals with the transaction processing in concurrent systems. Trace systems are special in that they enable both very detailed modelling and also offer opportunities for abstraction. In order to specify a trace system one has to specify an underlying spatial language and a dependency relation [28].

The trace system that represents the history of the map is maintained as a hierarchy of map objects and their ancestral dependency relationships [27]. Hence for the corresponding topological changes in the Quad-Edge data structure, a dependency relationship occurs when one atomic action uses the results of another: e.g. a Link action always follows a Split action, and it uses both the point that has been split and the point from which it occurred [32].

4. Map updates and map history

The user incrementally constructs or updates a map by giving map commands. A typical example could be updating a forest map to show the previous year's clear-cuts. Each map command is a sequence of atomic actions which will be executed on the Voronoi diagram, producing changes in spatial topology. All the state changes produced by map commands (induced by past map events) are permanently stored in the spatio-temporal data structure as levels of map updates [32].

The left side of Figure 9 shows the effects of a map update *"Add a Point P_6"* on a small portion of a map consisting of several line segments and points, illustrating the growth of the Voronoi diagram and of the Delaunay triangulation. The right side of Figure 9 shows the corresponding history structure for the map update shown in the left side, as described in detail below. The parent-child links (represented in Figure 9 by thick solid straight lines) allow us to know from which point a point has been split, or with which point a point has been merged. The connectivity links (represented in Figure 9 by solid curved lines) allow us to know if a given point has been linked to another point, and if it is the case, with which point, and through which half-line segments. The parent-child links associated with the creation of the two endpoints allow us to know the direction in which the digitizing of the line has been done between the two endpoints. The modification links (represented in Figure 9 by light solid straight lines) allow us to know if the Voronoi cell of a given object has been changed during a given map update (map command). The spatial adjacency links (corresponding to edges in the Delaunay quasi-triangulation) are drawn in dashed lines on both sides of Figure 9.

The map construction commands have a direct translation into a spatio-temporal topology, as the hierarchy of map objects and their corresponding Voronoi cells that have been added, moved, or inactivated over time.

Thus, the data structure encodes the history of map construction at two different levels: as spatial topology (or neighbouring relations), and as temporal topology (temporal adjacency relationships; see [32] also shown on the Figure 9 as the vertical links). Temporal topology is maintained through temporal or history links, which are the parent-child relationship links for the objects that have been split or merged, and modification links for the objects whose Voronoi regions have been changed. In Figure 9 the execution of the map command *"Add a Point"* is shown. We can see the addition of the Point P_6 to the Voronoi diagram. The

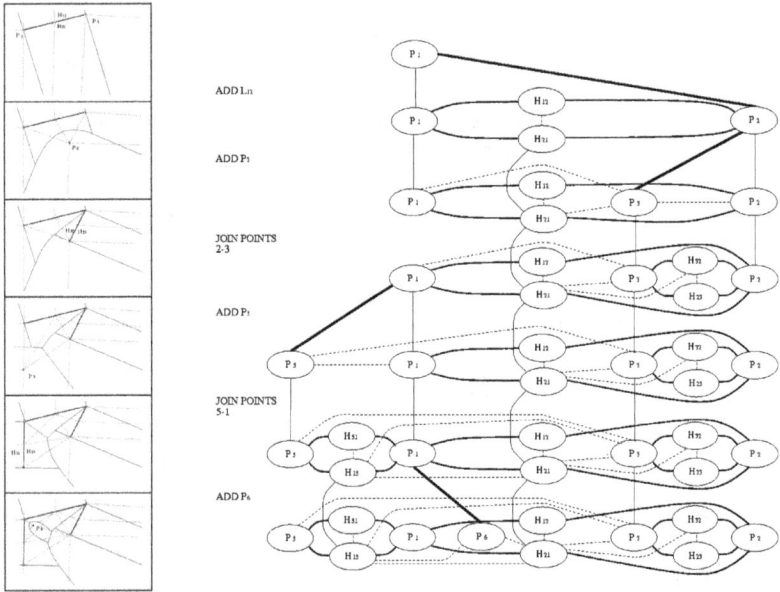

Figure 9. A sequence of map construction commands and the history of the map

Predecessor and the Successor map state are shown (as two update levels) together with the changes in the spatio-temporal topology. Each application of map commands is time-stamped and represented as an update level in our spatio-temporal structure. The hierarchical Quad-Edge data structure stores all the states of the Quad-Edge data structure (spatial topology states) and the connections between the update levels (the temporal topology in the Quad-Edge data structure). These connections between update levels correspond to the temporal topology, while the relationships in an update level correspond to spatial topology. These connections between update levels are the vertical component of the hierarchical Quad-Edge data structure.

5. Hierarchical Voronoi data structure

Ancestral dependency relationships are defined as timed division hierarchies of elemental map objects and their corresponding Voronoi regions, which are ordered by the dependency of one atomic action upon another.

In event-driven systems, such dependency relationships are isomorphic to event structures [48]. Event structures may be seen as a generalization of such structures [48]. We can say that the hierarchical Voronoi data structure is equivalent to an event structure. In event-driven systems the ordering of event occurrences is partial: there is no means to decide which of several independent events occurs first [28]. The only way to establish objective ordering of event occurrences is to find their mutual causal dependencies and to agree that a cause must always occur earlier than its effect [28]. Therefore, the dependency (or independency) of event occurrences should be a basis for the behaviour description of the event driven system [28].

Dependency graphs are a way of visualizing complex relationships [26]. In the following example, we can see update levels corresponding to the map updates shown in Figure 9, and their dependency links. Real-time sequencing of user-invoked events produces a temporal ordering of spatial objects inside the dynamic Voronoi data structure. Inside each map update level, the numerical order of the IDs of the objects corresponds to the temporal order of the commands that generated them. The temporal ordering in the hierarchical Voronoi data structure is maintained through history links (see Figure 9: parent-child links corresponding to the origin/destination of a Split/Merge or Link/Unlink action, and modification links representing a change in the Voronoi region of an object).

This temporal order has direct implications for the reversibility of the Voronoi data structure. Inside each map update level, it is possible to perform reverse execution without accessing a log file following the correspondence between the temporal ordering (maintained through the temporal topology links in Figure 9) of the atomic actions and the numerical ordering of the objects IDs (resulted from the decomposition sequences of Table 3).

Thus, the spatio-temporal model combines events and corresponding state changes in topology. Event structures together with "semantics of change" are essential for spatio-temporal reasoning and answering spatio-temporal queries.

This formal model for spatio-temporal change representation is used to develop the hierarchical Voronoi data structure (hierarchy of map objects ordered by their ancestral dependency relationships) suited for imprecise temporal data representation and spatio-temporal reasoning in the ordered event structures[10].

6. Spatio-temporal change representation

The lack of theory and formalisms for spatio-temporal change representation is a serious problem in research in spatio-temporal GIS [21]. One of the problems is related to the lack of the incremental map updates in current GISs [6]. In the Voronoi spatial data structures the clear specification of map updates (presented in the previous chapter) leads to a method for spatio-temporal change representation.

The formalism for representation of spatio-temporal changes in a dynamic Voronoi data structure and the method for map updates is based on the topological (or structural) properties of the line Voronoi diagram (see [16] and Figure 1). The map updates produce changes in spatio-temporal topology that are different for each map command and can be expressed using the theory of numbers as the structural topology changes.

The corresponding changes in topology are described in Tables 6 and 7. In Table 6, the corresponding "state changes" for each map command are represented as the number of newly created and inactivated Voronoi cells. We can observe that except for the "Move a Point" map command, the formulas for the numbers of inactivated Voronoi regions and newly created Voronoi regions generate couples of numbers, which pertain to different couples of residual classes modulo the prime number 5. The corresponding couple of numbers of

---

[10] There are significant implications of the temporal ordering of map construction events in the Voronoi data structure. The model has an implicit time ordering of events, visible through changes in topology. The dynamic Voronoi spatial data structure can support temporal data without precise temporal information. The changes in the spatio-temporal data structure capture the temporal and spatial semantics.

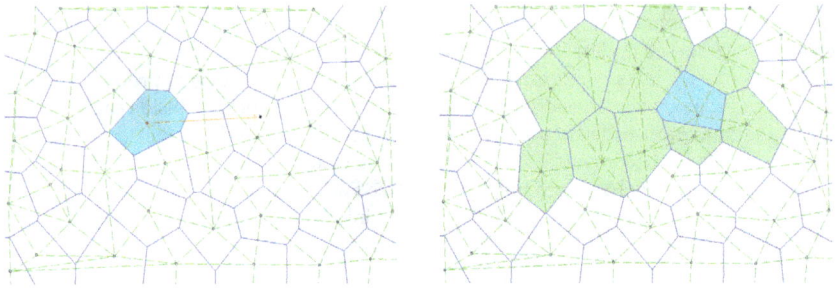

Figure 10. Predecessor and successor map states for a "Move a point" map command

inactivated Voronoi regions and newly created ones for the *"Move a point"* map command is $(0,0)$, which does not pertain to any other corresponding couple of numbers for other map commands. Therefore, we have an isomorphism between the set of map commands and the set of couples of numbers of inactivated Voronoi regions and of newly created ones. In the following figures the predecessor and successor map topology states for each map command are presented, together with the spatio-temporal changes that occurred, the appearance or disappearance of the Voronoi regions, as well as the modification of the neighbouring Voronoi cells. The newly inserted regions are displayed in dark gray, and the modified Voronoi regions in light gray. In Figure 10, the result of the map command *"Move a Point"* is shown. Figure 11 shows the effect of a map command *"Add a Point"*. Figure 12 shows the addition of a line segment to the Voronoi diagram, by an *"Add a Line"* map command. The last three sets of map commands show all possible combinations of the joining of different objects, points and line segments. Firstly, we can see the result of a map command *"Join two Points"* in Figure 13. Then, Figure 14 shows the effects of a map command *"Join Point and Line"*. Finally, Figure 15 shows the result of a map command *"Join two Lines"*.

Moreover, this isomorphism gives rise to a discrimination of map commands, that allows one to determine the number of line-line collisions that occurred in a given map update. The discrimination just described allows us to determine which map commands were applied just by knowing the predecessor and successor map topology states expressed in the number of newly created Voronoi regions and of inactivated ones (see Table 6). In formal terms, the connections between update levels are formally described by the surjective homomorphism from the Cartesian product (D) of the set of the numbers of new Voronoi regions by the set of the numbers of inactivated Voronoi regions, to the set of map commands.

In Table 7, the corresponding "state changes" for each map command are represented as the difference between the predecessor state and the successor state, expressed as the difference between the numbers of inactivated and newly created Voronoi edges. These state changes take into account the number c of intersections (between the newly created line segment and any existing objects) that occurred in the execution of the map command. When a moving point of a newly added line segment enters the last circumcircle of the line segment - this situation is named "the insertion context" after [31] and [30] the mutual splitting of the line segments occurs. The mutual splitting of line segments operation is visible as a replicating sequence in Table 3. Terms in parentheses are repeated for each line intersection (see Figure 16) detected in drawing the new line specified in the map command. When the first intersection

Map construction command	Inactivated Voronoi regions	Newly created Voronoi regions
	c = number of line intersections	
Move a Point	0	0
Add a Point	0	1
Delete a Point	1	0
Add a Line	0	$4 + 5c$
Delete a Line	$4 + 5c$	0
Join 2 Points	0	$2 + 5c$
Unjoin 2 Points	$2 + 5c$	0
Join Point & Line	0	$5 + 5c$
Unjoin Point & Line	$5 + 5c$	0
Join 2 Lines	0	$8 + 5c$
Unjoin 2 Lines	$8 + 5c$	0

Table 6. The changes induced by map commands in Voronoi regions

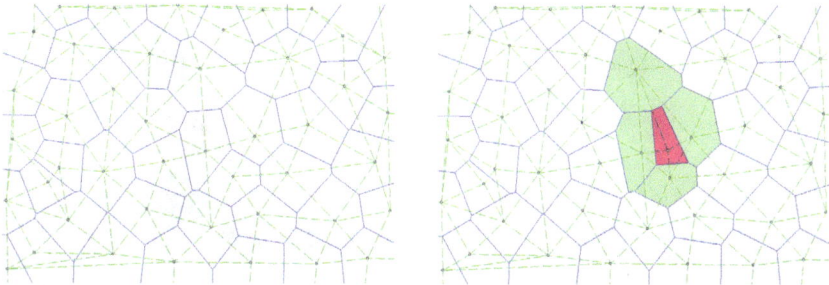

Figure 11. Predecessor and successor map states for an "Add a point" map command

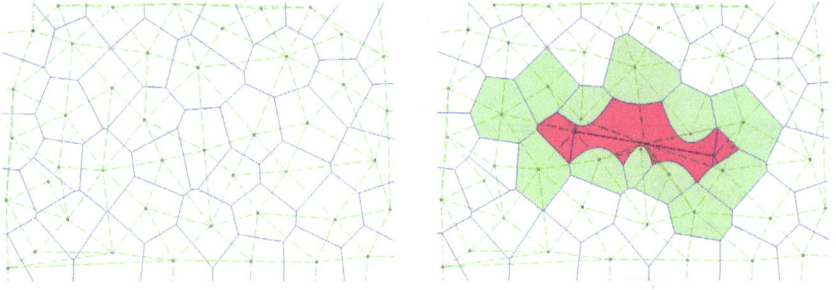

Figure 12. Predecessor and successor map states for an "Add a line" map command

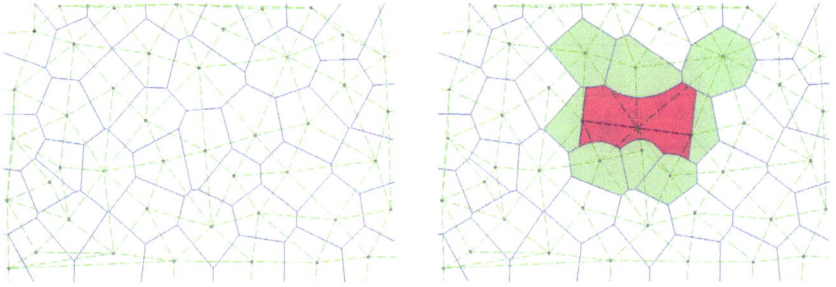

Figure 13. Predecessor and successor map states for a "Join two points" map command

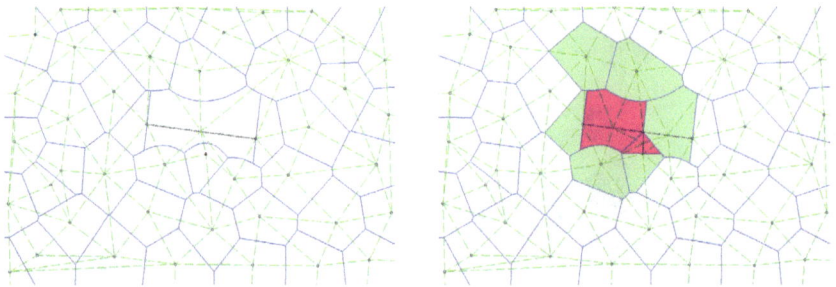

Figure 14. Predecessor and successor map states for a "Join point and line" map command

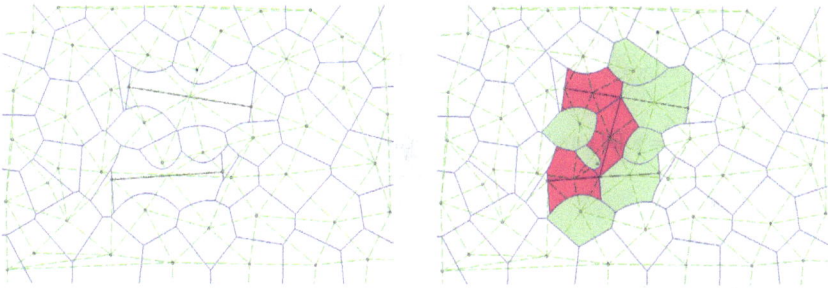

Figure 15. Predecessor and successor map states for "Join two lines" map command

(also called a collision) happens, the collision index is 1. Each time an intersection happens, the collision index is incremented. The intersections are computed incrementally, following the trajectory of the "moving point". In the "Add a Line" command, when the first intersection happens the sequence $(SLN^{t^{2i+1}} MSLN^{t^{2i+2}})=(SLN^{t^3} MSLN^{t^4})$ is added. Then, at the next intersection, the sequence $(SLN^{t^5} MSLN^{t^6})$ is added.

The language theoretic aspects of replicating sequences can be found in [31]. Longer and longer sequences [30] are produced by new map updates, only rearranging and replicating already existing information. Here, we can see that replication results in a growth of the sequences of the atomic actions. The sequence length is growing linearly. The replication mode is deterministic, therefore we have a growth function associated with a replicating system.

The map commands can be recognized by the changes between the predecessor and successor map topology states, expressed by the difference between the numbers of newly created Voronoi edges and of inactivated Voronoi edges (see Table 7), and vice versa. Indeed, the numbers generated by the formulas of differences (in the third column) pertain to sets (in the fourth column) which are mutually exclusive. Moreover, this discrimination allows us to determine the number of line-line collisions, and then the number of topological events that occurred in a map update (see example treated below). Mathematically speaking, there is an isomorphism between the set of map commands and the set of sets (in the fourth column) of possible corresponding changes between the predecessor and successor map topology states.

In the following example illustrated on the Figure 16, we will see the changes in topology induced by the map update. On the left Figure 16 the one line segment is shown and on the right Figure 16 the execution of the new map command is displayed. We can clearly see that the nine new Voronoi regions appeared in this map update level. From that result, we can determine (from Table 6) the number of line-line collisions that occurred in the map update: $9 \equiv 4 \mod (5)$. This implies that the update was *"Add a line"* and therefore we get the final result: $4 + 5c = 9 \Rightarrow c = 1$. We can see that the difference in the numbers of newly created and of inactivated Voronoi edges is 27. From this result ($27 \equiv 12 (15), 12 + 15c = 27 \Rightarrow c = 1$), we arrive (see Table 7) at the same conclusion: the map update corresponds to the addition of a new line segment intersecting one existing line segment.

Knowing the numbers of new Voronoi edges and of inactivated Voronoi edges, we can determine the exact number of topological events: $t + 23 + 37$ equals the number of new

Map construction command	New Voronoi Edges	Inactivated Voronoi Edges	Difference New - Inactiv.	Discrimination (set of the numbers corresponding to the difference New - Inactivated Voronoi Edges)
	t = total number of topological events			
Move a Point	t	t	0	$\{0\}$
Add a Point	$t+6$	$t+3$	3	$\{3\}$
Delete a Point	$t+3$	$t+6$	-3	$\{-3\}$
Add a Line	$t+23+37c$	$t+11+22c$	$12+15c$	$\left\{ \begin{array}{c} z \in \mathbb{Z}, z \equiv 12 \bmod 15 \\ \wedge z \geq 12 \end{array} \right\}$
Delete a Line	$t+11+22c$	$t+23+37c$	$-12-15c$	$\left\{ \begin{array}{c} z \in \mathbb{Z}, z \equiv 3 \bmod 15 \\ \wedge z < -12 \end{array} \right\}$
Join 2 Points	$t+20+37c$	$t+14+22c$	$6+15c$	$\left\{ \begin{array}{c} z \in \mathbb{Z}, z \equiv 6 \bmod 15 \\ \wedge z \geq 6 \end{array} \right\}$
Unjoin 2 Points	$t+14+22c$	$t+20+37c$	$-6-15c$	$\left\{ \begin{array}{c} z \in \mathbb{Z}, z \equiv 9 \bmod 15 \\ \wedge z < -6 \end{array} \right\}$
Join Pt & Line	$t+37+37c$	$t+22+22c$	$15+15c$	$\left\{ \begin{array}{c} z \in \mathbb{Z}, z \equiv 0 \bmod 15 \\ \wedge z \geq 15 \end{array} \right\}$
Unjoin Pt & Line	$t+22+22c$	$t+37+37c$	$-15-15c$	$\left\{ \begin{array}{c} z \in \mathbb{Z}, z \equiv 0 \bmod 15 \\ \wedge z < -15 \end{array} \right\}$
Join 2 Lines	$t+54+37c$	$t+30+22c$	$24+15c$	$\left\{ \begin{array}{c} z \in \mathbb{Z}, z \equiv 9 \bmod 15 \\ \wedge z \geq 24 \end{array} \right\}$
Unjoin 2 Lines	$t+30+22c$	$t+54+37c$	$-24-15c$	$\left\{ \begin{array}{c} z \in \mathbb{Z}, z \equiv 6 \bmod 15 \\ \wedge z < -24 \end{array} \right\}$

Table 7. The discrimination of map commands by means of their changes in topology, from [32]

Voronoi or Quad-Edge edges, (see Table 7). Therefore we know that the decomposition of the map command in atomic actions has the following form: $SN^{t_1}SLN^{t_2}\left(SLN^{t_3}MSLN^{t_4}\right)$ where $t_1 + t_2 + t_3 + t_4$ is the known total number of topological events.

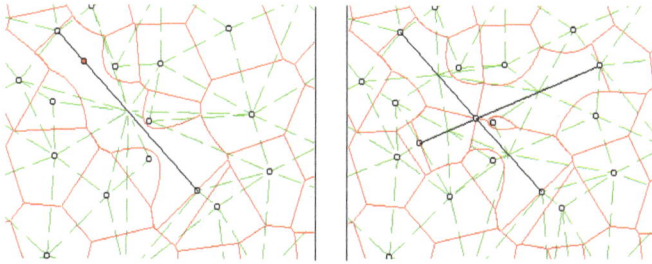

Figure 16. Mutual splitting of line segments

6.1. Retroactive map updates

For some cartographic applications users need to be able to access and update not only the current map, but also the past map states. In the case of retroactive map updates, all the map commands that happened after the date of the beginning of the retroactive map update are undone using reverse execution, returning to the corresponding starting state of the map.

Indeed, in the case of retroactive map updates, all the changes that occurred after the beginning of the retroactive map update are undone, and then the retroactive map update is performed, and the spatio-temporal data structure is updated. This is of particular interest in the cadastre due to the fact that land marking and division operations are rarely inserted in a cadastre information system at the same time as they are officially registered. In most cases the update time ordering does not match the official time ordering [7]. Moreover, the semantics of such operations (especially land marking) cannot be retrieved by comparing the cadastre maps before and after they occurred. Often, the documents and/or the landmarks disappear, making any operation much harder and more subjective. In this particular case of land marking, in order to retrieve the semantics we need to be able to reconstruct the sequence of operations that were performed together with their spatial adjacency relationships. In other words, we need to retrieve their spatio-temporal topology. The parent-child links and the modification links described above (see Figure 9) allow us to retrieve the temporal topology links between these operations, while the spatial adjacency links allow us to retrieve the spatial topology links between these operations.

Further benefits of the **formal model for spatio-temporal change representation** presented in the previous chapter include retroactive map updates. Even though there are some other methods, such as transaction logging, to keep track of the operations that have been applied, these methods have several drawbacks in the case of retroactive map updates. For example, for map updates in the past, an additional "audit file" needs to be maintained and updated with the exact sequence of atomic actions, while in our approach the "semantics of change" is simply maintained as the difference in the number of Voronoi edges between two update levels.

The theoretical work on the formalization of map update operations and the hierarchical Voronoi based spatio-temporal data structure has led to the three different methods for retroactive map updates presented here:

- Recomputation for a portion of the hierarchical data structure with retroactive updates. This method seems to be complex to implement, because it involves maintaining a large number of history links during the update operation. It could demand very complex algorithms for structure maintenance during the updates as well, due to the necessity for resolving potential spatio-temporal inconsistencies or conflicts[11]

- The second method for retroactive map updates is based on reversibility. The reversibility of map commands gives us a unique tool for another method for retroactive map updates. By exploiting reversibility the structure can be first undone until the moment in the past where the retroactive update has to take a place, and then redone up to the present. Here the efficiency of such retroactive update depends on how far in the past the user has to make the changes. Therefore for updates that need to access deep in the past this approach is not efficient enough. This gives rise to the third model for map updates that is based on the recomputation from the "log file".

- Within this kinematic, geometrical Voronoi structure [15], the map history is difficult to update locally. Indeed, for many updates, it could be a complex and time consuming task, because addition or removal of each point has to be recomputed up to the surface of the map. Therefore, for complex updates, the sequence (log file) could be updated, and the whole map with history recomputed! Therefore, the better solution is to rebuild the whole structure. Complete rebuild is also needed in the case of multiple updates and retroactive map updates from multiple sources.

The last method for retroactive map updates is the one that is most frequently applied in data structures. It is based on rebuilding the data structure from the list of entries.

7. Conclusions

In this research, a new spatio-temporal model based on a dynamic Voronoi data structure for points and line segments is presented. The approach is based on local changes in topology induced by spatio-temporal map updates. These map updates are performed through map construction commands that are composed of atomic actions on the dynamic Voronoi data structure. Even though previous research [13] on the dynamic Voronoi data structure for points and line segments describes the set of atomic actions and map commands used for the construction of the map based on the Voronoi diagram, those atomic actions and map commands were not formalized in a proper way. They were not deterministic and as a consequence not reversible.

This research succeeded in the formalization of the operations needed for constructing a Voronoi diagram for points and line segments, and of the corresponding topological changes. These operations are formalized at the lowest level, as the basic algorithms for addition, deletion and moving of spatial objects in the Quad-Edge data structure; defined as the atomic actions. Furthermore the map commands that are composed of these atomic actions are defined as well.

---

[11] **Spatio-temporal conflicts and inconsistencies caused by updates**
One of the problems arising from spatio-temporal updates in the Voronoi data model is that if we are deleting, creating or moving an object in the past, we are creating spatio-temporal inconsistencies in the present. This changed object is now inconsistent with the part of the map history which was previously rolled back, and now needs to be rolled forward to modify the present state accordingly.
The fact that the chronology of the events could be different after the change may produce spatio-temporal conflicts. Deletion of an object in the past do not remove its activities - there are a lot of links and interactions with other objects which have been recorded in the map history, and they should be resolved and recomputed up to the present state of map history and the log file.
The same conflict arises with the creation of a new spatial object - changes will affect its neighbours.

This research shows that the result of the formalization of the operations on the dynamic Voronoi data structure is a spatial language or a map grammar that is deterministic and reversible.

It was shown that the behaviour of the basic map operations is deterministic, and well defined in terms of topology changes.

Furthermore, a formal model for spatio-temporal change representation has been defined, where each map update is uniquely characterized by the number of newly created Voronoi regions. The recognition of map commands from the corresponding changes in spatial topology allows us to extend our data structure towards the hierarchical Voronoi data structure (hierarchy of map objects ordered by their ancestral dependency relationships). The hierarchical Voronoi data structure is well suited for spatio-temporal data representation even in the case of imprecise temporal data, and spatio-temporal reasoning in ordered event structures. The model has an implicit time ordering of events, visible through changes in topology. The changes in the spatio-temporal data structure capture the temporal and spatial semantics.

This research has shown that the hierarchical Voronoi data structure is equivalent to an event structure. The temporal ordering of past map events as well as states in a hierarchical Voronoi data structure provide a suitable model for the integration of Allen's temporal algebra, that is needed for reasoning about spatial objects and their temporal relationships [7]. Furthermore, a spatio-temporal structure that combines events and past map states is essential to answer queries about map changes over space and time.

This research has presented several applications of the hierarchical Voronoi data structure that are difficult to implement within the traditional GIS. The formal model of spatio-temporal change representation is currently applied to retroactive spatio-temporal map updates and visualization of map evolution.

The benefits of this approach reside in the possibility of reverting to previous states in order to visualize the evolution of the map or to perform spatio-temporal queries and analysis, as well as in performing reverse execution of the map commands previously applied to the Voronoi spatial data structure, to achieve retroactive map updates. The visualization of map changes enables fast comprehension of the events and processes that occurred in space and time. Visualization of map changes offers a powerful tool for spatio-temporal reasoning, which is needed in many GIS applications.

The map grammar described here allows us to build the deterministic spatio-temporal representations in which all the rules used in their construction are preserved. This is one of the main contribution of this paper, where for the first time in GIS research, a map grammar has been proposed as a method for handling map updates and building spatio-temporal representations.

Author details

Darka Mioc and François Anton
Technical University of Denmark, Denmark

Christopher M. Gold
University of Glamrogan, United Kingdom

Bernard Moulin
Université Laval, Canada

8. References

[1] Angel, E., 1997, *Interactive computer graphics: a top-down approach with OpenGL*, Addison-Wesley, Reading, MA.

[2] Anton, F., 1995, Le système de numérisation Voronoi intelligent, interactif et dynamique et ses applications à la foresterie, *Thèse de maîtrise, Université Laval, Faculté de Foresterie et Géomatique*, Québec, Canada.

[3] Anton, F., 1996, Reversible splitting of line segments within a Voronoi diagram for a set of points and straight line segments, *Internal document, Industrial Chair of Geomatics, Université Laval, Faculté de Foresterie et Géomatique*, Québec, Canada.

[4] Anton, F. and Gold, C. M., 1997, An iterative algorithm for the determination of Voronoi vertices in polygonal and non-polygonal domains, *Proceedings of the 9th Canadian Conference on Computational Geometry (CCCG'97)*, Kingston, Canada, pp. 257-262.

[5] Bedard, Y., Caron, C., Maamar, Z., Moulin, B. and Vallière, D., 1996, Adapting data models for the design of spatio-temporal databases, *Comput., Environ. and Urban Systems*, Vol. 20, No. 1, pp. 19-41.

[6] Chrisman, R. N., 1998, Beyond the Snapshot: Changing the Approach to Change, Error and Process, *Spatial and temporal reasoning in geographic information systems*, edited by Max J. Egenhofer, Reginald G. Golledge, Oxford University Press, New York, Chapter 6, pp. 85-93.

[7] Frank, A.U., 1994, Qualitative temporal reasoning in GIS-ordered time scales, *Proceedings of the Sixth International Symposium on Spatial Data Handling, Edinburgh, Scotland, In Advances in GIS Research, Proceedings*, Vol. 1, pp. 410-430.

[8] Frank, A.U., 1998, Different Types of "Times" in GIS, *Spatial and temporal reasoning in geographic information systems*, edited by Max J. Egenhofer, Reginald G. Golledge, Oxford University Press, New York, Chapter 3, pp. 40-62.

[9] Frank, M., Knight, T., Margolus, N., 1998, Reversibility in optimally scalable computer architectures, *The First International Conference on Unconventional Models of Computation*, January 1998, pp. 165-182.

[10] Gold, C. M., 1988, PAN graphs - An aid to GIS analysis, *International Journal of Geographical Information Systems*, Vol. 2, No. 1, pp. 29-41.
Proceedings of the Fourth International Symposium on Spatial Data Handling, Zurich, Switzerland, pp. 175-189.

[11] Gold, C. M., 1992, Dynamic spatial data structures - the Voronoi approach, *Proceedings of the Canadian Conference on GIS*, Ottawa, Canada, pp. 245-251.

[12] Gold, C. M., 1992, An object-based dynamic spatial data model, and its applications in the development of a user-friendly digitizing system, *Proceedings of the Fifth International Symposium on Spatial Data Handling*, Charleston, pp. 495-504.

[13] Gold, C. M., 1994, The Interactive map, In: *Advanced Geographic Data Modelling - Spatial Data Modelling and Query Languages for 2D and 3D Applications*, Eds. M. Molenaar and S. de Hoop, Netherlands Geodetic Commission Publications on Geodesy (New Series), No. 40, pp. 121-128.

[14] Gold., C. M., 1994, Three approaches to automated topology, and how computational geometry helps, *Proceedings of the Sixth International Seminar on Spatial Data Handling*, Edinburgh, Scotland, pp. 145-158.

[15] Gold, C. M., Remmele, P. R., Roos, T., 1995, Voronoi Diagrams of Line Segments Made Easy, *Proceedings of the Seventh Canadian Conference in Computational Geometry, (CCCG'95)*, Québec, Canada, pp. 223-228.

[16] Gold, C. M., 1996, An Event-Driven Approach to Spatio-Temporal Mapping, *Spatio-Temporal special issue, Geomatica*, Vol. 50, pp. 415-424.

[17] Gold, C. M., 1997, The Global GIS, *Proceedings of the International Workshop on Dynamic and Multi-Dimensional GIS*, Hong-Kong, China, 12 pp.

[18] Gold, C.M., Mioc, D. and Anton, F., 2008, Dynamic GIS, *ISPRS congress Book, Advances in Photogrammetry, Remote Sensing and spatial Information Sciences*, Eds: Z. Li, J. Chen, E. Baltsavias, Taylor & Francis Group, London, UK, pp. 289-303.

[19] Guibas, L. and Stolfi, J., 1985, Primitives for the Manipulation of General Subdivisions and the Computation of Voronoi Diagrams, *ACM Transactions on Graphics*, Vol. 4, No. 2, pp. 74-123.

[20] Hazelton, J. W. N., 1998, Some operational Requirements for a Multi-Temporal 4-D GIS, *Spatial and temporal reasoning in geographic information systems*, edited by Max J. Egenhofer, Reginald G. Golledge, Oxford University Press, New York, Chapter 4, pp. 63-73.

[21] Hirtle, S., 1998, Epilogue In *Spatial and Temporal Reasoning in Geographic Information Systems*, edited by Max J. Egenhofer, Reginald G. Golledge, Oxford University Press, New York.

[22] Hopcroft, J. and Ullman, J., 1979, *Introduction to Automata Theory*, Addison-Wesley, Reading, Mass.

[23] Kraak, M-J., MacEachren, A. M., 1994, Visualization of the Temporal component of Spatial data, *Proceedings of the Sixth International Symposium on Spatial Data Handling*, Edinborough, Scotland, pp. 391-409.

[24] Kraak, M-J., Edsall, R., MacEachren, A. M., 1998, Cartographic Animation and Legends for Temporal Maps: Exploration and/or Interaction, *Proceedings of the Seventh Annual Conference of Polish Spatial Information Association*, Warsaw, Poland, May 1998, pp. 287-296.

[25] Langran, G., 1992, Time in Geographic Information Systems, *Technical Issues in Geographic Information Systems*, Taylor & Francis, London, UK.

[26] Linz, P., 1996, *An introduction to formal languages and automata*, D. C. Heath and Company, Lexington, Massachusetts.

[27] Lück, H. B. and Lück, J., 1976, Cell number and cell size in filamentous organisms in relation to ancestrally and positionally dependent generation times, In *Automata, Languages, Development* , North-Holland, Amsterdam, Netherlands, pp. 109-124.

[28] Mazurkiewicz, A., 1989, Basic Notions of Trace Theory, *Proceedings of REX Workshop and Summer School "Linear Time, Branching Time, and Partial Order in Logics and Models for Concurrency"*, Noordwijkerhout, May/June1987, In *Lecture Notes in Computer Science* Vol. 354, Springer-Verlag, Berlin, pp. 285-363.

[29] McFarland, G. and Rudmik, A., 1993, Object-Oriented Database Management systems, *A Critical Review/ Technology Assessment*, Source: http://www.utica.koman.com/tech/oodbms.ToC.html

[30] Mihalache, V., Paun, G., Rozenberg, G. and Salomaa, A., 1996, Generating strings by replication: a simple case, *TUCS Technical Report No. 17*, Turku Centre for Computer Science, Turku, Finland.

[31] Mihalache, V. and Salomaa, A., 1998, Language-Theoretic Aspects of String Replication, *International Journal for Computer Mathematics* 66, pp. 163–177.

[32] Mioc, D., Anton, F., Gold, C. M. and Moulin, B., 1998, Spatio-temporal change representation and map updates in a dynamic Voronoi data structure, *Proceedings of the Eight International Symposium on Spatial Data Handling*, Vancouver, Canada, pp. 441-452.

[33] Mioc, D., Anton, F., Gold, C.M., and Moulin B., 2006, Map updates in a dynamic Voronoi data structure, *Proceedings of ISVD'06*, Banf, Alberta, pp. 264-269.

[34] Mioc, D., Anton, F., Gold, C.M., and Moulin B., 2007, Reversibility of the Quad-Edge operations in the Voronoi data structure, *Proceedings of ISVD'07*, Glamorgan, UK, pp. 135-144.

[35] Mioc, D., Anton, F., Gold, C.M., and Moulin B., 2009, On Kinetic Line Voronoi Operations and Finite Fields, *Proceedings of ISVD'09*, Copenhagen, Denmark, pp. 65-70.

[36] Mioc, D., Anton, F., Gold, C.M., and Moulin B., 2010, Kinetic Line Voronoi Operations and Their Reversibility, *Transactions on Computational Science*, pp. 139-165.

[37] NCGIA, 1988, Proposal to the geography regional science program at the National Science Foundation, *Technical Paper 88-1, National Center for Geographic Information and Analysis*, University of California, Santa Barbara, CA.

[38] Newell, R. and Batty, M., 1994, GIS databases are different, *AM/FM'94*, pp. 279-288.

[39] Okabe, A., Boots, B., Sugihara, K., 1992, *Spatial Tessellations - Concepts and Applications of Voronoi Diagrams*, Wiley & Sons, Chichester.

[40] Pequet, D. and Wentz, E., 1994, An Approach for Time-Based Spatial Analysis of Spatio-Temporal Data, *Advances in GIS Research, Proceedings 1*, pp. 489-504.

[41] Proulx, M-J., 1995, Développement d'un nouveau langage d'interrogation de bases de données spatio-temporelles, *Thèse de Maîtrise, Département des Sciences Géomatiques, Université Laval*, Québec, Canada.

[42] Prusinkiewicz, P., Lindenmayer, A., 1990, *The Algorithmic Beauty of Plants*, Springer-Verlag, New York.

[43] Rekers, J., 1995, A Parsing Algorithm for Context-Sensitive Graph Grammars, *Technical report 95-05 of Leiden University*, Leiden, The Netherlands. Source: ftp.wi.leidenuniv.nl, file /pub/CS/TechnicalReports/1995/tr-05.ps.gz.

[44] Roos, T., 1991, Dynamic Voronoi diagrams, *Ph.D. Thesis, University of Würzburg*, Germany.

[45] Van Oosterom, P. J. M., 1993, *Reactive Data Structures for Geographic Information Systems*, Oxford University, Bookcraft Ltd.

[46] Van Oosterom, P. J. M., 1997, Maintaining Consistent Topology including Historical Data in a Large Spatial Database, In *1997 ACSM/ASPRS Annual Convention & Exposition, Seattle, Washington, Auto-Carto 13*, Vol. 5, pp. 327-336.

[47] Vaario, J., 1993, An Emergent Modeling Method for Artificial Neural Networks, *Doctoral dissertation, University of Tokyo*, Japan.

[48] Winskel, G., 1989, An introduction to event structures, *Proceedings of REX Workshop "Linear Time, Branching Time, and Partial Order in Logics and Models for Concurrency"*, *Lecture Notes in Computer Science*, Vol. 354, Springer, Berlin, pp. 365-397.

Chapter 5

Feng-Shui Theory and Practice Investigated by Spatial Regression Modeling

Jung-Sup Um

Additional information is available at the end of the chapter

1. Introduction

Feng-Shui (meaning wind and water) has a strong indication for the placement of a physical object in space since it determines propitious locations by investigating the vital energy of the earth (Qi) throughout the lay of the land. Feng-Shui is the product of ancient Chinese cosmological beliefs and their responses to the land which were directly linked to survival and healthy living since it was believed that it could be used to harmonize people with their environment. Feng-Shui practices could be divided into landform Feng-Shui and living-space Feng-Shui (Sung, 2001). Landform Feng-Shui practices have been conducted with emphasis on grave, residential sites, village, town, and city sites while living-space Feng-Shui focused on the physical layout of such elements as the front gate, bedroom and kitchen. Today, living-space Feng-Shui is being widely used in the United States and Western Europe (Kim, 2003). In some respects, these two phenomena are quite different. However, they both involve patterns in siting that seem to express cosmological beliefs, numinous feelings and aesthetic experience.

Feng-Shui has been practiced in China and many other areas of Asia over the past five thousand years and it is becoming increasingly popular in western countries, such as the United States and Western Europe, thus gaining popularity world-wide (Chris, 2002; Diana 2004). Feng-shui theory and practice were based on considerable anecdotal evidence rather than on scientific proof. It should be noted that conflicts and debates exist when applying Feng-Shui rules because of the flexible practices and ambiguous literature (Andrew, 1968; Zetlin, 1995). Even Asians, who are familiar with ancient oriental literature and philosophy, have difficulty understanding the basic Feng-Shui concepts and methods of application.

So far the spatial variation of Feng-Shui locations is poorly defined despite abundant interest in this problem. Previous research for living space Feng-Shui has been based on the small number of in-situ observations reflecting the specific attributes of experimental sites such as the window, mirror and bed room (Hobson, 1994; Verlyn, 1995; Lu & Jones, 2000). Field practices for living-space Feng-Shui have been researched with a focus on questionnaires (June, 2002). Field observations have the disadvantage that they provide only limited information on the area-wide distribution of the Feng-Shui locations.

Although area-wide data such as a topographic map for villages, towns and city sites in landform Feng-Shui have been used, it is rare to find enough experimental locations to prioritize the relative importance of spatial variables for Feng-Shui practice in those applications since just one sample used to occupy a huge area (e.g city site) (Hammond, 1995,; Xu, 1998; Zitao, 2000; Jun, 2003; Lynch, 2003; Mak & Ng, 2005). Spatially prioritized parameters cannot, therefore, be adequately identified on the basis of such a few Feng-Shui locations. Until recently, the investigations of an area-wide distribution pattern of Feng-Shui locations and their spatial characteristics remained largely theoretical because the non-GIS survey technique had difficulty in assembling multi-thematic factors simultaneously (Earle, 1992). In the case of burial mounds, there is area-wide spatial evidence to document them. The actual grave distribution is a result of the complex interaction of historic and recent environmental and human factors. Tomb footprint shows a tenacious vision of what constitutes a proper mound location. The footprint of a grave characterizes Feng-Shui's spatial structure that varies with a multitude of factors such as the land use pattern and surface roughness. However, the field practices of tomb Feng-Shui have seldom been empirically examined in terms of area-wide spatial arrangements. It is necessary for the complicated belief system of Feng-Shui to be explained by describing how tomb footprint is distributed in space or with respect to each other. In Feng-Shui, every piece of information has to be looked at individually and put it together with others and look at the whole (or bigger) picture. Everything has its own place in every different situation and position. Concentrating on one point to the exclusion of all others does not lead to a balanced point of view. GIS is an ideal technology capable of integrating, merging and analyzing simultaneously multiple data layers to explore spatially prioritized relationships that are necessary when studying Feng-Shui footprints.

The multivariate spatial regression modeling compares multiple maps at one time and answers questions such as "how much more important is one map over the other". It is expected to reveal a consensus on how the Feng-Shui practice may have a universal sense of what is the most important variable for any place. If the importance or preference of each factor relative to other factors is identified in the context of a spatial multi-criteria framework, high priority areas in Feng-Shui will be identified. Therefore, the aim of this research was to evaluate major controlling factors in locating Feng-Shui tombs, using the spatial prioritization modelling techniques of GIS, based on the area-wide grave footprints in South Korea.

2. General methodology

2.1. Model parameters

Feng-Shui, by studying the configuration of mountains and rivers, selects a location where the vital energy (Qi) that flows throughout the earth is connected by water and not scattered by the wind. There is often a lack of understanding as to where this Qi is coming from. An important question is how to measure or represent criteria that are important to Feng-Shui practices and should be included in a spatial regression model. Preference variables perceived according to different cultures and geographical differences vary significantly across individuals and across the groups these individuals represent. Each grave location in a cluster, or isolated, takes into consideration the astrological conditions at the site for its unique occupant. It is important that the occupants of the sites are rich or poor, or what clans they represent. Facing a mountain at great distance is also considered as one of the important variables of a propitious location. The grave density is subject to complex, interacting influences, several of which, notably competition for favorable landscape and micro-site variability.

It is impossible to measure and quantify many of them for the purpose of spatial modeling since they need to be measured based on people's perceptions and feelings. It would clearly be impossible to model the behavior of every subjective variable involved in the burial decision. This method is extremely subjective since the model has to be run under vast numbers of combinations of rules, its approach is conceptually non-spatial. In such circumstances, there is no objective means of ascertaining the data quality and extent of the model. Such variables are too much subjective to be of great use in the modeling exercise.

Landform is frequently recognised as the most important factor in the grave location since it includes indicative information such as water features, vegetation, landslide susceptibility and soil moisture, etc. for the site investigated (Choi, 2001; Yoon 2006). Water moves through and over the landform, so the spatial distribution of soil and vegetation properties should be related to topographic attributes such as slope, elevation, and aspect and solar radiation in a specific grave location. The arrangement of meaningful Feng-Shui landscape "mountain in the back, water in the front" is closely related to landform factors such as the direction of maximum slope and transverse to the slope. In this regard, landform is believed to characterize changes in the flow of energy direction (Qi), flow velocity and Qi transport processes based on the orientation of mountains and spatial arrangements of topographic attributes such as slope.

In general, models of physical systems are forced to adopt aggregate approaches because of the enormous number of individual objects involved. It was necessary to focus on the variables which can be quantified to ensure the quality of the model, compared to the traditional use of subjective rules of thumb. These studies have employed only the most readily available and accessible data that are directly integrated with GIS to focus on representative and typical features in the experimental design and in order to evaluate the field Fung-Shui practices in an objective manner.

The main focus has been given to demonstrating the influence of the hillshade, elevation, slope and aspect as the most important factors controlling the spatial location of a grave. Landform multi-criteria models may provide an equitable and efficient means for exploring the spatial structure of Feng-Shui in terms of only stable physical geography. However, new grave location procedures could be influenced by various human factors beside landform parameters. In order for these tools to be effective, they should include criteria that are locally relevant and measurable in a human framework. To examine land-use effects and proximity to population on grave densities, several proximity variables calculated in accordance with the distance from roads were included as the model parameters (Table 2).

2.2. Study area and data analysis

Feng-Shui is responsible for many aspects of the cultural landscape in Korea, especially the location of tombs. The process of tomb location is extremely important in South Korea, since it is a common belief accepted by the Korean public that the location of the tomb would affect the descendant's well-being of the tomb's occupant. It is well known that a strong belief in Feng-Shui led kings in past dynasties to exhume and rebury their ancestors. "Even former and current presidents, and leaders of major political parties in South Korea, have moved their ancestors' graves or sought out propitious locations. Also, it was thought that burying one's parents in a place deemed propitious under Feng-Shui principles was a way to demonstrate filial piety (Kim, 2002)." The direction and position of grave were carefully selected by geomancers. "The spirit of Feng-Shui is still visible and strongly felt in the countryside, tightly integrated into the rural way of life " (Choi, 2001).

The burial areas are well maintained by a clan and memorial ceremonies are conducted on the traditional ancestor worship days such as Thanksgiving day and New Year's day, etc. at the tomb (Hough 1999). In this process, "Feng-Shui plays the role of a chain of life that connects an individual with his ancestors as well as descendants" (Yoon, 2006). "This is why Feng-Shui has continued to be influential despite being constantly criticized as a superstition (Kim 2002)". Such practices make the Korea the best place to explore the principles of Feng-Shui in the entire world.

The study area is situated in the south eastern part of South Korea between latitude 35.55°N and 36°N and longitude 128.38°E and 128.44°E. It belongs to the Gongsan-Dong (administrative districts) in the city of Daegu, a metropolitan city recognized as the third most important city in South Korea, covering approximately 56.5 km2 (Fig. 1). It possesses several advantages that make it an appropriate choice for such a study. Its elevation ranges from approximately 60–834 m above sea level and shows a hilly character, frequently observed as the topography of typical grave locations in South Korea (1). The area is located on the south eastern edge of Palgong Mountain which is well known for geomantic ideals as well as one of the most spiritual mountains in the country. In Korea, Buddist monks appear as geomancer-monks who choose most of the auspicious places and thus the locations of many Buddhist temples have geomantic legends (Yoon, 1975). The study area was located in a place where it is easy to find many old temples in every valley and Buddhist heritages are scattered (Korean Heritage Online, 2006). The area is relatively isolated from the urban center, thus grave locations in the area still keep the appearance of the old dynasty period. The study area is located at the place where a cemetery is not available to investigate the natural location pattern of tombs by individual clans.

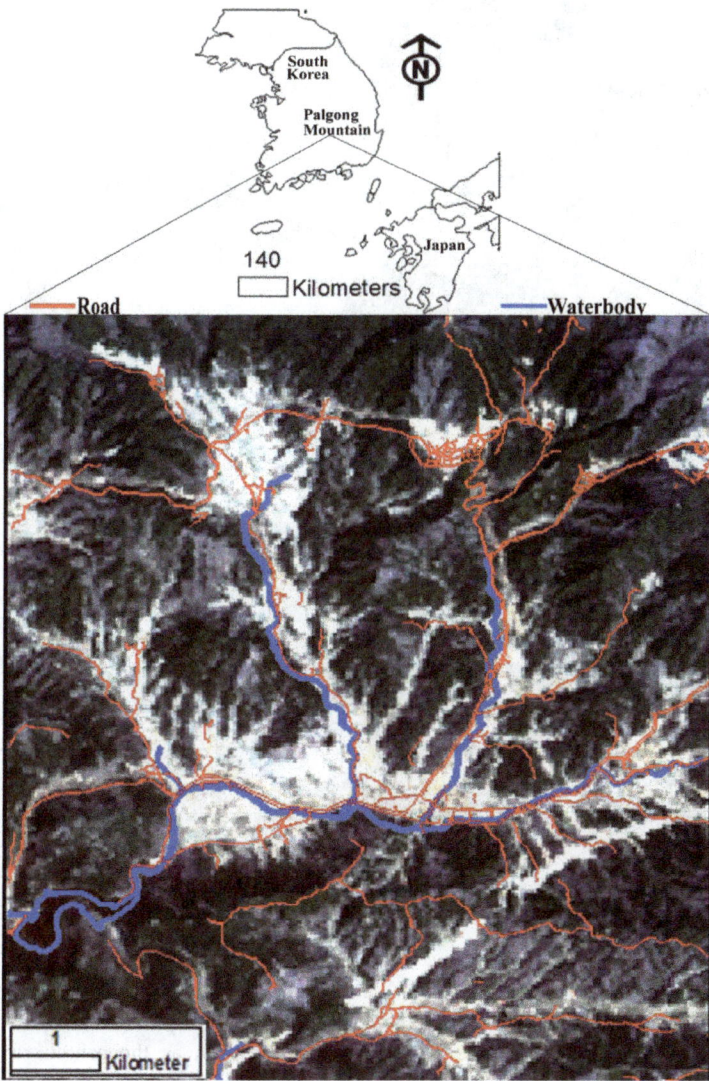

Figure 1. Location map of experimental site presented by a Thematic Mapper satellite image (acquired in April 2002)

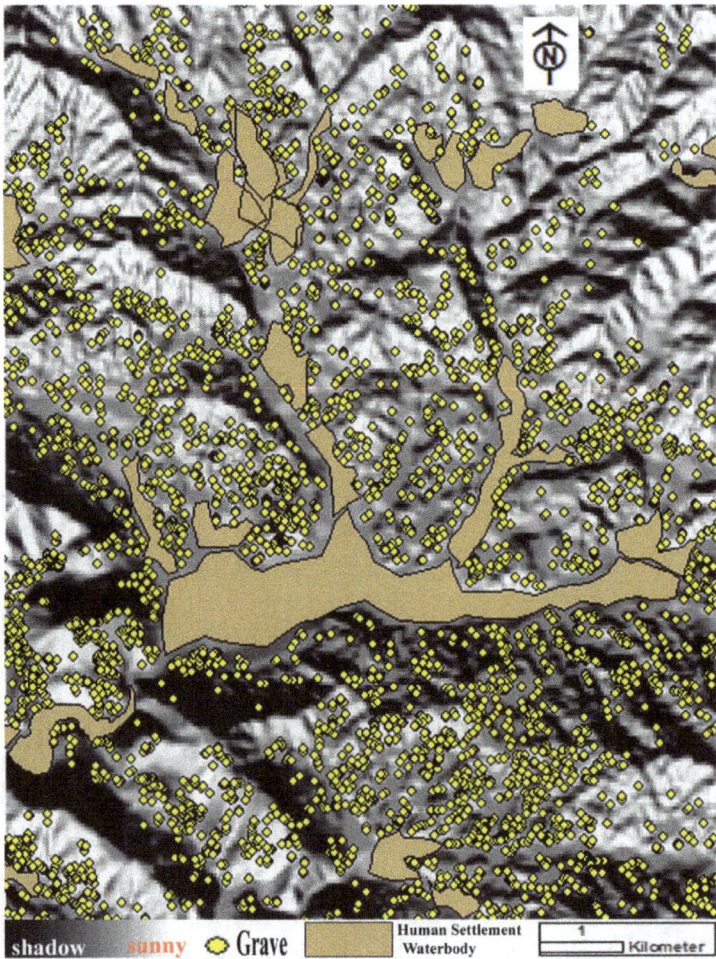

Figure 2. Solar radiation map combined with grave locations.
A shadow area at the south eastern end and northern part of the study area shows a very small number of graves.

A 1:5000 digital topographic map (produced in 1996 from aerial photography taken during the same year) was accessed via the national mapping authority of South Korea. The base map has been used to extract a Feng-Shui database such as the contour with 25 m equidistance, administrative boundary and road and tomb available in the study site. The human settlements, golf course and water-body have been excluded for further analysis since graves could not be located there (Fig. 2). The analysis in the ArcGIS was performed to identify and calculate spatial variables to be used in a regression model. A localized grave density map was created by defining the mean number of points per unit area in the ArcGIS to pinpoint spatially intensified distribution areas and examine the surrounding causal factors such as landforms. The search radius for the calculation of each pixel (30 m) was set to 300 m by an initial option suggested by the density slicing function of ArcGIS. The distributions of the concentration levels and predictor variables are presented by percentile scale as shown in the Fig. 2.

From the contours and spot heights, a DEM (Digital Elevation Model) was generated with pixel size 30 m × 30 m. Hillshade, slope and aspect layers were created as derivatives of the DEM. Computing the hillshade requires one to enter values that identify the sun's height and location in the sky. The hillshade grids were produced with an azimuth of 195 degrees, an altitude of 60 degrees based on the Korean holiday (Hansik) falling on the 105th day after the winter solstice (13:05 April 6, 1996) of the study site. It is a day for the family to move their ancestors' tombs and to hold their own memorial services. Each hillshade cell contains a value which indicates the sun intensity, ranging from 1 to 255. It was used to determine what was in shadows and what was is in sunlight at that time in 1996.

The road-related spatial database, population density and land prices were already available for most areas of Korea. The population register (1996) was acquired from the Korean central statistical Office and land prices (1996) from the Korean Association of Property Appraisers (1996). The population and land price statistics were added to each administrative unit by means of joining the data table to a theme's attribute table. Feng-Shui databases were built as integrated GIS layers and were overlaid accordingly. The data acquired from the spatial analysis were stored in MS Excel and basic statistical parameters were calculated to acquire the overall feature of the data sets. Multivariate spatial regression was undertaken by the SPSS statistical package to examine the relationships between grave concentrations and spatial variables potentially associated with grave density (Tables 2). The fit of the spatial models was assessed by the percentage of explained variation (R^2), slope estimates of the parameters and their VIF (Variance Inflation Factor) were given.

3. Result

Summary data for the grave density in relation to proximity to roads is presented in Table 1. The grave density data does not show a significant degree of variability for different proximity conditions from roads. The nearby places to the road are not more related than distant places in terms of grave concentration. Similarly the results of the regression analyses indicate that proximity to a road is not an important predictor of grave density. The population density did not make a significant contribution to explaining the variance in R^2 since the owners of tombs are mostly residents of urbanized areas far away from the study area. The proximity or human variables were the least significant variables in the regression model. Furthermore, when the human variables were excluded, the R^2 values for the models decreased by just 0.009 (0.754 to 0.745), showing the lower importance of road, population, and land price variables. By contrast, grave concentrations in landform parameters showed relatively high variability (Table 2).

Road (6 m width)	density	Road (4m width)	density	Road (2 m width)	density
50 (meter)	84	50 (meter)	83.9	50 (meter)	78.9
100	108.9	100	109.4	100	78.7
200	118.8	200	124.6	200	76.6
400	110.4	400	104.1	400	80.2
600	83.6	600	102.4		
1000	98.9	1000	96.5		

Table 1. Area-wide grave density (number of graves per km²) in relation to buffer distance from road

	Model 1 (with proximity and human variables)			Model II (with landform variables alone)		
Summary of model	Adjusted R²=0.754, number of observations (65523), F=16719.154 (P= 0.000)			Adjusted R²=0.745, number of observations (65523), F= 27405.752 (P= 0.000)		
Explanatory variables	Standardized coefficients	t-value	VIF (Significance levels*)	Standardized coefficients	t-value	VIF (Significance levels*)
solar	1.590	178.293	21.173	1.557	289.977	7.417
elevation	-.858	-148.941	8.832z	-.806	-152.509	7.189
slope	-.191	-34.111	8.329	-.116	-22.696	6.771
south	.049	18.413	1.894	.054	20.469	1.182
east	.063	23.922	1.818	.069	26.662	1.746
west	.018	8.135	1.275	.015	6.595	1.262
north	.001	.472	1.101(.637)	-.002	-1.145	1.09 (.252)
road (6 m)	.114	33.563	3.086			
road (4 m)	.098	28.204	3.242			
road (2 m)	-.003	-.817	2.776(.414)			
population	-.006	.609	26.13 (.543)			
land price	-.080	-8.695	22.36			

* Significance levels are not presented in the case of p < 0.001.

Table 2. Multiple regression results including human variables and landform variables alone (dependent: grave density)

The R² values for the models indicate that 74.5% of the variability of grave densities observed in the site was explained by the landform variables (Table 2). Initial observation of the grave map overlaid on the solar radiation clearly disclosed strong evidence for the influence of the sun index as a derivative of the integrated landform parameters at the grave distribution, as shown in Figs. 2-3. Solar illumination made the most significant contribution (b: 1.557) to explaining the variance in grave presence or absence (Fig. 2-3). Solar radiation is affected by various factors such as the latitude of the site investigated, slope and aspect of a site, sun's height and location in the sky. There was a more distinct preference for low altitude sites (Figs. 4-5) and a stronger association (b: -0.806) with decreasing elevation than those of slope (b: -0.116) and aspect. The relationship between the slope and the grave density variable is less distinct and a clear trend cannot be observed. A scatter plot shows that graves occur more frequently where the slope is intermediate (Fig. 6).

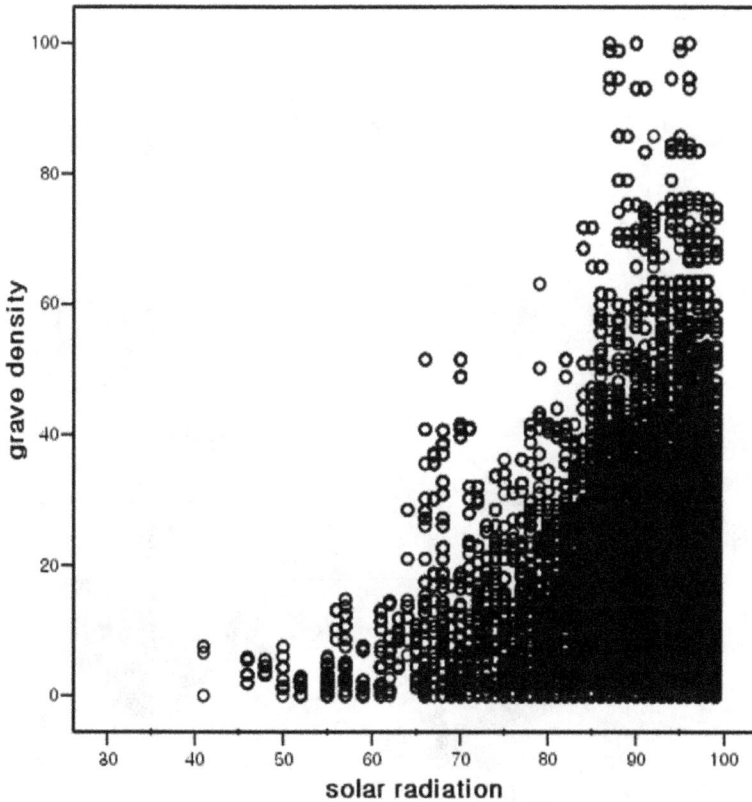

Figure 3. Plot of grave density vs. solar radiation
(percentile scale, One hundred percent represents the highest grave density in the experimental site). Local grave density displays distinctively increasing trends in respect to the occurrence of a sunny site condition. The plot shows that the sun index is a crucial control factor on grave density.

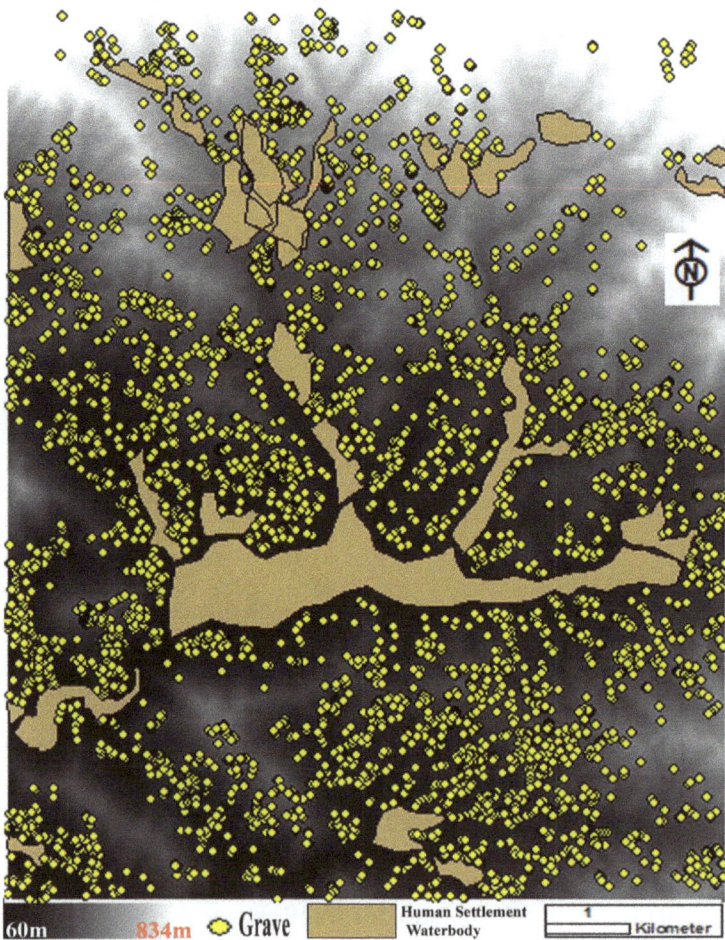

Figure 4. Digital Elevation Model combined with grave locations.
A highly elevated landform condition at the northern edge of the study area shows a very small number of graves.

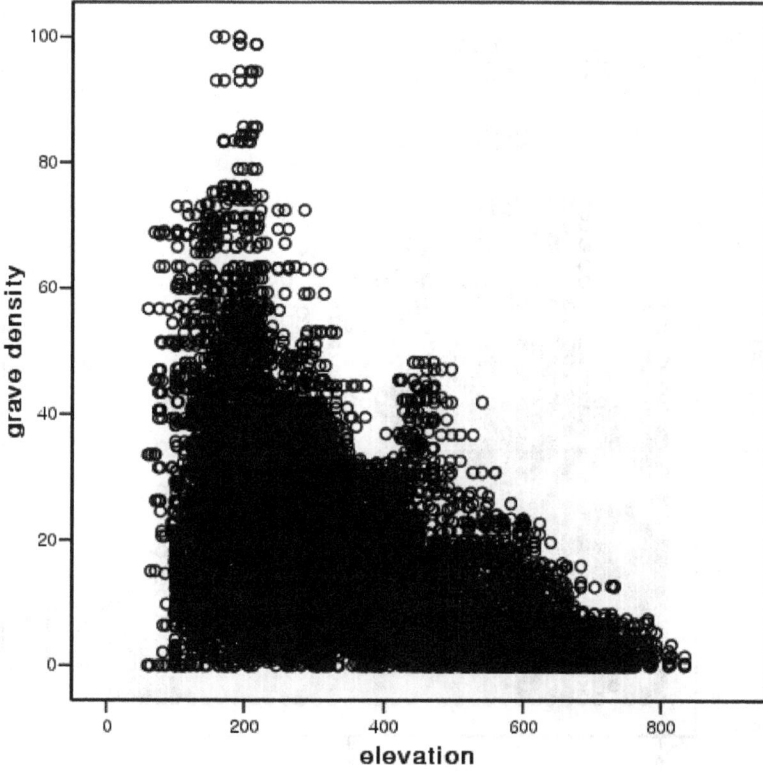

Figure 5. Plot of grave density (percentile scale) vs. elevation (m). Local grave density displays decreasing trends in respect to the occurrence of a high altitude condition.

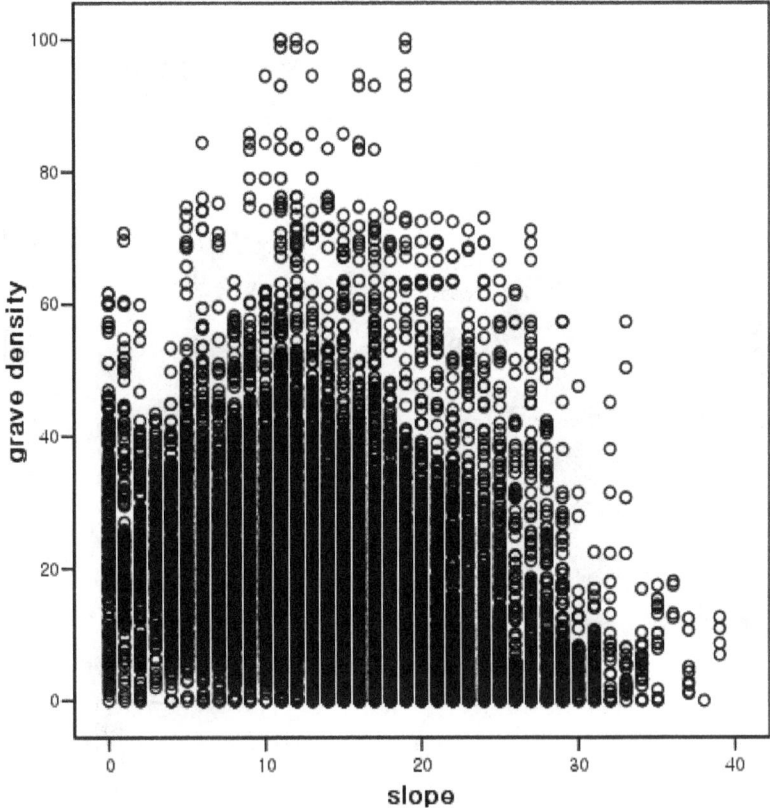

Figure 6. Plot of grave density (percentile scale) vs. slope (°). There is considerable scattering in the individual data points, but a higher grave density is generally found on the intermediate slope range.

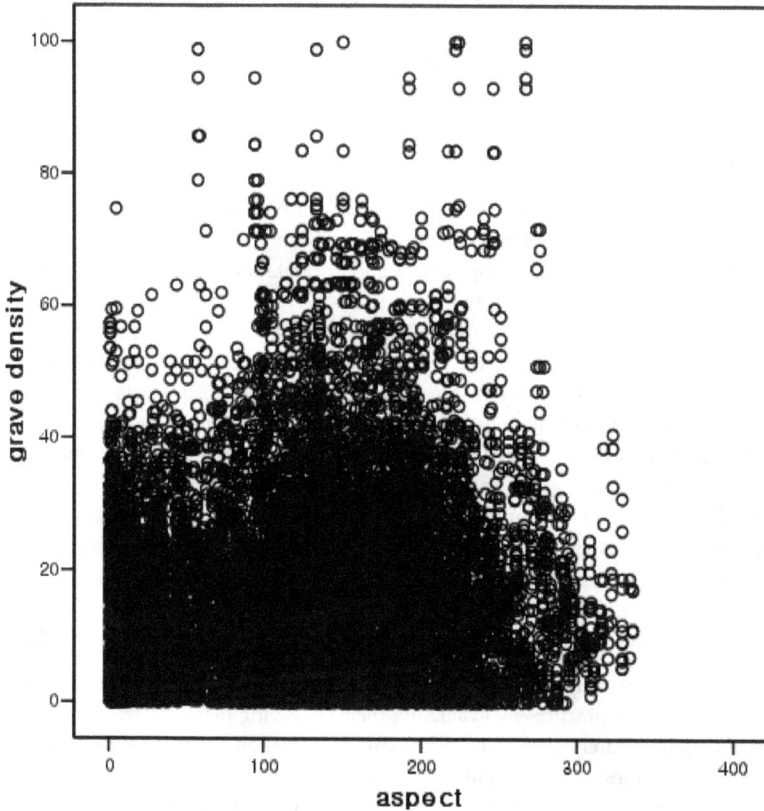

Figure 7. Plot of grave density (percentile scale) vs. aspect (º). Grave density is generally higher on south (136º-225º) and east-facing (46º-135º) slopes than on north (0º-45º, 316º-360º) and west-facing slopes (226º-315º). A distinct preference for south direction is not observed.

Grave density was generally higher on south and east-facing slopes than on north, and west-facing slopes (Table 2). The number of graves in the southern direction is not significantly higher than that of the east or west, different from the common belief that a south-facing is most favorable. The aspect parameter showed a much weaker influence (Fig. 7) on the grave distribution as a whole in the regression model than those of the sun's index (b: 1.557) and elevation (b: -0.806).

A comparison of scatter plots for the sun index with other landform variables indicated that grave density is much more abruptly decreased in a shadow condition than those with other explanatory variables such as elevation or slope. The sudden drop of grave density in accordance with the sun index bands indicates that the grave density is strongly associated to the site condition within the sun index 92 percentile. There were high joint effects for

several variables, reflecting the colinearity between the variables. The most significant colinearity was among these three, the sun index 99 percentile, elevation 251 m and elevation
423 m. Despite this colinearity, these variables were retained in the models because they
could display their differing relative importance in the regression trees.

4. Discussion

Several studies have shown that Feng-Shui locations tend to be dependent on landform
parameters, and increase if favorable conditions are observed, like when a site is located in
low elevated hills, has a smooth slope and a good orientation facing south (Andrew, 1968;
Yoon, 2006). The results of this study extend previous findings, specifically focused on the
associations with the landform surface. The present study has utilized area-wide and
pertinent information in the analysis of a Feng-Shui location to identify the spatially
prioritized variables. In this project, the number of samples (5549) was large enough to
extract reliable information relating to the existing state of the grave density in question, as
substantial evidence of the spatially prioritized relationship caused by the Fung-Shui
application. The area-wide grave data acquired from the topographic map produced by the
interpretation of aerial photography provided sufficiently reliable information by replacing
the questionable accuracy of a conventional field survey by a permanent record. This study
has also demonstrated that a Geographic user friendly interface can be utilized for a factual
assessment for Feng-Shui distribution by overcoming any serious subjective judgment
suffered from the lack of a cartographic representation in the text based statistical data.

The spatial regression analysis allowed for assessing the level of consistency among the spatial
variables in Feng-Shui practices by evaluating their stratifying hierarchy simultaneously. The
grave locations were rarely affected by their degree of isolation or their degree of connectivity
to roads, the dynamics of local populations and the price of the land segments. The results
show that grave locations generally are determined only by the ideal Feng-Shui model
established more than two thousand years ago without considering the surrounding
environmental changes caused by urbanization. The spatial prioritization model clearly
demonstrated that the sun index is perceived as extremely important while the elevation is
considered moderately important and the aspect and slope are less important.

The actual importance (regression coefficient) value of the sun index presented by the
spatial prioritization model can answer such a question as "how much more important is the
sun index over the others?". In usual site selection cases, selecting one site over the other
requires a tradeoff among spatial variables, provided the variables have a similar
importance on site selection. The key issue in the spatial prioritization model would be a
further tradeoff among first, the sun index, then the second most important and a
combination of the rest. The magnitude of solar radiation and its relative importance is great
enough to offset the difference in other landform parameters. It is believed that the Feng-
Shui experts used to take the most important factor, the sun index and make tradeoffs
between the second important factor, elevation and the reduced combination set of the last
two factors (slope and aspect).

Feng-Shui emerged in the context of agricultural societies located in a temperate zone. It was essential to find a sunny location for farming and to survive in a chilly climate. In this regard, the relative merits of other landform factors in the overall Feng-Shui siting were very weak in comparison with the solar radiation in the spatial model. The sun index is much more influential a factor in the model's predictions since it is jointly determined by various landform parameters and the solar system in different locations. The findings of this study have important implications for the application of Feng-Shui and for future research. Also, the findings of this paper may assist in the more accurate targeting of Feng-Shui applications at their locations. It is believed that the postulates of the classical Feng-Shui theory are applicable to a specific sun light deficient area and not to the general case. In a tropical area, the sun can be quite intensive. It is not necessary for the sunlight to be shining into the sites all day (Tan, 2006). That is the reason why it is so hard to find real evidence of Feng-Shui applications in tropical regions such as South America and Africa.

This research has been carried out in one of the best country to explore the principles of Feng-Shui in the entire world. The experimental site is located on the site well known for geomantic ideals as well as one of the most spiritual mountains in the country. The spatial prioritization model derived in the study area may reflect Feng-Shui practices under ideal circumstances. The spatial dimensions of Feng-Shui locations vary according to the way different societies construct different shared cultural spaces. Moreover, the characteristics of the special case assumed by the classical Feng-Shui theory happen not to be those of the economic society in which we actually live. The perceptions about the surrounding environment generally vary with geographical differences. However, it is expected that the result of a Feng-Shui application will almost always parallel with the modern solar siting techniques and theories as the outcomes of these inspections have illustrated if the ancient Feng-Shui principles have been applied properly. Therefore the spatial prioritization model of this research will provide a standardized form against which to compare the real phenomenon of Feng-Shui. Although the cultural and geographical differences between the Western and Eastern Feng-Shui rarely behave this way, it is still useful to know what would happen if they did, as a basis for comparison.

5. Conclusions

Analyzing spatially prioritized parameters in relation to the Feng-Shui surface represents a critical step toward understanding how a local condition contributes to Feng-Shui landscape and will help to evaluate Feng-Shui locating strategies. This is one of a few studies that have looked at the relative importance of the differences of spatial variables within Feng-Shui locations and the first to formalize a solar illumination priority in a Feng-Shui tomb with area-wide empirical evidence. The spatial prioritization model showed that it could play a key role in evaluating Feng-Shui tomb distribution in which different kinds of landform mosaic mixes could be accommodated in the same location and could also occur in different locations. Therefore, this study made a valuable first step toward reducing its aura of superstitious mystery by investigating a major contribution factor for Feng-Shui locations through more rigorous and systematic approaches.

Solar radiation was the most important predictor of grave density in the Feng-Shui locations. Similarly, spatial clustering technology identified the fact that high concentrations of grave necessarily accompany the significantly increasing trends of solar radiation. The results of the regression analyses indicate that the grave density could be explained by the four landform parameters alone yielding R^2 values of 0.751. In contrast to the typical theory, slope and aspect were not a dominant determining factor upon the dependent variable of grave density. Also, the significantly increasing trends of grave density were not observed in line with a southern direction. A clear verification has been made for the hidden assumptions in Feng-Shui 's long history that its approach is found to be more appropriate in avoiding shadow conditions, rather than exploring the ideal landform location

However, this research endeavor leaves many questions for further discussion and investigation. This study only finds statistically significant association between grave density and certain variables describing landscape and it does not attempt to elucidate what this association means in cultural terms with a more structured code of practice for Feng Shui application. One possibility is that dispersed patterns of graves could be the typical result of competition for a particularly auspicious piece of land. The presented approach to identify spatially prioritized parameters of Feng-Shui has yet to be tested in more diverse situations (agriculturally, environmentally and socio-economically) and the parameters of the analysis have to be more thoroughly validated through further research. This type of information generally requires extensive cross-cultural and comparative studies. There is a need to compare a variety of cultures equally from an objective viewpoint, to explore the spatially integrative relationship between the grave density and their surrounding landscape. This work constitutes an early step in the process of identification and characterization of the full suite of spatially prioritized parameters of Feng-Shui.

Author details

Jung-Sup Um
Kyungpook National University,
South Korea

Acknowledgement

Thanks are extended to the different national agencies of South Korea (Spatial Data Warehouse of National Geographic Information Institute, Central Statistical Office, Korean Association of Property Appraisers) for sharing their data for this project. I am grateful to Prof. Michael Goodchild of the National Science Foundation-funded National Center for Geographic Information and Analysis (NCGIA) at the University of California, Santa Barbara (UCSB) and Prof. Andrzej Weber, Department of Anthropology, University of Alberta, Canada for their critical remarks that helped in improving the manuscript. I would like to thank Shin Joon-Ho for the statistical advice and for the English corrections suggested by Paedar.

Notes

This chapter was revised from the paper initially published in International Journal of Geographical Information Science (Taylor & Francis), Vol. 23, No 3-4, March-April 2009, 513-529.

6. References

Andrew, L.M., (1968). An appreciation of Chinese geomancy. *The Journal of Asian Studies*, 2. pp. 253-267.

Carre, F. & Giard M.C., (2002). Quantitative mapping of soil types based on regression kriging of taxonomic distances with landform and land cover attributes. *Geoderma*, 3-4, pp. 241-263.

Choi, W.S., (2001). A study of geomantic auxiliary temples and pagodas in the Yongnam region. *Study of Korean History of Thought*, 17, pp. 169-204.

Chris, T., (2002). The role of Feng-Shui: Haymarket priority project [Sydney, Australia] *Landscape Australia* 1, pp. 10-12.

Diana, L., (2004). California assemblyman proposes resolution that promotes building with Feng-Shui. *Architectural Record*, 4. pp. 48.

Earle, M.J., (1992). Spiritual landscapes: A comparative study of burial mound sites in the Upper Mississippi River Basin and the practice of' Feng-Shui' in East Asia. Ph.D Thesis, University of Minnesota, 270.

ESRI (Environmental System Research Institute), (2006). ArcGIS software help menu (spatial analysis toolbox).

Hammond, J., (1995). Ecological and cultural anatomy of Taishan villages. *Modern Asian Studies*, 3, pp. 555-572.

Hobson, J.S.P.1., (1994). Feng-Shui: its impacts on the Asian hospitality industry. *International Journal of Contemporary Hospitality Management*, 6, pp. 21-26.

Hough, W., (1999). Korean clan organization. *American Anthropologist*, 1, pp. 150-154.

Jun, X.A., (2003). Framework for site analysis with emphasis on Feng-Shui and contemporary environmental design principles. Ph.D Thesis, Virginia Polytechnic Institute and State University.

June, M.B., (2002). Feng-Shui: Implications of selected principles for holistic nursing care of the open heart patient. Master Thesis, University of South Africa.

Kim, D.G., (2002). Feng-Shui (Pungsu): chain of life that connects ancestors with descendants. *Koreana*, 4, pp. 24-31.

Korean Association of Property Appraisers, (2006). http://member.kapanet.co.kr/cgi-bin/gsv/ (accessed 10 February 2008).

Korean Heritage Online, (2006). Buddhist culture in Palgong Mountain, http://www.heritage.go.kr/eng/tou/tou_the_12_sce.jsp.(accessed 10 February 2008).

Lu, S.J. & Jones, P.B., (2000). House design by surname in Feng-Shui Source. *The Journal of Architecture*, 4, pp. 355-367.

Lunch, E. (2003). Feng-Shui as a site design tool: Assessing conditions of human comfort in urban places. Master Thesis, University of Arizona.

Mak M.Y.& Ng.S.T., (2005). The art and science of Feng-Shui - a study on architects' perception. *Building and Environment*, 3, pp. 427-434.

Sung, D. H. (2001). Removing an illustrious imperial garden and the establishment of Hwasong according to Chongjo's view of geography and geomancy, *Study of Korean History of Thought* [South Korea] 17, 121-167.

Tan, M.L., (2006). Filipino 'feng shui'?
http://www.inq7.net/globalnation/col_pik/2003/aug14.htm (accessed 10 February 2008).

Verlyn, K., (1995). Bold illusion for California: unexpected ideas transform a San Jose tract house. *Architectural Digest*, 5, pp. 156-161.

Xu,P., (1998)."Feng-shui" models: structured traditional Beijing courtyard houses, *Journal of Architectural and Planning Research*, 4, pp. 271-282.

Yoon, H. K., (1975). An analysis of Korean geomancy tales. *Asian Folklore Studies*, 1, pp. 21-34.

Yoon, H. K., (2006). The Nature and Culture of Feng-Shui in Korea.Lexington Book

Zetlin N, M., (1995). FENG-SHUI: Smart business or supersition? *Management Review*, 8, pp. 26

Zitao, F., (2000). Feng-Shui in site planning and design: A new perspective for sustainable development. Master Thesis, Arizona State University.

Chapter 6

Behavioural Maps and GIS in Place Evaluation and Design

Barbara Goličnik Marušić and Damjan Marušić

Additional information is available at the end of the chapter

1. Introduction

This chapter addresses the subject of behavioural maps, their characteristics and significances; the ways they can be created or produced as well as their applicability in place analysis, its evaluation and design, using GIS as working and analytical milieu. It argues that data immanent to behavioural maps enable various simulations of places, exploring their characteristics and qualities, checking potentials of places for certain development, post occupancy evaluations etc., but also checking on skills of designers in their achievement towards user friendly places.

Traditionally, models and simplified symbols which match certain requirements regarding metric dimensions of places or spatial elements that form places have always been used in place planning and design. The question is how much freedom in place creation do such models actually allow? Saying with other words, how well do they reflect real situations and relations between functionality and physical characteristics of places they represent and their actual occupancies? In former times such templates were available in plastic models for layouts and cross-sections in different scales. Nowadays they are available within computer-aided design software packages. A critical eye must realise that such templates serve only as indicators for recognition of the main purposes of places and usually do not reflect actual shapes and do not respond to actual usage of places. For example, pre-designed symbols for basic elements such as furniture, and shapes of places or rooms derived from such basic elements are limited by their forms. When they are used in a place design process, the process itself and the final results are limited within the given framework as well. In such an approach a goal to create user friendly places is often overlooked. Attractive layouts and cross-sections are produced to fascinate clients. In terms of responsible planning and design, this should not be acceptable. Planners and designers must strive for more accurate and refined reading of relationships between places and their users.

The main assumption is, that behavioural maps are fairly dynamic means for place planning and design, as opposed to pre-designed models or templates; used either for single elements or the entire spatial systems such as rooms in buildings, groups of rooms in buildings or even buildings and patterns of open spaces between them in cities and towns. Beside the fact, that IT development enables the production of pre-designed elements and templates for design and therefore speeds up the production process, it limits the responsive part of the process. However, at the same time some other aspects of this same IT development bring advantages towards responsive design. It opens new possibilities in place design and decision-making. For example, usage of merely static templates and symbols can be replaced with dynamic responsive templates based on dynamic patterns of spatial occupancies. Further assumption is that such spatial templates which are based on behavioural maps can help to produce better place design and more user friendly solutions.

The aim of this chapter is to exceed schematic annotation of places. It takes a point of view that places differ from each other. Addressing places via behavioural maps seems an optimal scanning process which can lead towards successful decision-making and design. Behavioural maps are seen as direct links between users in places and physicality and functionality of places themselves.

The chapter discusses behavioural maps as means to addresses usability and the spatial capacity of places, via several examples and from different viewpoints. Firstly, it comments on selected relevant case studies where behavioural mapping was applied to as the appropriate method to address the original research question of each study. Mostly, the discussion is based on a research applied to urban squares and parks in two European cities, Edinburgh (UK) and Ljubljana (Slovenia), to reveal common patterns of behaviour that appear to be correlated with particular layouts and details (Goličnik, 2005). Furthermore, a research study addressing behavioural patterns of urban cyclists in Ljubljana (Slovenia), is commented mostly regarding questions relevant for mapping and tracing (Goličnik Marušić et al., 2010). Different types of behavioural maps are discussed. Secondly, the chapter discusses possibilities for various simulations of place assessment and design using principles of behavioural mapping. It shows how existing behavioural maps can be used to evaluate the quality of existing environments as well as the quality of environments to be developed in a desired way. Additionally, it shows how simulations of uses can be arranged for checking the quality of (new) proposals, using principles and characteristics of usage-spatial relationships learnt from previously observed places.

At the same time the chapter promotes GIS as a successful practical tool to build, develop and maintain a body of empirical knowledge gained from different types of behavioural maps. In this context, GIS database offers a transparent examination of places through different combinations of behaviour pattern attributes e.g. the type of activity, gender, age etc. (Goličnik Marušić, 2011). It enables a designer to look at places from any desired viewpoint of attribute combinations, which may most intrigue him or her. In addition to providing information about different elementary peculiarities of patterns of occupancies, GIS based behavioural maps can show the results which have arisen from deeper investigation, e.g. how often a certain activity has happened, how intensively it has occurred per certain temporal unit, how the patterns of each certain activity were differentiated with regard to the presence of others, and so on.

2. How to produce a behavioural map?

Behavioural map is a product of observation and a tool for place analysis and design at the same time. It was developed by Ittelson et al. (1970) to record behaviour as it occurs in a designed setting. Accordingly, spatial features and behaviour are then linked in both time and space. There are some fundamental conditions which need to be met before any recording of behaviour can start. It is necessary to obtain an accurate scale map of the area to be observed, to clearly define the types of activities and details about behaviours to be observed, to schedule specific times and their repetitions for observation, and to provide a system of recording, coding, counting and analysing.

Chronologically, some of the most common ways are systematically writing notes and filling formatted tables. The development of photo-video techniques has influenced the latter methods, and nowadays computer-oriented and supported techniques are forthcoming. Environment-behaviour studies take as their basis the inseparable duality of the behavioural phenomenon and its environmental context, especially in the outdoor environment often recorded through photography and/or video. Both these media clearly show the evidence and temporal consequences of events within a place but they cannot directly give either very detailed micro-scale relationships or the overall synoptic situation of the place itself. However, for the purpose of this chapter, the latter objectification is especially necessary. Thus, the medium of (urban) designers and planners - the map, a physical layout of a place, is adhered to. This medium can convey basic and exhaustive information about an environment and enables the researcher to record changes, suggestions, decisions or any other movement in/about a place, too.

However, although an accurate scale map is a first pre-condition of behavioural mapping, the literature review on the application and development of this method/technique shows that it does not always serve as a recording board. It could have a role of an informative object in the whole process and can result in some general notes, while, for example, behavioural records are collected in a carefully developed table. This shows that there are different techniques available in recording observations, depending on the scale and the nature of the research problem. Typologically different manners in which the behaviour can be recorded are possible. Ittelson et al. (1970) stress that other possible ways of presentation include graphs, pictures, and combinations in which tables may be superimposed; and that floor plan, table, and any other methods of data presentation are all equivalent behavioural maps. This chapter discusses behavioural mapping matrixes, behavioural maps in a narrow sense and a type which combine both the manners. The latter is discussed in relation to GIS.

2.1. Behavioural mapping matrix

The literature review shows (e.g. Bechtel et al., 1987) that behavioural tables were quite often used in recording behaviour in indoor environments. Such a table usually consists of rows representing physical locations and columns representing behaviour. An index mark at the intersection of a row and column then indicates whether the behaviour had occurred at that location. Goličnik (2005) redesigns and upgrades some sort of table for recording observations of outdoor places in Edinburgh (UK) and Ljubljana (Slovenia).

Data on the use of parks and squares in each city were collected during the month of May (May 2002 – Edinburgh; May 2003 – Ljubljana). May was chosen as the time when the weather was likely to be warm and the outdoor activity pleasant. A day observation unit represents four sections: morning (10am-12pm), early afternoon (12-2pm), afternoon (2-4pm) and late afternoon (4-7pm); during the week as well as weekend. Observations were usually conducted from one location in a setting, from where a good overview across a place was provided. As some places were too big to be observed with one overview across the entire place, they were divided into more sub-areas, usually three or four. Each spatial unit was observed for 10 minutes. Altogether two parks, six squares and one square-like street were observed in Edinburgh; and two parks, five squares and a park-like square in Ljubljana; 105 observations were made in Edinburgh and 106 in Ljubljana.

Because of the complex feature of the table and because each cell could collect more than one information, a table is addressed with the behavioural matrix. Every cell in the matrix can collect quantitative and qualitative data. The completed matrix from the field observation is exemplified in Figure 1.

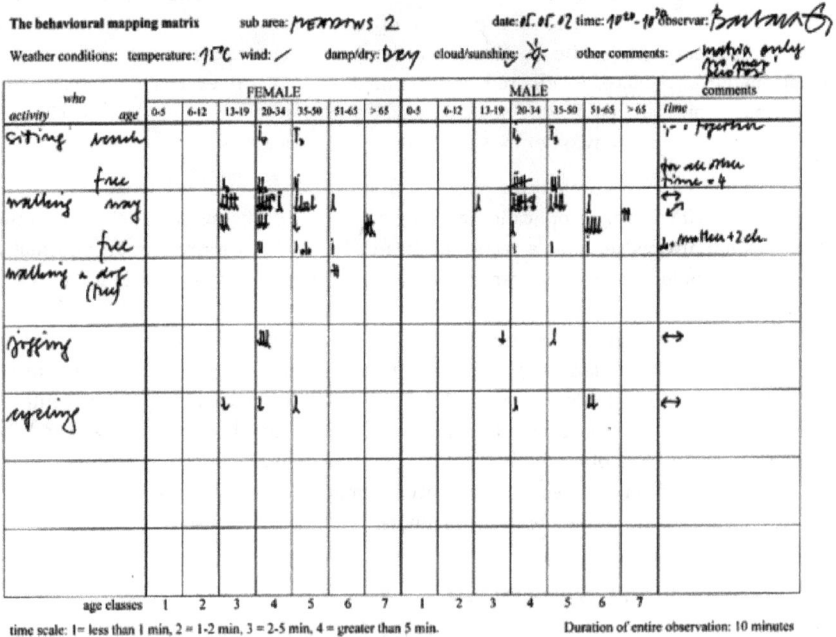

Figure 1. The behavioural matrix with the recorded data (Goličnik, 2005).

From methodological point of view Goličnik (2005) tested two types of behavioural maps: behavioural matrixes and drawn behavioural maps, so called behavioural maps in a narrow sense. The results showed that there were only a couple of situations where mapping as behavioural matrix was appropriate. This was the case for quite simple places such as square-like street with outdoor settings of tables and chairs, where beside those sitting at tables, people were only moving in transition through the place, usually alongside. In order to be able to comment on usage-spatial relationships as thoroughly as possible, drawn behavioural maps were introduced in places, where more complex situations occurred.

2.2. Behavioural map in a narrow sense

In order to create a database as informative as possible, it is important to organise the entire process of mapping very thoroughly. The process of recording behaviour itself needs to be as condensed and inclusive as possible. Accordingly, attributes such as the type of activity, the users' gender and age, duration of the activity, time of the day of occupancy, time of the week of occupancy, movement direction and weather conditions at the presence of the activity; all describe an observed activity in a place. Consequently, a coding and counting system needs to be selected before a technique's implementation is addressed. It is important that the system is designed in a way, which minimises the effort required for recording observations. Thus mapping preparation includes a list of some anticipated activities, their assigned symbols and additional coding (e.g. duration, age group). However, it is of key importance that the list of anticipated activities stays open-ended for any possible new activities to be added, and attached symbols for these unexpected or infrequent activities to be developed in the course of the observation.

Behavioural maps produced at sites were manually drawn on paper prints. The list of symbols for expected behaviours in places distinguishes symbols for male and female participants (see Figure 2). The appropriate symbols were then drawn on prepared prints of site plans. They were accompanied by some qualitative information as well; such as the duration of an activity, the estimated age of the participant (age group), the direction of movement, as well as the time and date, and the description of the weather conditions when the activity occurred. Equally the areas occupied by certain activities or behaviours were documented on a map in scale 1:1000 (see Figure 2).

2.3. Behavioural map in a narrow sense combined with behavioural matrix

In principle, the usage of both behavioural matrix and behavioural map in a narrow sense combined together at the same time at the observation field is hardly ever reasonable. Practically, it is suggested to select the most appropriate technique, depending on the nature of the research problem. However, as field observation usually requires some repetition, i.e. there are always several maps produced for the observed place. If for any grounded reason the observer mixes both recording techniques for the same place during the observation period, all attributes of coding systems must be previously synchronised. It is more likely to find a combination of both types of recording when manually gathered data is digitalised and databases are created, or when any direct recordings, e.g. GPS tracing, are introduced and data are organised into databases. Following subchapters exemplify some of such situations.

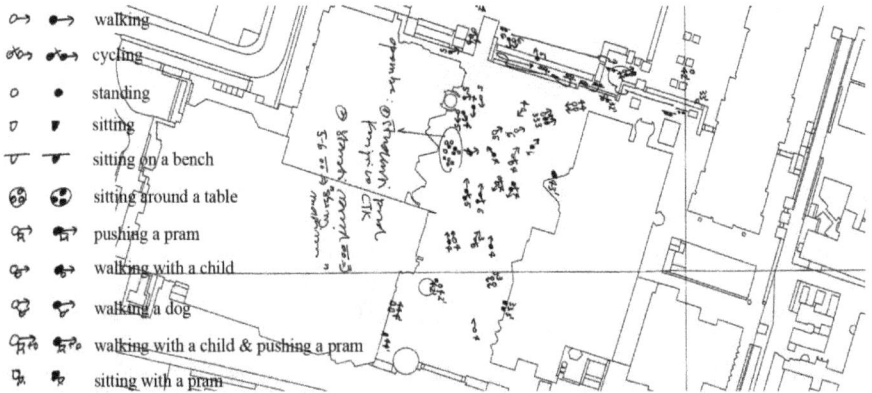

Figure 2. A set of activities from the entire list of expected activities including their attached symbols, specifying male and female users, used for recording activities in squares and parks of city centres of Edinburgh and Ljubljana (Goličnik, 2005), applied in Trg republike, Ljubljana within one observation session.

2.3.1. Digitalisation of data collected via behavioural matrixes or behavioural maps in a narrow sense

The process described at this point is relevant to the research conducted in Edinburgh and Ljubljana (Goličnik, 2005). Every symbol recorded manually was transmitted into its digital version in the same way as recorded in the original map. A map of the observed area was projected on the computer screen. The location of the re-recorded activity was identified with the cursor and clicked when the location was verified. All the attributes of each re-mapped symbol were described in the attached table under its given serial number. Following such a procedure, every database of each place consisted of information layers, based on a day-order structure. Point symbols within the layers represent single users originally recorded in the place. Properties of an activity included in symbols developed for manual behaviour mapping and the characteristics of other circumstances, such as weather conditions, time of day and day of the week, captured within symbols and matrixes of original records, were described in the table attached to those point symbols visualised on the map. So GIS behavioural map is composed of a map of geo-located activities and the attached table with detailed descriptions of these activities (see Figure 3).

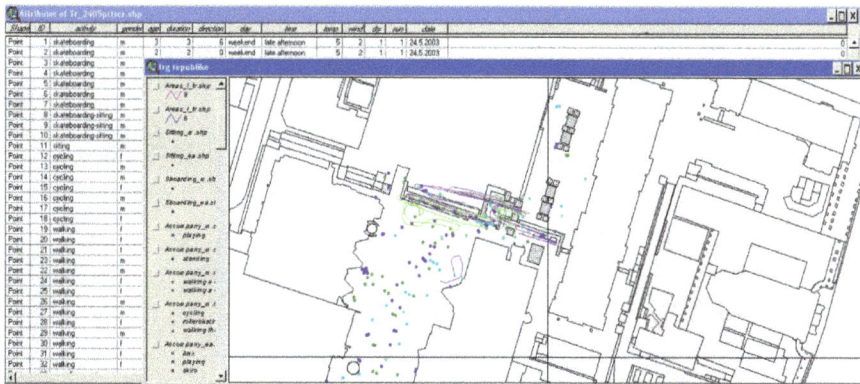

Figure 3. Representation of a GIS supported map transformed from a manually produced behavioural map of Trg republike, Ljubljana (Goličnik Marušić, 2011).

In some cases, where behavioural matrix technique was used to record behavioural patterns in places, re-rendering followed qualitative notifications in the matrix's cells and picked up graphical information from spatial sketches and schemes along with the original matrix.

2.3.2. GPS device as a data source

Usage of GPS devices is increasing in popularity in transportation studies, from traffic flow to cyclists and pedestrian tracking (e.g. Nielsen, 2005; van Schaick and van der Speck, 2007). Visualisation of collected tracks is some sort of a behavioural map, although recording process itself does not reflect all the key elements immanent to classic behavioural mapping. However, usage of tracking technologies such as GPS makes it possible to collect large datasets on human behaviour with a high level of accuracy, combining directly temporal and spatial data. The important advantage of behavioural maps conducted with the use of a GPS device is that collected information is well compatible with any GIS application. Nevertheless, the biggest disadvantage is that this approach might lead to intervention effects and inhibit natural behaviour, as users are aware of their participation in a research project. However, there are situations when usage of GPS to produce behavioural maps is credible.

Goličnik Marušić et al., (2010) studied Ljubljana cyclists' behaviour implementing several approaches, including the use of a simple GPS device. Cyclists, who were interested in participating in the research, contacted the researchers and were given a time period when a device was available. Each participant kept the device for several days and collected data of her/his journeys. Besides the GPS device, each participant collected a blank behavioural matrix designed especially for the study to collect other qualitative information relevant for each recorded trip (e.g. gender, age group, purpose of the journey etc.) (Figure 4, part 2).

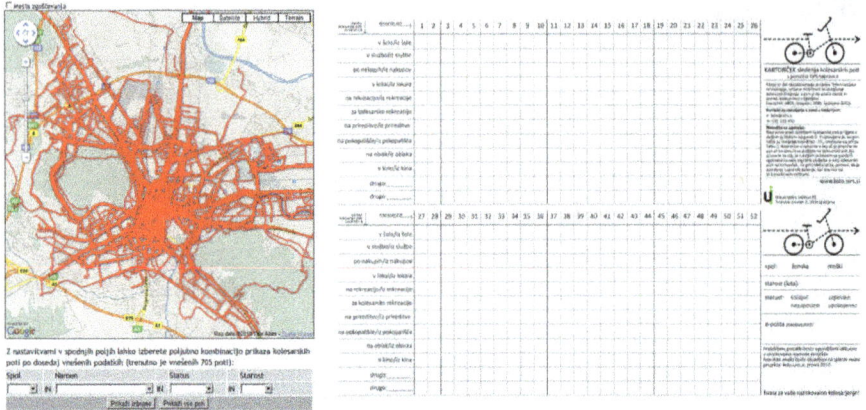

Figure 4. Behavioural map of cyclists' tracks resulted from the use of GPS devices and two sided blank behavioural matrix where each cyclist collected other required data (Goličnik Marušić et al., 2010; http://kolo.uirs.si/).

Data collected from a GPS device and those recorded into a matrix were merged together to get the whole picture about users and their journeys. Monitoring and review of an emerging behavioural map was performed on Google Maps API (Figure 4, part 1). The approach combined both recoding techniques: behavioural mapping in a narrow sense was automatic; other relevant data was collected manually. Although the approach combined both recording techniques, it represents a deviation from classical behavioural mapping as spontaneity of observation is excluded and because it focuses on specific users; all users except cyclists were excluded in advance.

2.3.3. Behavioural mapping in a virtual environment

The study of cyclists' paths in Ljubljana (Goličnik Marušić et al., 2010) developed a web based GIS portal for production of cycling behavioural maps of Ljubljana. This web based portal offers 3D and 2D virtual environments in which cyclists digitalise their own cycle-tracks on the map (Figure 5). Beside this GIS interface, the portal http://kolo.uirs.si/ includes user registration module and a secured online survey to assure basic descriptive information about the cyclist and the purpose of his/her journeys. Thus, it was assured that physical behavioural patterns with linked descriptive data were collected at the same time, so that finally a comprehensive behavioural map was composed.

Figure 5. Examples of the established GIS portal showing steps in data collection and data representation (Goličnik Marušić et al., 2010).

The example clearly shows that the accuracy of collected data may be an issue and that some corrections and adjustments of collected data must be provided before its use for further analysis (e.g. Lachance-Bernard et al., 2011).

2.4. Implications and limitations

Commenting on the purpose of behavioural mapping, Bechtel et al. summarise that: "it is to locate behaviour on the map itself, to identify kinds and frequencies of behaviour, and to demonstrate their association with a particular site. By associating a behaviour with a certain environment it is then possible to both ask questions and draw conclusions about the behaviour and its relationship to a place" (Bechtel et al., 1987: 23). Discussed findings in research practice show possible different manners in which behaviour can be recorded, and suggest two general principles: behavioural tables/matrixes and behavioural maps in a narrow sense.

Every technique used claims a different approach and has some advantages and disadvantages. The developed behavioural matrix enables simple and complex collection, both of quantitative and qualitative data. Observing and recording using the matrix considers especially the number of people involved, noting their age, gender and many times the duration of an activity as well. Drawing a map emphasises the particular location, certain uses and their duration. Age and gender are easily recorded variables in each approach. It seems to be most suitable to use a matrix to record the activities in which the main characteristic is movement such as walking, jogging or cycling; and to use a map technique to record more static activities such as sitting, playing within a certain area, waiting and similar. The maps were found especially appropriate for observing areas with frequent changes. Furthermore, their overlapping allowed the researcher to get a brief intermediate review.

The combined use of both recording techniques, especially in GIS supported environments and various IT related means for recordings, assures a qualitative database for further comparisons, raises new challenges and offers new possibilities in data collecting, their further use and analysis. In this way this chapter can contribute some innovations to the methodology in the practical and theoretical field of environment-behaviour studies.

To sum it up, one must be aware of limitations of literary behavioural mappings. The accuracy of recording the location of observed activities on a map may have some degrees of error, even when all researchers are trained and tested for inter-observer reliability. Recording the location of activities by the use of a GPS device might appear to offer a more accurate way of locating individuals but it may inhibit natural behaviour. However, it would derive a considerable volume of data for analysis. Therefore, such approach might have less disturbing moments if a particular type of behaviour is in focus, e.g. cyclists. GPS is also proven to be quite good to study behaviours representing transition (van Schaick and van der Speck, 2007).

However, either of the behavioural maps provides a shorthand description of the distribution of behaviours throughout a place. They are useful if sufficient repeated observation in a place is done. Examples shown compare the use of spaces on the settings by male and female users, by activities engaged, duration of the activities and their distribution. The major value of behavioural maps as a research tool, lies in the possibility of developing general principles regarding the use of space that apply in a variety of settings. Overlapping individual drawn behaviour maps can show some characteristics and changes in using spaces in terms of activities, number of people engaged, gender, and all the other variables that were explored.

3. Applicability of behavioural maps

Behavioural maps record people's behaviour in real spatial settings and, by that, talk the language of research in a design manner. They offer great potential to represent behavioural patterns as visual data, and as such act towards the reconciliation between design and research in the field of planning and place design. At this point applicability of behavioural mapping and its contribution in spatial planning, place design and decision-making is addressed. Accordingly, the chapter is divided into sub-chapters discussing the roles and potentials of behavioural maps for quality of places, for refinement of designers' notions and knowledge about usage-spatial relations and their inclusion in place design, and for comprehensive simulations which enable modelling of liveable environments.

Firstly, the chapter discusses behavioural maps as scripts of the actual uses mapped in places, using repeated observation at different days, times and weather conditions. Such value of behavioural maps is represented in empirical knowledge about dimensions and spatial requirements, especially for some long-stay active uses, such as ball games in parks and skateboarding in squares, and how long-stay passive uses such as sitting, might relate to them, as well as how transitory activities relate to both long-stay engagements. In addition, it illustrates how some activities can be contiguous, while some others require 'buffer' zones between them for effective use.

Secondly, the chapter uses behavioural maps to address activities imagined in parks and squares by urban designers, using two approaches: mapping likely uses in detailed maps of selected places, and revealing a physical structure of a particular place by knowing its behavioural patterns. On this basis, the chapter examines designers' tacit knowledge about the usage-spatial relationships and highlights potential applicability, the role and value of behavioural maps and empirically gained knowledge in the design of parks and squares. This shows that designers' beliefs and awareness about uses in places, in some aspects, differ from actual use. From this point of view, the chapter reveals a need for effective design-research integration and stresses the importance of behavioural mapping as a source of empirical knowledge and its incorporation in design.

All examples are related to databases and analyses from Goličnik (2005). The data collected by mapping is reliable because of repeated observations on different days, times and weather conditions. Conditions regarding the reliability of the data collected at the workshops with urban designers were met by asking experts about places that were unfamiliar to them.

Finally, possibilities for simulations as results based from knowledge stored in behavioural maps are discussed. This sub-chapter assumes that any dynamic system can be characterised as behaviour in a space. Therefore the notion of behavioural mapping is understood as wide as possible.

3.1. Behavioural map as a check-list for quality of places

Repeated behavioural observation resulted in some common patterns of occupancies that appear to be correlated with particular spatial layouts and details. Behavioural maps analysis show actual dimensions of effective environments for one or more uses and show how design guidance can be arrived at, based on the particularities of the case study sites and cities. Here lies the potential for using information derived from behavioural maps analysis for assessment and evaluation of quality of places.

Analysis of different parks shows that a certain spatial definition such as a corner or a path with different degrees of transparency are not the ultimate clues to spatial occupancy per se. Groups of trees, some prominent single trees or any other objects can play a crucial role. What matters is a spatial articulation and a placement of uses in a place relying on a certain distance from it. It is reflected, for example, in occupancies, distanced at least 5 metres away from transparent edges such as tree lines along pathways of the patches, predominantly without trees, congregations right up against a solid edge, whether a steep slope or a bank, and in the areas of smaller groups of trees or solitaires. Figure 6 summarises and illustrates some of these situations.

Different spatial qualities of settings and their conduciveness to passive usage, such as sitting or lying down in a park, are exemplified on the basis of empirical evidence represented on the assembly behavioural maps for Tivoli, Ljubljana, Princes Street Gardens, Edinburgh and the Meadows, Edinburgh. The upper set of Figure 6 shows sitting right next to a solid edge such as a slight slope, and people in the areas of smaller groups of trees or solitaires. The lower set exemplifies sitting further away from transparent edges, such as tree lines along pathways, especially near their intersections, and no sitting along any of the broader zones along the path with no other spatial definition.

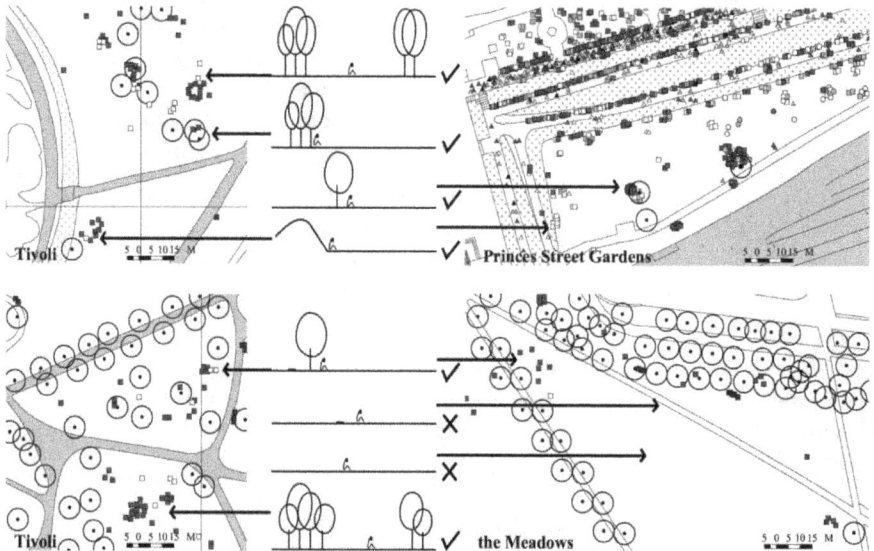

Figure 6. Spatial qualities of settings and their correlations to passive usage (Goličnik, 2005).

The results also show that, even if the lawn patch is huge, if it is not articulated, unless any temporary articulation is available, uses such as sitting or lying down are less likely to occur. The importance of spatial articulation reveals, especially in places where there are not very many different elements of spatial definition, that it is not only physical spatial definitions that might direct uses in a certain spatial occupancy, but also that the presence of other uses, to a certain degree, can perform this function as well. Mainly larger groups of active participants can articulate places and, in doing so, create room for themselves and for others (see A and P in Figure 7). Goličnik (2005) has found that the size and the shape of lawns in parks are not particularly crucial for any passive occupancy; but they can be of greater importance for informal ball games, especially playing football.

Analysis of the parks showed (Goličnik and Ward Thompson, 2010) that spatial articulation is the clue to spatial occupancy. Activities, especially those significant for active group games, form patterns buffered by voids, in several quite predictable ways. There are two significant types of buffer zones that different active, long-stay users need: the buffer between an edge, whether solid or transparent, and active users (e.g. informal football); and buffers between a number of adjacent active groups occupying different territories.

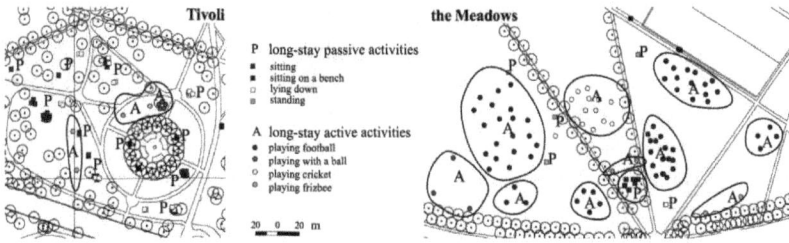

Figure 7. Examples of a situation where, when more uses take place in a park and make new boundaries, how sitting and lying, 'move in' with regard to them (Goličnik, 2005).

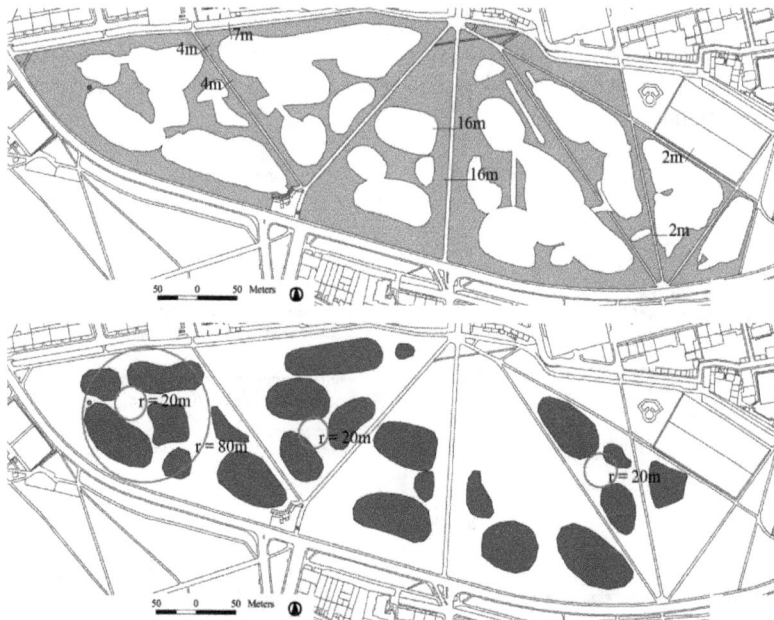

Figure 8. Cumulative minimum buffer zone (above) and relationships between size and shape of a patch and its occupancy by long-stay active uses based on records from a daily pattern (below) (Goličnik Marušić, 2011).

To illustrate the first type, compact groups of informal football players are likely to require a distance of at least 4m from an inner transparent edge, such as a tree-lined path. Activities forming looser and smaller groups, such as a couple of frisbee players, are likely to occupy a space closer to an inner edge, e.g., at least 2m from a path. For the second type the minimum 'common open' area between activity spaces is quite difficult to define precisely in terms of a surface area, since the activities taking place depend on the size, shape and edge qualities of a green patch. However, an abstract form which can describe the minimum activity buffer space commonly needed between groups of activities is a circle of 20m radius.

Addressing conduciveness and quality of places from the usage point of view, some observations of squares in city centres of Edinburgh and Ljubljana (Goličnik, 2005) have shown that even though none of the squares examined was planned as a skateboarders' platform, a certain articulation of those places has stimulated its users to be there and to use it for their pastime. However, this certain articulation in itself did not ensure optimal use. The size, shape and vertical articulation of the available space are of key importance.

For one of the usual skateboarder's actions, which consists of approaching an elevated spatial element such as a step, in order to jump on it, slide along, and then jump off it, the necessary full length of a step required needs to be at least 15m. The adjacent area before such a step should allow a skateboarder to approach it along a curve of a circular line of at least 20m in diameter, and to undertake virtually a jump-preparation journey on it of at least 5m. Thus, a platform at least 3m wide, attached to a long step, would allow such a minimum jump-preparation journey.

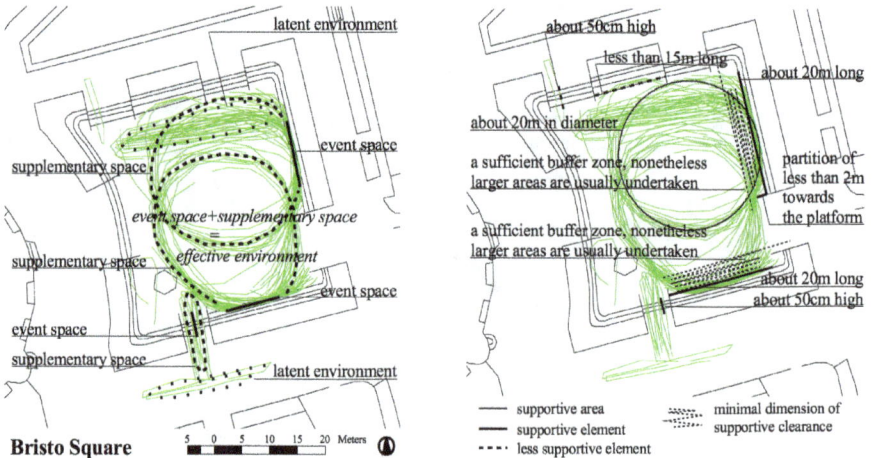

Figure 9. Effective environments, their structure and dimensions exemplified in Bristo Square, Edinburgh (Goličnik, 2005).

When addressing usage-spatial relationships in more detail, a spatial configuration of places becomes important. The examined cases show that steps which merge into a flat platform, are essential elements that attract skateboarders; but the merged, flat area is crucial to enable their actual use. Physical traces of actual activities, represented as graphical information on the map, elucidate the inner structure of the effective space, reflect usability and in this way, address its spatial capacity. Equipment such as boxes and some other light structures, which skateboarders brought to the stage, evokes latent environments. Bristo Square in Edinburgh and Trg republike in Ljubljana were the most illustrated examples of these skateboarders' performances (Figures 2, 3, 9,10)

Further analysis shows (Goličnik, 2005) that neither of those squares (Bristo Square, Trg republike) has any benches but participation in sitting is remarkable. Compositions of steps are attractive as sitting places as well. This examination showed also that sitters' and skateboarders' actual effective environments do not overlap. Sitters are searching for sheltered, back-covered, less exposed areas, and places with a view of either attractive landscape or actions (see figures 9 and 10).

Although the intensities in participation at Bristo Square, Edinburgh in any long-stay occupancy on a cold, windy and cloudy afternoon (first map in Figure 10), in comparison to a warm, sunny weekend afternoon (second map in Figure 10) is lower, both maps show a similar behavioural pattern of occupancy. People sitting were mostly occupying the upper steps in the parts where broad planting beds enclose the square. The distribution of uses recorded on a nice early afternoon during the week (third map in Figure 10), when there were no skateboarders around and people sitting there could have occupied any square inch of the staircases, it shows a similar pattern of passive occupancy with the other days. One may speculate that sitting along the eastern edge was not evident as the skateboarders' performance on the stage was missing.

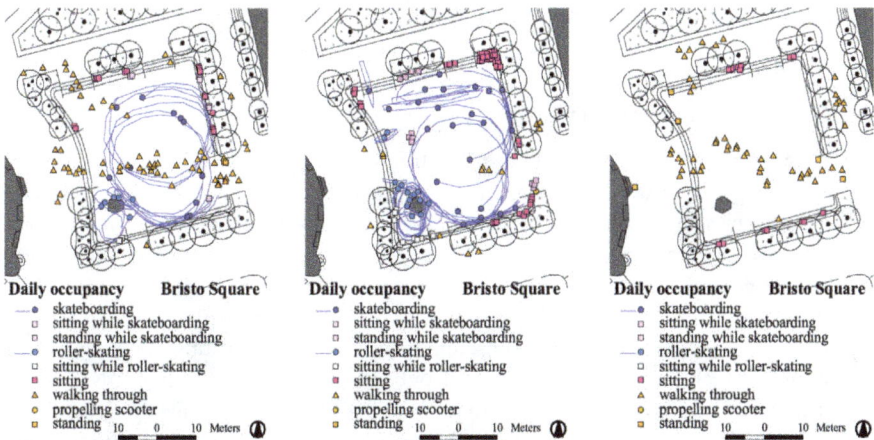

Figure 10. Maps of typical occupancies in Bristo Square, Edinburgh, represent from left to right the daily occupancy on a weekday afternoon of poor weather, the daily occupancy on a weekend afternoon of very good weather, and the daily occupancy on a weekday early afternoon of good weather (Goličnik, 2005).

These examples show that, spatially speaking, optimal settings for sitting and skateboarding are different, and that Bristo Square successfully serves both activities at the same time. They also show different concerns addressing time aspects of occupancies. According to a weekly occupancy, transitory activities are more common to occur on weekdays than during the weekend. Time differences recorded for different times of the day, reflect that active long-stay activities such as roller-skating and skateboarding, are usually participated in on an afternoon, no matter what the weather, whereas they might not be seen earlier in the day.

The common act of observation and behavioural mapping shows up as an effective way of searching for empirical knowledge about usage-spatial relationship. It is a comprehensive way of collecting the evidence about 'where, how and what is going on' in a place. Moreover, mapped physical dimensions of uses are seen as a potential which could inform about the design of places and could become a way of negotiating landscape forms.

3.2. Behavioural map as a check-list of tacit knowledge of designers

The examination of the variety and richness of information and empirical knowledge, which may be gained from observation and GIS mapping, raises the question, what type of information about usage-spatial relationships, if any, would be most required or would be highly demanded to inform designers about better design? A search for an answer might reflect on the meaning, the value and the role that empirical knowledge may play in the design (process). In other words, what are designers' perceptions and imagination about public places and the activities that are likely to occur in them? To illustrate some of these challenges, Goličnik (2005) conducted several workshops with urban landscape designers in Edinburgh, a sample of 35 participants in total. Participants carried out the tasks individually. This inquiry was intended as a pilot study, to look for some basic insights about such issues, rather than a fully-fledged investigation.

The examination of the use of public spaces from both angles, what activities do users actually engage in, and what designers' perceptions and beliefs about them are, has shown that often there is a lack of knowledge about the nature of occupancy of one particular use or a certain combination of them. A brief conclusion showed that despite the fact that generally there had been a good recognition and listing of likely activities in specific places, concerns about the spatiality of uses and conduciveness of places to support them were not always fully considered. For example, a scrutiny of sitting, as one of the more usual activities in urban open public spaces was not sufficiently precise, especially when concerns relate to selectivity in usage's placement. Some spatial entities, which the workshop participants suggested were most likely to be used for sitting, were revealed by the observation-based evidence to be marginally in use or even unused (see Figure 11).

Further analysis of various overall assembly maps shows that, in many cases, the likely settings proposed by the designers for some activities such as skateboarding, playing or even sitting, differ in their rates and intensity of use from that recorded by surveying the actual use of settings which usually facilitate any such activity. It shows that designers were not only generous when placing certain activities such as sitting, practically everywhere possible, but also general in graphical expression using a 'bubble-diagrammatic' technique.

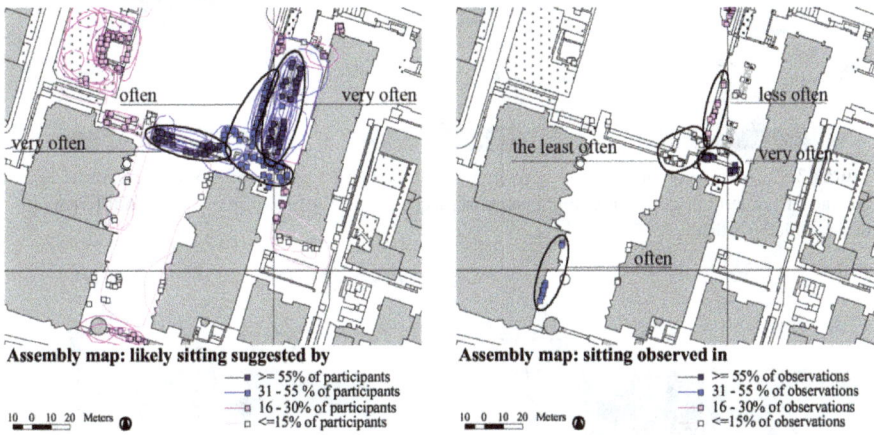

Figure 11. Locations of likely occupancy by sitting in Trg republike, Ljubljana as they resulted from the workshops and actual occupancy recorded in the entire observation period (Goličnik, 2005).

This implies that designers are not used to express their visions of places via behavioural patterns of their imagination of actual activities in places, which also indicated that they do not operate with behavioural maps. Such findings argue for the importance of getting information from empirical evidence that might inform a better understanding of the usage-spatial relationship, as well as the design itself. Accordingly, knowing about physical dimensions, spatial peculiarities of activities and their interrelations, as well as about physical conduciveness of places to occupancy, and by this recognising the potential and effective environments, is of key importance when aiming for responsible design. Behavioural maps are effective sources of such knowledge.

The examination of design proposals, which were produced by workshop participants based on known behavioural patterns of places, not only shows sometimes partly considered responses but also other characteristics of these proposals. They are reflected in solutions which, many times, are driven by a spatial context itself, a formal structural message from the area or the ideas beyond that, addressing a place's integration in a broader context. In fact, in all such cases, there was little response to a behavioural pattern as such in the first place. Goličnik (2005) reveals three categories reflecting the basic design layouts: 'quite responsive' (13 cases), 'mostly indifferent' (14 cases) and 'non-classified' (8 cases). These latter usually refer to proposals which actually do not address design as such, but which stay in a level of zoning and bubble diagram expression. Figure 12 exemplifies typical representative drawings for each category as described above. Although the examples which were classified as 'mostly indifferent' were significant for hardly responded at all, there were also examples which were overwhelmed by the design proposal. These latest are exemplified in Figure 13.

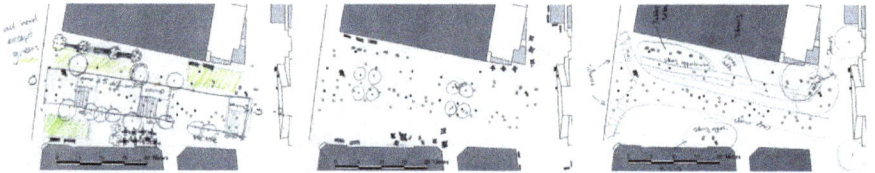

Figure 12. Categorisation of drawings in relation to comprehensive design proposals for Dvorni Trg, Ljubljana, from left to right exemplifying responses as follows: quite responsive, mostly indifferent, non-classified (Goličnik, 2005).

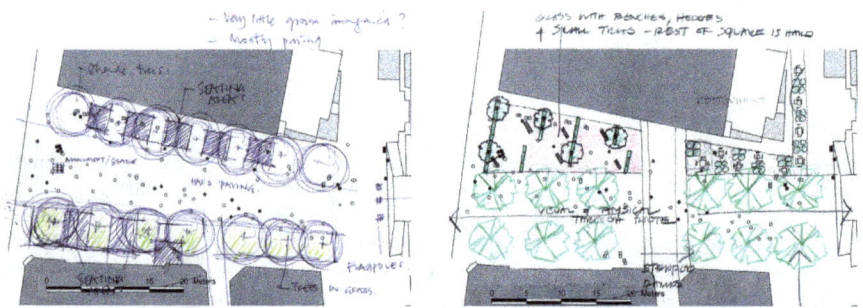

Figure 13. Left example shows indifferent design proposal for Dvorni Trg, Ljubljana, where except for certain respect given to a main transitory flow, no other behavioural pattern has been taken into account at all. Right example shows indifferent design proposal for Dvorni Trg, Ljubljana, overwhelmed by the repetitive, geometric pattern, disregarding some particular behavioural patterns or their combinations (Goličnik, 2005).

Figure 13 shows the responses, where the initial indications to a reading of the behavioural patterns are quite good; but then a sort of 'make up' process spoilt them. It is about the production of usually geometrical and repetitive patterns of spatial definition elements such as benches, trees, terraces and the like, with less and less connection to the existing behavioural patterns (see left example in Figure 13). The example of Figure 13 on the right shows the representation where the attention is paid to the spatial composition and its rhythmic order, and also how this emerging composition fits into a broader context. Thus, the proposed designs resonate more with the spatial properties themselves, rather than focusing more on a response to the behavioural patterns themselves.

Speaking about designers' tacit knowledge and their responses to usage of places, these examples hint at the importance of design as a composition of spatial elements with regard to their inner order and a response to its structural spatial context, rather than a combined response to applied creativity and consideration of the 'social structure' of a place.

3.3. Behavioural map as the key input data for comprehensive spatial simulations

Behavioural maps are applicable in a variety of different scales, from the small scale of a single room, to a medium scale of a street, a square or a park to the large scale of a whole region. How behavioural maps can be created and applicable in the detailed scale of urban design has been already discussed so far. However, considering these different scales, they can for example, represent movement of individual users in particular rooms, from the living room in a house to a classroom in a school, as well as migrations in regions. In this relation it is important to consider that behavioural maps cannot be limited exclusively to individuals. Accordingly, regarding the scale and the nature of the research problem it is possible to address behaviour as a variable, which varies from a single person, group of persons or even other subjects than people relevant in spatial planning and design (Batty, 2005; Porta et al., 2009; Chaker et al., 2010). Actually, the focus on the subject depends on the viewpoint chosen and the scale preferred. When paying attention to individuals on the street, then commuters are in focus and the street represents their physical environment or context. On the other hand, when the street as a cumulative result of single commuters is in focus, its surrounding becomes recognised as the context of the studied phenomena. Both views have been addressed in a study of cyclists in Ljubljana (Goličnik Marušić et al., 2010).

Beside sliding over different scales and by this accommodating the focus on the subject to be observed, it is also possible to discuss how behavioural maps at one level or certain scale can help to interpret behaviour of subjects from another related layer. For example, behavioural patterns in squares as discussed before, can help to interpret liveability of (local) businesses in the influential area of the square. Similarly, the liveability rhythm map of such businesses is also one of the behaviours of the square; i.e. cafés open first, followed by shops, crafts and restaurants. Such mapping can be applicable on the scale of a block, a quarter, or even an entire city or a town. Via such behavioural mapping it is possible to address the capacity of places by documenting timetables of working hours, frequencies of peak occupancies, and the like, for both, spontaneous or programmed uses.

Combination of different observations in different scales and accuracies (e.g. patterns of people, rhythm of activities in businesses and services etc.) can lead to a comprehensive simulation of a place. Data or information that influences behaviour of a certain place is not always directly visible. Non-spatial data, which backgrounds a certain place, such as money flows behind the business in a place, can also be mapped and therefore spatially expressed. However, such abstract descriptions of places reflect some physical characteristics of places which can be expressed or measured by behavioural maps. The point is that the behavioural mapping approach enables us to visualise primarily non-spatial data (e.g. detailed parameters of users in places, economic efficiency, frequency of cultural events, etc.) on the maps. Thus such data are placed into the contexts of other data of spatial realities. In an urban environment for example, they may be suitable for models and simulations of revitalisation and regeneration processes. (e.g. Goličnik and Nikšič, 2009)

The discussion showed that it is possible to look at different relevant spatial parameters in similar ways. Behavioural mapping is recognised as an approach, which can be beneficial due to the way of documenting and organising data which describe the observed phenomena. However, when an approach enables different and diverse data to be gathered in compatible ways; they are more likely to be put onto a common denominator, set into a compatible database and compared between each other. Such database or network of databases can serve as a base for comprehensive spatial models. Sufficient amount of data allow also simulations of behaviour in places. Such simulations showed that by successful simulation the amount of actual observation and data collection may be reduced. And vice versa, sufficient amount of observation can offer a foundation for better design practice or simulations.

On one hand, a collection of data from comparable sites enables reliable reasoning about the site which has not been sufficiently observed, or has not been observed at all. However, checking the site (e.g. one or two observations) is required afterwards to assure reliability of the simulation. On the other hand, results gained from repeated observations such as sizes and shapes of effective environments for certain activity or more of them can be used for simulation of new places. This is not applicable only in the physical design of places. It may also be helpful in the phase of programmatic and economic efficiency planning within the entire design and decision-making process.

When using data for simulation in relevant or comparable places, it is of key importance to select a set of criteria which enable us to define compatibilities of places. For example, data collected in public open places in Edinburgh and Ljubljana was reliable because a sufficient number of repeated observations was done, but most of all as sites were proven to be comparable. Both cities are middle-sized European capital cities with a relatively small population in total (Edinburgh about 450,000 inhabitants, Ljubljana about 300,000). Both are important national and international cultural, educational, as well as political centres; in this respect, they have a similar atmosphere in terms of their daily routine. Both cities belong to the mid-latitude temperate climate zone, Edinburgh to the oceanic, Ljubljana to the continental; which causes some differences during different times of the year. In mid spring, a popular season for outdoor activities, they are quite similar, especially when conditions are dry, no matter if it is sunny or windy. By contrast, the frequency, duration, and volume of rainfall are more likely to be different. In Ljubljana's continental climate, heavy rain is usually a downfall, whereas in Edinburgh, types of downfall can vary from mist to mild showers or heavy rain. A comparable number and typology of selected places representing popular, central public open spaces of different sizes and micro-spatial contexts, were selected for study within an area of about 2 km² in each city.

Behavioural maps as scripts of behaviour of any studied spatial phenomena are especially effective within the GIS environment as it is a tool which can convey data referenced to different scales and enables organisation of data, its visualisation and analysis. These characteristics of GIS place it as a highly valuable source and environment for spatial simulations.

3.4. Implications and limitations

Addressing spatial characteristics of places by their usability and by reflecting from that on the conduciveness of places to occupancy, this chapter and the results of commented researches have shown in some detail the nature of effective environments and have suggested some vocabulary for their descriptions (see Figure 9). By examining the effective environments for skateboarding in more detail, it has shown that it consists of two adjacent spaces: the 'event space' and the 'supplementary space'. The event space is the actual space through which the activity is 'installed' in the place. It represents a position which a person or a group of people engaged in a particular activity occupies in a place. The supplementary space is the available space at hand to this person or a group of people, which actually enables the complete activity to happen fully. As the event space is necessary for that activity to be invited into the place, its supplementary space addresses its satisfactory staying in a place. Both spaces together form the effective space of an activity. The same is true about spaces for playing football and spaces for sitting. Therefore, it is important to understand the spatial articulation as a necessary but not sufficient condition for some kinds of use. In this respect, the examination of places through the distribution and physical dimensions of behavioural patterns in them, has enabled a discussion about what the effective environments are and how to imply their importance and relevance to design practice.

Attempts to find generalisations from findings related to dimensions of usage-spatial relationships can offer important contributions to our knowledge and understanding, even if they are still rather speculative. That is so, firstly because as the sub-chapter dedicated to designers' tacit knowledge showed, that designers' beliefs and awareness about uses in places currently often differ from actual use, and secondly, because previous theoretical stands and guidance for the built environment have now been supported by data for large open space occupancies, as well.

The discussion also showed that behavioural mapping has potential that can be used not only in studying actual patterns of occupancies of places but that the same approach can be applied to different subjects of spatial reality. However, one must bear in mind also the limitations related to this.

4. The role of GIS in relation to behaviour mapping and place design

The main advantage of GIS in relation to behaviour mapping is that the system can be updated practically with any information. Results showed that GIS databases can effectively serve as an inventory tool, providing basic descriptions and information about activities in places. They offer an understanding of those places by patterns of spatial occupancy with regard to their elementary characteristics, those that describe their peculiarities when being carried out. Maps, as products of visualisation, can represent the spatial data of behaviour patterns as patterns reflecting occupancies at different times of a day, or days of a week; as patterns structured by the duration, nature or type of occupancy; as patterns showing the occupancies only under pleasant weather conditions and the like.

Therefore, if the demand in practice for better designs calls for the importance of empirical knowledge, the technical possibilities performed by GIS (Jiang and Yao, 2010) can show and reflect on richness of its contents. In this chapter, the initial contents of the empirical knowledge directly reflect the information recorded through the observation. A GIS application upon it elucidates different aspects of this basic information about the usage-spatial relationship and provides a variety of different information derived further from this original collection. From this point of view, especially because the knowledge is visualised on maps, it also reveals the possibility for more effective design-research integration and stresses the effective incorporation of empirical knowledge in design.

A specific value of an empirical knowledge gained by GIS behaviour mapping lies in the notion about the effective environment, that what happens in any particular environment depends on those who use it. Hence, while urban designers might create potential environments, people create effective environments. The challenge is not only to see to what degree or how much of a potential environment can be transformed into an effective one, but also to discover its inner structure. Empirical knowledge, stored in digital interactive GIS databases and shown on maps after a selection process, can provide some insights into different dimensions of the usage-spatial relationship, such as gender or age differentiations and the like.

On the other hand GIS enables observation of hidden dimensions of subjects relevant and related to the dynamics of spaces. Qualitative descriptions of variables are immanent to such a system. Especially, as GIS (e.g. Ratti et al., 2005; Porta et al., 2005; van Schaik and van der Speck, 2008; Jiang and Yao, 2010) is a system which can be sourced in various ways, from actual data collection in the field, to data collection from a virtual environment, such as the internet. Moreover, GIS itself represents the initial platform which may offer some basic observation for a certain level of subject observed. Thus, no actual field observation is needed at all and data input is automatic. However, quality control of such data input is necessary.

4.1. Implications and limitations

This chapter found that GIS, as an analytical and evaluation tool, draws the closest approximation to meeting the challenge of 'talking about the physicality of spaces, using the language of patterns of uses'. GIS maps are recognised as an effective tool to represent and interpret behavioural patterns as visual data. They also translate recorded evidence into a body of empirical knowledge and preserve the connection of related non-spatial data to the material place. By such an association of behaviour with a certain environment, it is possible to ask questions and draw conclusions about the behaviour and its relationship to a place, and from such reasoning, move towards a reconciliation between design, decision-making and research.

Although the main advantage recognised within the topic of this chapter is openness of GIS as a system for updating the system with any relevant data, one must bear in mind that there are limitations in data collection which can significantly influence the system as such.

Quality of data depends on the quality of the observer. Generally, credibility of data can be relatively high if the observer or group of observers is trusted. Some errors may appear when data is transferred from the manual to the digital version. The accuracy in locating studied subjects is better by direct input to digital media, especially if addressing inhibition of natural behaviour; but the practical difficulties of implementing this are considerable. When using GIS web portals to collect the data of users' experiences in place usage (e.g. cycling in Ljubljana, Goličnik Marušić et al., 2010), the skills, seriousness and accuracy of person providing the information directly influence the overall quality of data provided. Besides, when using GIS web portals to collect data, another limitation must be born in mind. Although such approach can assure quite big amount of data, it does not include user groups which are not familiar with the internet and IT. When web portals are used as data sources reliability of the data itself may be questionable. Therefore, no matter the way in which data are collected, quality control must be provided by cross-checking for example, so credibility of database can be assured, including the 'smoothness' of data.

Another very important issue when using GIS either as a source of data or as an environment to produce data, is checking on compatibility and comparability of data offered within the system. For example, the socio-economic context, the functions and density of the surrounding area may vary and are certainly likely to influence the activities and level of use within a space. In this discussion, the sites chosen were roughly comparable with regard to such considerations but this potential limitation must be recognised before generalising to other parks in different (e.g. suburban) parts of other towns and cities. Well-used (and well-maintained) city parks are likely to be perceived as safe places to visit, sit on the grass, etc., but this may not be the case for emptier or poorly maintained spaces, or where there is no surrounding land use that provides informal policing of the area. However, the context of each site may also be analysed and compared with other sites using GIS.

To sum it up, when using behavioural maps of any kind, the context of the studied behaviour is always important. Various characteristics of this context are described in GIS with different levels of accuracy. Therefore it is important to develop mechanisms for detecting comparable information, and sites as GIS as a system can be quite reliable for comparative spatial studies.

5. Conclusion

This chapter discusses the actual uses observed in certain places to illustrate the role of behavioural mapping (based on observation and the use of GIS databases) in place design, monitoring and decision-making. The major value of the use of behavioural maps as a research tool, lies in the possibility of developing general principles regarding the use of space that apply in a variety of settings. GIS based behaviour maps extract behaviour evidence into layers of spatial information to give a better understanding of the individual and collective patterns of use that emerge in a place. The overlap of behaviour maps can show some characteristics and changes in using places in terms of activities, number of people engaged, gender, time of day, duration of activity and similar.

The empirical knowledge gained by behaviour mapping is seen as an addition and a complement to other research approaches and tacit designers' knowledge. Such empirical knowledge brings a good template and/or a starting point for further post-occupancy evaluation analysis, as well as benefits to public participatory processes in planning and design decision-making. It is especially important when addressing user-groups such as youngsters or elderly people, even homeless people, who may not respond to participation, but whose preference and existence in a place is important, especially when talking about democratic, all-inclusive design.

Whatever technique is used or can be expected to be used as the most efficient one in future, taking into account IT development, GIS as a tool, with the ability to produce and use databases, remains of key importance. Another practical challenge in the future lies in the technical field of computer software. Having the appropriate and affordable equipment for recording digital data directly (in the field), a programme which could support a simultaneous coding of all sorts of behaviour attributes such as gender, type of activity, its duration and similar; as well as any other conditions regarding the weather, time of the day or any other relevant aspects, would be very helpful.

The combination of GIS and activity mapping provides a powerful tool to support designers with empirical evidence of the relationship between environmental design and the use of open space that is spatially explicit and therefore presented in a spatial and visual language familiar to designers. This makes it more likely to be useful and useable.

Another contribution of behavioural maps which has been brought up within this chapter is the notion that they are not necessary to be limited to the examination of behavioural patterns of activities in places. Behavioural maps and behavioural mapping can be interpreted according to various relevant aspects related to place design and research. Such viewpoints enhance the role of GIS in behavioural mapping, as GIS as a system does not serve only as a tool for visualisation and interpretation within the context of studied phenomena, but represents a common comprehensive database and works as a generator of simulations. One of the key aspects which can contribute to building such a comprehensive database is the ability of GIS to perform linkages between different scales of any data stored in such database originally may belong to. Another crucial characteristic in terms of GIS efficiency for behavioural maps is the ability to compare compatible patterns or phenomena.

Author details

Barbara Goličnik Marušić
Urban Planning Institute of the Republic of Slovenia, Slovenia

Damjan Marušić
The Surveying and Mapping Authority of the Republic of Slovenia, Slovenia

6. References

Batty, M. (2005). Agents, cells and cities: new representational models for simulating multiscale urban dynamics, Environment and Planning A, vol. 37, no. 8., pp. 1373–1394 ISSN 0308-518X

Bechtel, R.B., Marans, R., & Michelson, W. (1987). Methods in Environmental and Behavioural Research, Van Nostrand Reinhold, New York.

Chaker, W, Moulin, B & Theriault, M. (2010). Multiscale Modeling of Virtual Urbn Environments and Asociated Populations, In Geospatial Analysis and Modeling of Urban Structure and Dynamics, Jiang, B. & Yao, X. (Eds.), pp. 139-162, Springer, ISBN 978-90-481-8572-6, London.

Goličnik Marušić, B. (2011). Analysis of patterns of spatial occupancy in urban open space using behaviour maps and GIS. Urban design International, vol. 16, no. 1, (Spring, 2011), pp. 36-50, ISSN 1357-5317

Goličnik Marušić, B., Tominc B., Nikšič, M., Bizjak, I. & Mladenovič, L. (2010). Informacijska tehnologija, urbana mobilnost in izboljšanje kakovosti življenja: z GSM-i do analiz stanja in potreb kolesarstva v Ljubljani. Information technology, urban mobility and improvement of quality of life: state of the art and needs regarding cycling in Ljubljana using GIS, Urban Planning Institute of the Republic of Slovenia, Ljubljana.

Goličnik, B. & Nikšič, M. (2009). Geographic information system behavioural and cognitive mapping in the city centre revitalisation process. Journal of urban regeneration and renewal, vol. 3, no. 2, (October – December 2009), pp.161-179, ISSN 1752-9638

Goličnik, B. & Ward Thompson, C. (2010). Emerging relationships between design and use of urban park spaces. Landscape and Urban Planning, vol. 94, no. 1, (January 2010), pp. 38–53, ISSN 0169-2046

Goličnik, B. (2005). People in Place: A Configuration of Physical Form and the Dynamic Patterns of Spatial Occupancy in Urban Open Public Space. PhD thesis. Edinburgh College of Art. Heriot Watt University, Edinburgh.

Ittelson, W.H., Rivlin, L.G. & Prohansky, H.M. (1970). The Use of Behavioural Maps in Environmental Psychology, In Environmental Psychology: Man and his Physical Setting Prohansky, H.M., Ittelson, W.H. & Rivlin, L.G. (Eds.), pp. (658-668). Holt, Rinehart & Winston, New York.

Jiang, B. & Yao, X. (Eds.) (2010) Geospatial Analysis and Modeling of Urban Structure and Dynamics, Springer, ISBN 978-90-481-8572-6, London.

Lachance Bernard, N., Produit, T., Tominc, B., Nikšič, M. & Goličnik Marušić, B. (2011). Network Based Kernel Density Estimation for Cycling Facilities Optimal Location Applied to Ljubljana, Computational Science and its Applications, Procedings of ICCSA 2011, Part II, ISBN 978-3-642-21886-2, Springer, Santander, Spain, June, 2011

Nielsen, T.S. (2005). The Potential for the Exploration of Activity Patterns in the Urban Landscape with GPS-positioning and Electronic Activity Diaries, Life in the Urban Landscape: International Conference for Integrating Urban Knowledge and Practice, Gothenburg, Sweden, May 2005.

Porta, S., Crucciti, P., & Latora. V. (2008). Multiple centrality assessment in Parma: A network analysis of paths and open spaces. Urban Design International vol. 13, no. 1 (Spring, 2008), pp.41-51, ISSN 1357-5317

Porta, S., Strano, E., Iacoviello, V., Messora, R., Latora, V., Cardillo, A., Wang, F. & Scellato, S. (2009). Street centrality and densities of retail and services in Bologna, Environment and Planning B: Planning and Design, vol. 36, pp. 450-465.

van Schaick, J. & van der Spek S. (2007). Application of Tracking Technologies in Spatial Planning Processes: An Exploration of Possibilities, REAL CORP 007: To Plan is Not Enough: Strategies, Plans, Concepts, Projects and their successful implementation in Urban Regional and Real Estate Development, ISBN 978-39502139-3-5 , May 2007

van Schaick, J. & van der Spek, S. (Eds.). (2008). Urbanism onTrack: Application of Tracing Technologies in Urbanism. IOS Press, ISBN, Amsterdam, the Netherlands.

Chapter 7

Probabilistic Evaluation of the Extent of the Aquifer – Case Study

Marek Kachnic

Additional information is available at the end of the chapter

1. Introduction

Environmental researchers often analyze phenomena and objects which can be determined as "poorly-defined" (Fisher 1999). Due to mathematical rules of sets these are the objects which are difficult to be assigned to a specific class of objects in compliance with dichotomic rules of binary (Aristotelian) logic. The extents of lithofacial, stratygraphical and tectonic units are represented in the cartographic studies based on the point or local reconnaissance performed in the field. With respect to distances, those limits are of probable course, more or less similar to the natural boundary. Error assessment of graphic presentation of the geological units has not been expressed in values yet. There is even a lack of approximate estimation of probability to define the unit borders. A similar problem can be identified in hydrogeology. There we have to evaluate homogenic areas and units with similar properties of groundwaters or aquifers. Hydrogeologic cartography offers diversified studies, due to the credibility of used data. It directly finds reflection in the accuracy and likelihood of estimation of the extent of groundwater bodies, their amounts and quality.

In environmental researches, the proper use of the information (or the lack of the information) leads to searching for way to represent this kind of data. It is argued (Leung & Leung, 1993) that the application of Boolean logic (the all-or-nothing system) in the GIS design causes the following problems: a) it imposes artificial precision on intrinsically imprecise information, graded spatial phenomena and processes, b) it fails to determine and communicate to users the extent of imprecision and error, c) it is inappropriate to human cognition, perception and thinking processes, which are generally embedded with imprecision (Leung & Leung 1993).

The aquifer is a good illustration of a "poorly-defined" object (fig. 1). This result from the lack of information on its extent (especially for the confined aquifer), facies changes within the aquifer and various definitions of the aquifer.

Figure 1. Example of "poorly-defined" objects: limits of forest, valley and aquifer

In order to describe correctly "poorly-defined" objects in modeling proper methods should be found. They should allow intermediate values to be defined between conventional evaluations like 1 and 0, true or false. For describing "poorly-defined" objects we can use one of the multi-valued logic such as "fuzzy logic" (Zadeh, 1965), kernel-based probability density function estimation (Brundson, 1995) or other probability methods such as Bayesian theory or Dempster – Shafer theory (Shafer, 1976; Klir and Yuan 1997; Eastman 1999b). This paper attempts to evaluate the extent of the unconfined aquifer in a nonparametric – probabilistic scale **with help of last one**.

The main study objective was to evaluate the probability that an unconfined aquifer may be found in each pixel location in a surface represented in the studied area. Due to a large amount of data IDRISI software was used to achieve the aim.

2. The area of research

The research area of 1300 km^2 in the east part of the Pomeranian Lakeland in Poland was chosen for a testing procedure. This area lies completely within the limits of the last (Veichselian) glaciation. Along with the relatively slight hypsometric differentiation, the relief of the studied area is characterized by a few forms of fluvioglacial and glacial source. The main form is outwash sediments (the Wda sandur) and a moraine plateau (Fig. 2).

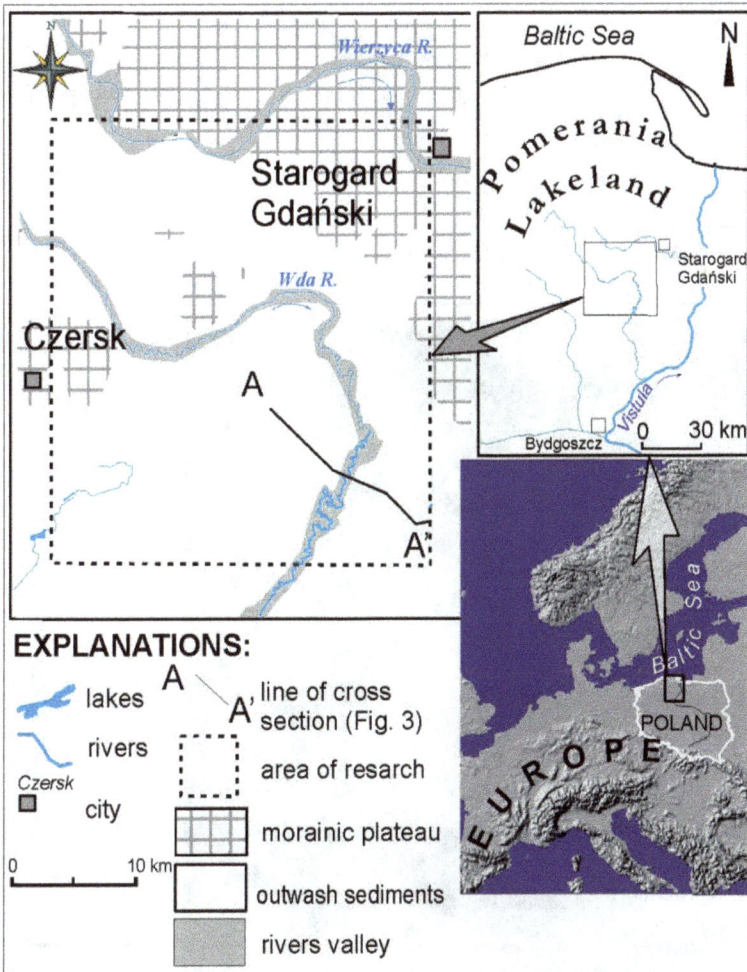

Figure 2. Location of the study site

There are only Cenozoic water bearing strata recognized within the log wells. The Pleistocene water bearing layers are the major aquifer for the studied area. It consists from of the unconfined aquifer and a few confined aquifers (Fig. 3).

The (hydro)geologic recognition of research area is rather shallow and diversified due to uninhabited area. These is shown on the Fig. 4

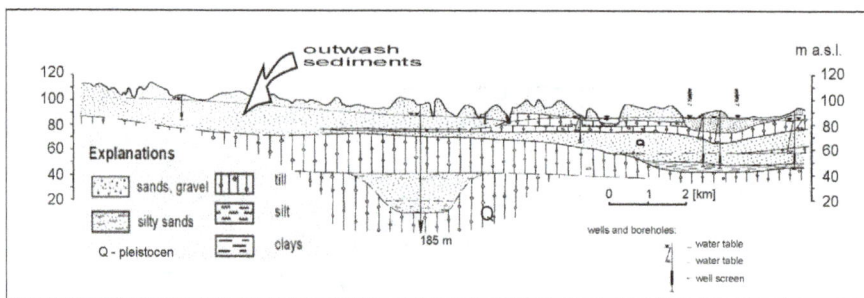

Figure 3. Geologic cross-section along line A-A' in figure 4.

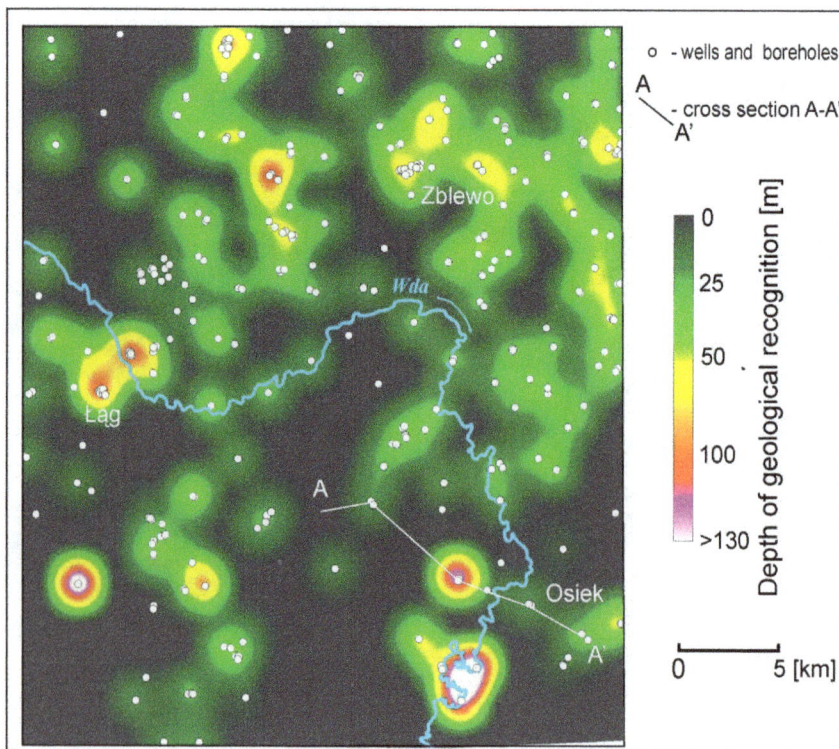

Figure 4. Map of depth of geologic recognition of the research area.

3. Methodology

The Dempster-Shafer theory (Shafer, 1976) is an extension of Bayesian probability theory. This theory makes a distinction between probability and ignorance and allows for the expression of ignorance in uncertainty management (Lee et al., 1987; Klir & Yuan, 1995). The basic assumption of Dempster-Shafer theory are that ignorance exists in the body of knowledge, and that the belief for hypothesis is not necessary to the complement of the belief for its negation. By using the "belief functions" to represent the uncertainty of hypothesis, the theory releases some of the axioms of probability theory. The resulting system becomes a superclass of probability theory. However, it suffers from the need for large numbers of probability assignments and from the need for independence assumptions (Malczewski, 1999). Unlike Bayesian probability analysis, D-S theory explicitly recognizes the possibility of ignorance in the evaluation, i.e. the incompleteness of knowledge or evidence in the hypothesis (Eastman, 1999).

The research objective was performed on IDRISI (ver. Andes) raster based software program. In IDRISI, the BELIEF module (Fig. 8) can be used to implement the Dempster-Shafer theory. BELIEF constructs and stores the current state of knowledge for the full hierarchy of hypotheses formed from a frame of discernment (also called state space). BELIEF first requires that the basic elements in the frame of discernment be defined. As soon as the basic elements are entered, all hypotheses in the hierarchical structure will be created in the hypothesis list. For each line of evidence entered, basic probability assignment images (in the form of real number images with a 0 – 1 range) are required with an indication of their supported hypothesis.

4. The development of knowledge base

The research question guides us to define the frame of discernment – it includes two elements [present] and [absent]. The hierarchical combination of all possible hypotheses, therefore, includes [present], [absent] and [present, absent]. We are most interested in the result generated for the hypothesis [present]. The final results produced for the hypothesis [present] are dependent on how all evidences are related together in the process of aggregation.

Given knowledge about existing wells and given expert knowledge about the occurrences of aquifers, each evidence is transformed into a layer representing likelihood that a aquifer exists. The aggregated evidence produces results that are used to predict the presence of an aquifer and evaluate the impact of each line of evidence to the total body of knowledge.

For study several bitmaps and pixel maps were prepared. At the beginning each map included separately: point, line or area data which all confirm or deny the occurrence of the aquifer in a **dichotomy scale 0 and 1**. In the next stage, the information on each map was changed due to the prepared membership function. As a result, the **pixel map** with values **from 0 to 1** was obtained. Finally, all the maps (information layers) were put to the BELIEF module **and probability map was compute**.

4.1. Data input for the unconfined aquifer

There is significant difference between analyzing the extent of the unconfined or confined aquifer with the use of the GIS methods. With the exception of the wells as the best indicator of existing aquifers, there is far more indirect evidence of occurrences for the unconfined aquifer than the first one. For example, they are: springs, rivers, lakes, the area of extent of alluvial or outwash deposits. There is high probability that the unconfined aquifer will be close to these forms (Kachnic, 2010).

The author focused here only on the unconfined aquifer. Still, GIS methods are a tool for, by and large, two dimensional data. And there is no advanced GIS raster program for analyzing three dimensional data yet, required for analyzing the confined aquifer.

For estimating the extent of the unconfined aquifer in a probabilistic scale the following data was selected:

a. locations of wells and boreholes,
b. area of the extent of the outwash and moraine plateau,
c. course of main rivers and lakes,
d. map of depth to the water table in the area where there are no impermeable sediments on the terrain surface.

4.2. Creating probability maps (fuzzyfication)

The stage of fuzzyfication is a procedure, which allows for converting a discrete image (bitmap) into images with a probabilistic (nonparametric) scale. The reliability of the obtained maps depends on the applied parameters of fuzzyfication controlled by a membership function. For researches the following assumptions were taken:

4.2.1. Probability for background

Initially, for the whole research area the background value was assumed as constant 0.5. That means there is no proof for the **existing of** unconfined aquifer and there isn't evidence for the lack of the aquifer in research area.

4.2.2. Membership function for wells

Wells are the best point markers of the aquifer. For these features, the area in the close vicinity of the wells obtained high likehood. The map with locations of the wells with unconfined conditions were rasterised and all the pixels where there were wells, obtained the value "one". The pixel values are high in the area calculated by means of empirical formula, and finally, the pixel value decreased down to the level of the background (Fig. 5).

The empirical formula was applied as one of the assumptions for the extent of the unconfined aquifer. That was a formula:

$$R = 2s\sqrt{k * H} \tag{1}$$

known as the Kusakin formula (Bear, 1979; Hölting, 1996).

Where R – is a radius in [m] of depression cone; s – the maximum depression observed in a well [m], k – the coefficient of permeability, in [m/24h] and H – thickness of the aquifer in the well log, [m].

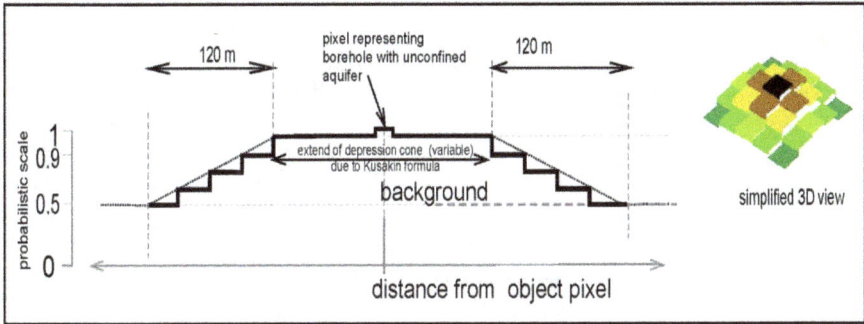

Figure 5. Graph of the membership function for wells in an unconfined aquifer.

The Kusakin formula isn't good assumption for extend of aquifer and should be established the better one.

4.2.3. Membership function for wells and boreholes with lack of unconfined aquifer

The value "0" was assigned for the pixels where boreholes exist and there is not unconfined aquifer noticed. In the vicinity of those pixels, probability increases from "0" to the value of background for the range 300 [m] (Fig. 6). The above distance was established subjectively as the optimal one after analyzing the geological and hydrogeological cross-section from the research area.

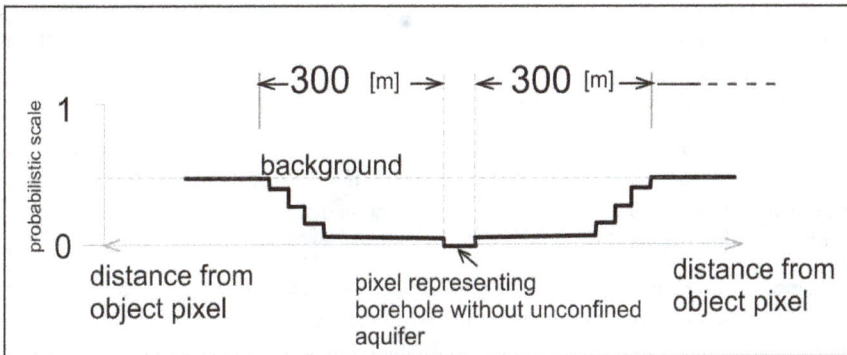

Figure 6. Graph of the membership function for boreholes and wells without unconfined aquifer.

4.2.4. Membership function for boreholes with observed an unconfined aquifer

The value "1" was assigned for the pixels where boreholes exist and there is an unconfined aquifer notice. In the vicinity of those pixels, probability increases from "1" to the value of background for the range 300 [m].

4.2.5. Membership function for the area of outwash sediments and moraine plateau

In the research area the main body of the unconfined aquifer is associated with fluvioglacial outwash (a sandur). The area of the outwash extent was digitized from the Geological Map of Poland in a scale 1:200 000 (Butrymowicz et al., 1978). The rest of the area was classified as a logic negation, which means the area without sand sediments on the terrain surface (i.e. moraine plateau). The value "0.8" was assigned to all the pixels which represent the area of outwash sediments and the river valley (Fig. 2). For the remaining area a constant value "0.3" was established *a priori*.

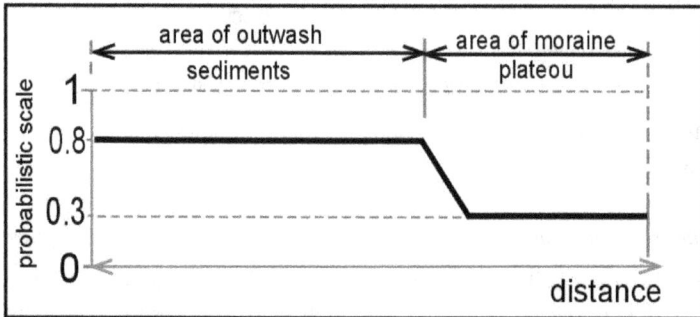

Figure 7. Graph of the membership function for area of outwash and moraine plateau.

4.2.6. Membership function for area in the vicinity rivers and lakes

Rivers and lakes are hydrologic objects with frequent connection to the aquifer, especially the unconfined aquifer. Close to a river or a lake there are often sand sediments with the aquifer, therefore, this closeness to water indicates the plausibility for the aquifer. Only rivers that are longer than 5 km and lakes with the area bigger than 1 ha were analyzed.

Simple statistical methods were used in order to develop the relationship between the distance to water and the locations of the wells. On the basis of that procedure the author found that there should be higher likelihood (the value of 0.8) in the zone 200 [m] from the river banks or lake shores (Fig. 8).

The information about the depth to the water table and the extent of the unit "a" where there are no impermeable sediments on the terrain surface, was taken from the computer Hydrogeological Map of Poland in a scale 1:50 000 (HMofP). HMofP is a new kind of a map, prepared and stored in GIS (Geomedia) system as a multisheet map (Paczyński et al., 1999). From 1994 to 2004, 1069 sheets covering the whole of Poland were made. The map is based on the concept of the main usable aquifer which is a productive aquifer meeting the following criteria: thickness at least 5 m, transmissivity at least 50 m^3/24h, and potential discharge of a well at least 10 m^3/h. All data is kept in 19 information vector layers, which contain among others: topographic situation, well and spring locations, type of the aquifer, water quality classes, aquifers pollution risk classes, land use and hydrodynamic information, e.g. hydraulic head, groundwater flow directions and transmissivity distribution (Herbich, 2005; Fert et al., 2005).

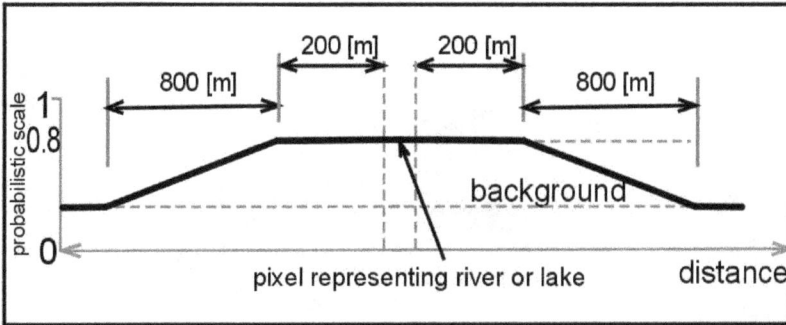

Figure 8. Graph of the membership function for hydrological objects.

5. Stage of calculating

All prepared pixel maps were put to the BELIEF module (Fig. 9).

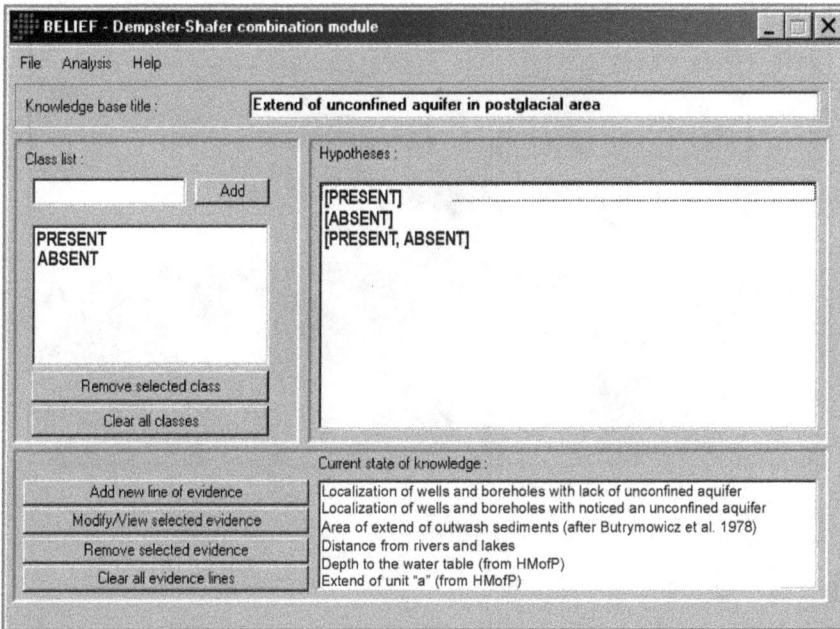

Figure 9. The window of BELIEF module of IDRISI program.

After processing in the BELIEF module a set of maps was generated. These were maps of the degree to which evidence provided concrete support for the hypothesis (belief) (Fig. 9) and the degree to which that evidence did not refute the hypothesis (plausibility).

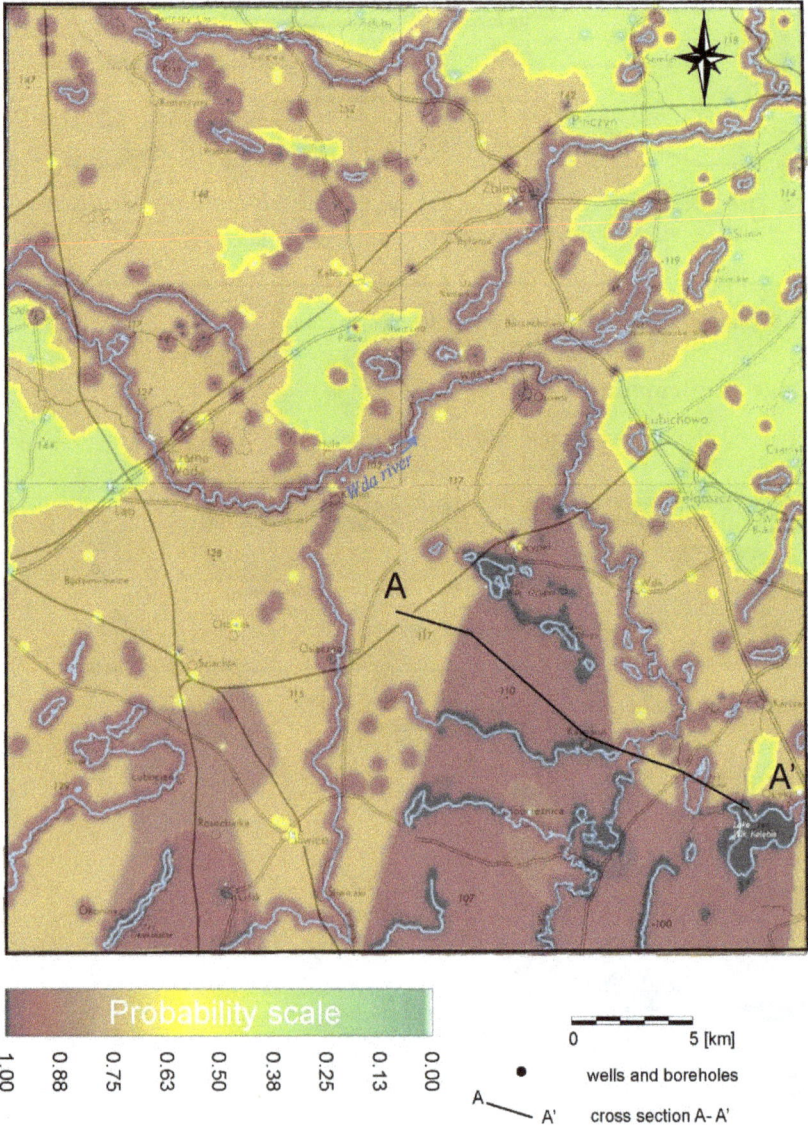

Figure 10. The probabilistic map of the extent of the unconfined aquifer in a research area.

6. Conclusion

The limits of geological and hydrogeological units (structure) presented on the cartographic studies include often significant errors due to poor of recognition. Those errors are highest in the area recognized by drilling holes, i.e. point recognition.

The purpose of the methodology presented herein is to produce a probabilistic information layer of the extent of the unconfined aquifer in the studied area (Fig. 10). It is an attempt to use Dempster-Shafer theory in hydrogeology. Taking into account the fuzzy set theory, the author proposed the assessment of hydrogeological structure extent based upon hydrogeological boreholes where pumping test had been carried out.

Additional information for the probabilistic map come from hydrological, geomorphological investigations and data from other geological reconnaissance point. The accuracy of such map is largely determined by the reliability of database information and established membership functions.

The author treat the generated maps as a supplement to a classic set of information concerning hydrogeology and which provides a new form of a map layer which can be useful for decision-makers. The statistical description of the pixel value on the result map may be used for the assessment of reliability of groundwater model and as decision support for sustainable groundwater management.

Author details

Marek Kachnic
Nicolas Copernicus University in Toruń, Departure of Geology and Hydrogeology, Poland

7. References

Bear, J. (1979). *Hydraulics of Groundwater*, McGraw-Hill Book Company, Inc., New York, pp. 320–347

Brundson, C. (1995). *Estimating probability surfaces for geographical point data: an adoptive kernel algorithm*. Computer and Geosciences, 21: 877-894

Butrymowicz, N. & Murawski, T. & Pasierbski, M. (1978). *Geological map of Poland in a scale 1:200000 - Chojnice sheet*. Wyd. Geol. Warsaw [in Polish]

Eastman, J. R. (1999b). *IDRISI guide to GIS Image Processing*, Vol. 1: 126

Eastman, J.R. (1999). *IDRISI guide to GIS Image Processing*, Vol. 2: 41

Fert, M. & Mordzonek, G. & Węglarz, D. (2005). *The management and data distribution system of the Hydrogeological Map of Poland 1 : 50,000*, Prz. Geol., 53, 10/2: 917-920. Available from www.pgi.gov.pl/pdf/pg_2005_10_2_15.pdf

Fisher, P.F. (1999). *Models of uncertainty in spatial data*. In: *Geographical Information Systems: Principles and Technical Issues*, Longley P.A., (Ed.), 191-205. New York: John Wiley and Sons

Herbich, P. (2005). *Hydrogeological Map of Poland 1: 50000 – Present state and development of computer data base.* Prz. Geol., 53, 10/2: 924-929, Available from www.pgi.gov.pl/pdf/pg_2005_10_2_12.pdf

Hölting, B. (1996). *Hydrogeologie.* 3rd ed. Stuttgart: Enke Verlag [in German].

Kachnic, M. (2010). *The extent of the unconfined aquifer based on the Dempster-Shafer theory on the example of postglacial sandur area.* Biuletyn Państwowego Instytutu Geologicznego 441: 55-62 [in English] Available from

www.pgi.gov.pl/images/stories/NW/Biuletyny_PIG/441/b441_kachnic.pdf

Klir, G.J. & Yuan, B. (1995). Fuzzy sets and fuzzy logic: theory and applications. In: *GIS and multicriteria decision analysis.* Malczewski J. 1999. p. 129. New York: John Wiley and Sons

Lee, N.S. & Grize, Y.L. & Dehnad K. (1987). *Quantitative Models for Reasoning Under Uncertainty in Knowledge-Based Expert Systems.* International Journal of Intelligent Systems, no. 2: 15-38

Leung, Y. & Leung, K.S. (1993). *An intelligent expert system shell for knowledge-based geographical information system.* International Journal of Geographical Information Systems 7 no. 3: 189-213

Paczyński, B. & Razowska L.& Sadurski A. (1999). Aquifers risk assessment by considering their vulnerability presented on the Hydrogeological map of Poland. In: *Proc. XXIX IAH Congress,* Fendeková M. (ed.) and Fendek M. - Bratislava, p 93-99

Shafer, G. (1976). *Mathematical theory of evidence.* Princeton University Press. London

Zadeh, L. (1965). Fuzzy sets. *Information and Control,* 8 no 3: 338-353

Chapter 8

Assessing Agricultural Potential in South Sudan – A Spatial Analysis Method

Xinshen Diao, Liangzhi You, Vida Alpuerto and Renato Folledo

Additional information is available at the end of the chapter

1. Introduction

After almost five decades of war and armed conflict, South Sudan achieved its independence in July 2011. Expectations are high that the independence will bring peace, food security, improved health, and prosperity to its people. The world's newest nation, South Sudan is naturally endowed with agricultural potential given its favourable soil, water, and climatic conditions. It is estimated that about 70 percent of total land area is suitable for producing a wide range of agricultural products, including annual crops such as grains, vegetables, tree crops such as coffee, tea, and fruits, livestock, fishery, and various forest products. To realize such agricultural potential and achieve economic development and broad-based improvements in the nation's living standards, a realistic understanding of the country's initial conditions is required such that appropriate policy measures and agricultural growth strategy can be designed in the near future.

This chapter focuses on analyzing a more realistic agricultural potential in South Sudan in five to ten year horizon. While such analysis seems to be straightforward in most other countries, it is a monumental task in South Sudan given its protracted history of violence. A functional government statistics system that regularly collects socio-economic data literally did not exist during the turmoil years. Hence, our analysis needed to put together different spatial data from several available sources. The key GIS datasets that we used are the 2009 Land Cover data which provides land use information for South Sudan, the Oak Ridge National Laboratory's 2001 LandScan population data, and the most recently updated road condition surveys conducted by World Food Program (WFP). We combine these GIS datasets with the 2008 population census and 2009 National Baseline Household Survey (NBHS) carried out by the country's National Bureau of Statistics (formerly known as Southern Sudan Centre for Census, Statistics and Evaluation). While the agricultural potential is analyzed spatially, the socio-economic datasets, which are both nationally representative, allow the statistical analysis to be carried out at subnational levels such as at the state and livelihood zone levels.

In the next section, we estimate the size and distribution of the different types of land use, as well as the association between agricultural potential and population density in South Sudan. Based on the agricultural consumption and production patterns, the current agricultural values in monetary terms are calculated in Section 3. In the same section, we then estimate the agricultural potential value in the next five to ten years by simulating an increase in cultivated area though cropland expansion and improvements in agricultural productivity. Section 4 concludes.

2. Spatial distribution of different types of land use

The country's current land use and coverage in the different states[1] and livelihood zones[2] is described in this section. Then, we use the length of growing period (LGP)[3] as proxy for determining typologies of agricultural production potential and describe the relationship between such potential and population density.

Current land use

We use a two-step process to derive South Sudan's land use from almost 300 types based on Land Cover data obtained from FAO in 2009. First, the land use types were resampled and aggregated into 18 classes as depicted in Map 1. In the second step, we further aggregated the land use types into 8 categories (Table 1). For agricultural production potential, we use LGP equal to or more than 180 days as an indicator for sufficient moisture and temperature conditions that permit crop growth. Using this threshold, about 80 percent of the country's territory is under climatic conditions that are considered suitable for agriculture. However, the aggregation of the land use types indicates that most of the land that is suitable for agriculture is still under natural vegetation. As shown in Table 1, land that is currently under crop cultivation, most of which are rainfed, accounts for less than 4 percent of total land. Conversely, the largest part of the country is still under trees and shrubs (62.6 percent). Given the country's favorable agricultural climate condition, this ratio is clearly very low as the crop areas account for more than 28 percent of national land in Kenya and 8 in Uganda. Before South Sudan became an independent country, crop areas in Sudan as a whole accounts for 7 percent of total land. Given that the agro-climate conditions are less favorable in the northern Sudan than that in South Sudan, it is obvious that South Sudan is

---

[1] South Sudan has ten states: Upper Nile, Jonglei, Unity, Warrap, Northern Bahr el Ghazal, Western Bahr el Ghazal, Lakes, Western Equatoria, Central Equatoria, and Eastern Equatoria.

[2] The country is divided into seven livelihood zones that are identified under the country's livelihood profile project and defined based on climate conditions and farming systems (SSCCSE, 2006): Eastern Flood Plains, Greenbelt, Hills and Mountains, Ironstone Plateau, Nile-Sobat Rivers, Pastoral, and Western Flood Plains.

[3] The concept length of growing period is used in the Global Agro-Ecological Zone Project led by the International Institute for Applied Systems Analysis and the UN Food and Agriculture Organization. For more detailed information, see Fisher et al. (2002).

significantly underdeveloped in agricultural production. While the large land areas under natural vegetation definitely indicate huge agricultural potential in the country, the challenges to develop them into agricultural land, including required large physical investments and difficulty in identifying suitable farming systems and crop patterns, are huge.

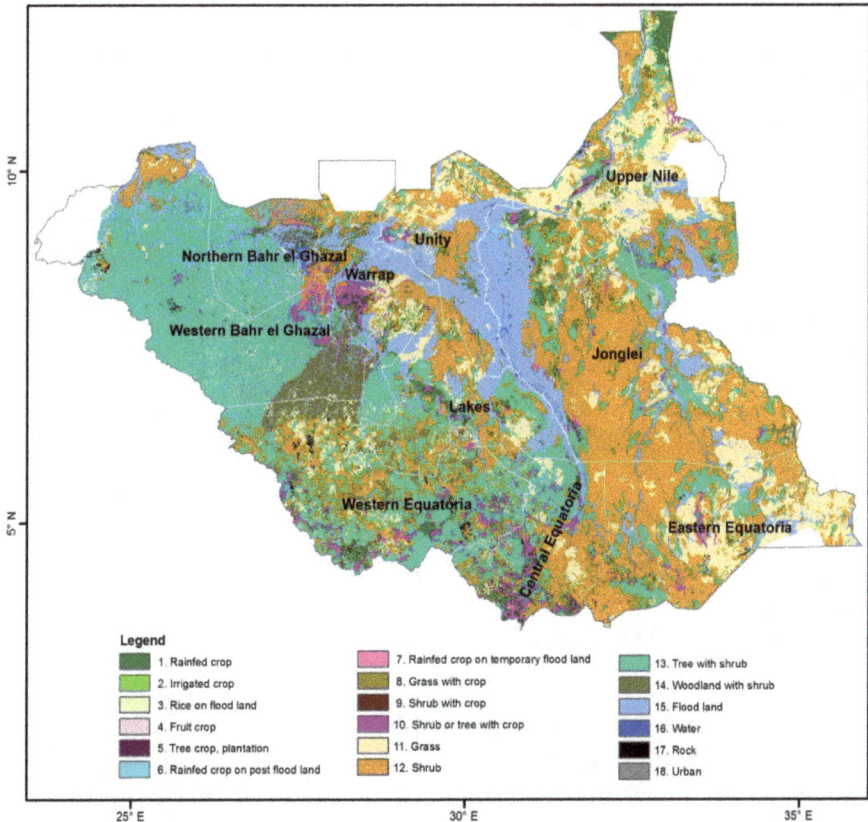

Source: Authors' aggregation using Land Cover database (FAO 2009).

Map 1. Spatial distribution of aggregated types of land use

We further consider the extent of land use types at both state and livelihood zone levels to understand its distribution (Map 2). In terms of cropland distribution, Western Flood Plains, which covers parts of Northern Bahr el Ghazal, Warrap, Unity and Lakes, is the most important livelihood zone, providing 34.2 percent of national cropland and 24.2 percent of national cropland mixed with grass and trees. Moreover, this zone has the highest ratio of cropland over total land, as cropland and cropland mixed with grasses/trees account for 8.5 and 5.4 percent of zonal territorial area, respectively. Greenbelt (spanning parts of Western

Equatoria and Central Equatoria) and Eastern Flood Plains (encompassing Upper Nile and parts of Jonglei) are the two other major crop producing regions, accounting for respectively, 17.6 percent and 26.2 percent of national cropland, and 25.7 percent and 14.6 percent of the country's land mixed crops with grasses/trees. Both zones also have high ratio of cropland to total land as lands with crops and crops mixed with grasses/trees account for 11.4 percent of total land in Greenbelt and 6.8 percent of total land in Eastern Flood Plains. In total, these three agricultural zones provide 78 percent of national cropland and 64.6 percent of national cropland mixed with grass/tree, but only covers about 47 percent of national territorial area.

	Area (in 1000 ha)	Share of total land (%)		Area (in 1000 ha)	Share of total land (%)
A: By 18 types of land use categories			*B: By 8 aggregated categories*		
Rainfed crop	2,379.3	3.7	Cropland	2,477.7	3.8
Irrigated crop	32.1	0.0	Grass with crop	325.1	0.5
Rice on flood land	6.0	0.0	Trees with crop	1,707.3	2.6
Fruit crop	0.1	0.0	Grass	9,633.8	14.9
Tree crop, plantation	6.2	0.0	Shrub and tree	40,526.9	62.6
Rainfed crop on post flood land	25.4	0.0	Trees, shrubs and other vegetation on flood land	9,497.6	14.7
Rainfed crop on temporary flood land	28.5	0.0	Water and rock	482.7	0.7
Grass with crop	325.1	0.5	Urban	37.0	0.1
Shrub with crop	4.3	0.0	**Total**	**64,688.3**	**100.0**
Shrub or tree with crop	1,703.0	2.6			
Grass	9,633.8	14.9			
Shrubs	20,506.6	31.7			
Tree with shrub	17,694.9	27.4			
Woodland with shrub	2,325.4	3.6			
Tree, shrub, and other vegetation on flood land	9,497.6	14.7			
Water	350.1	0.5			
Rock	132.6	0.2			
Urban	37.0	0.1			
Total	**64,688.3**	**100.0**			

Source: Authors' aggregation from 2009 Land Cover.

Table 1. Area and share of total land, by aggregated types of land use

Agricultural potential and population density

Based on the LGP classification, about 27.3 percent of cropland in South Sudan is located in areas with high agricultural potential (LGP of more than 220 days) and another 41.5 percent in the medium potential areas (LGP between 180 to 220 days) (Table 2). To some extent, population determines the current crop production, as well as fulfilling crop system's potential for intensive farming in the short to medium term. Roughly 34 percent and 46 percent of population lives in such areas of high and medium agricultural potential, respectively.

The majority of South Sudanese (85 percent) lives in rural areas, which we classify into two categories: "low density" areas with population less than 10 per square kilometer (10/km²) and "medium to high density" areas with population above that threshold. With 13 people per km², the average population density is very low in South Sudan compared to other countries in the region. The low average is driven by the fact that only 25 percent of the population lives in 83.4 percent of the total territorial lands in South Sudan (Table 2). Accordingly, the population density averages 4/km² in these areas. In contrast, the remaining 75 percent of the population resides in "medium to high density" areas representing just 16.6 percent of country's total land, thereby resulting to density of 57/km².We combine the LGP and population density categories that results in six agricultural potential typologies (Table 2; Map 3).

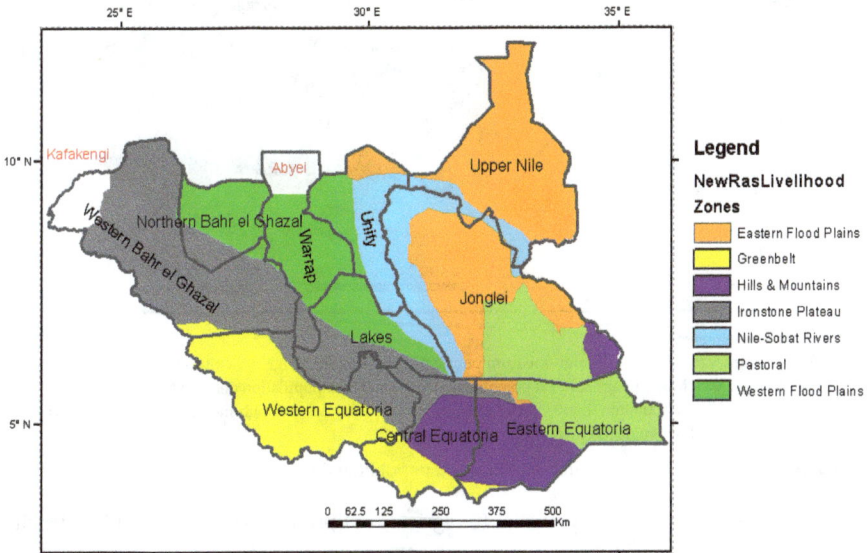

Source: Authors' estimates.

Map 2. The ten states and seven livelihood zones

Our analysis indicates that Type HH, HL, and MH, which are the three typologies of high agricultural potential areas, collectively cover 54 percent of total crop land. This is mostly driven by large areas of MH in Warrap and Lakes representing 26.7 percent of total cropland area (Map 3). This is followed by Type HH (15.3 percent) which can be attributed to the similarly large areas of high population density-high agricultural potential in Western Equatoria and Central Equatoria. Among crop production zones, Greenbelt has the highest share of cropland distinguished as Type HH, while Western Flood Plains dominates the MH category (Map 3). On the other hand, half of the cropland areas in the Eastern Flood Plains are characterized as LL primarily because of the large contribution of Upper Nile region that falls under this category.

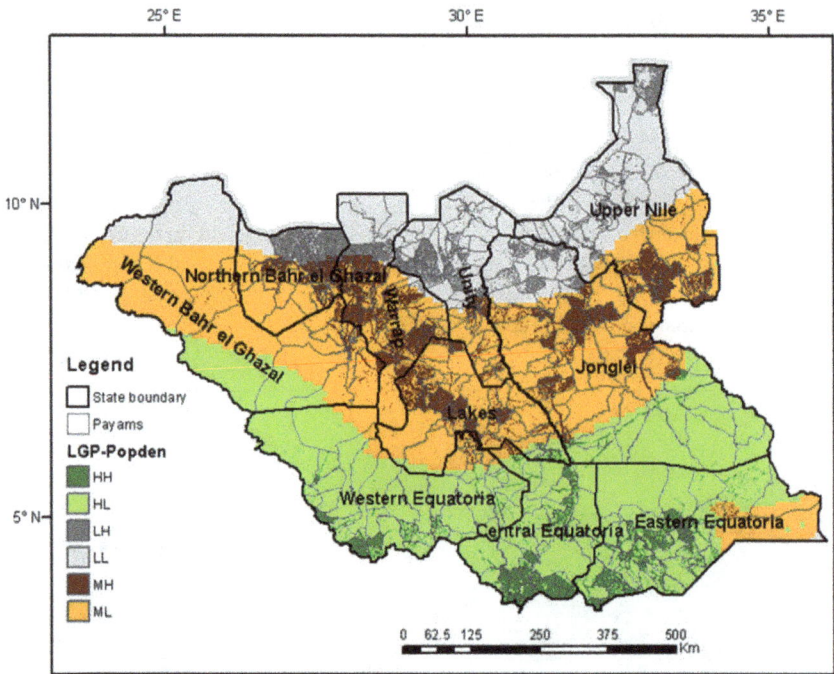

Source: Authors' estimates.

Note: **HH**: High agricultural potential /high-medium population density; **HL**: High agricultural potential/low population density; **MH**: Medium agricultural potential and high-medium population density; **ML**: Medium agricultural potential/low population density; **LH**: Low agricultural potential and high-medium population density; and **LL**: Low agricultural potential and low population density.

Map 3. Spatial patterns of agricultural potential and population density

The results also show that the potential for agricultural production and population density are spatially correlated. The areas classified as having "high" and "medium" potential have the highest population density at 66/km² and 54/km², respectively (Table 2). Both are greater than the 50/km² threshold that is often used to identify the possibility for promoting intensive farming system in an area (Boserup1965; 1981). However, some areas in "high" potential Western and Central Equatoria that are parts of the Eastern Flood Plains have population densities that are low (e.g. these areas are Type HL). This indicates the difficulty of developing an intensive smallholder farming system even in areas with high agricultural potential. Moreover, because the cropland area under "high" potential is almost equally split between "medium to high" and "low" population density, the area of cropland at 0.18 hectare/per capita in the highest agricultural potential areas is extremely small compared with the national average of 0.30 ha/per capita. Nonetheless, among the six typologies, the ones that are best positioned to generate high returns from investments are HH, HL, and MH. Given that more than half of the cropland areas fall under these categories, these areas should be prioritized for agricultural development programs.

		Agricultural potential defined by LGP			
		High LGP>220 days	Medium 180-220 days	Low <180 days	Total
Population density High-medium Population>10/km²	Population (%)	25.4	33.8	15.8	75.1
	Population density	66	54	51	57
	Land (%)	4.8	7.8	3.9	16.6
	Cropland area (%)	15.3	26.7	17.9	59.9
	Cropland ha per capita	0.18	0.23	0.33	0.24
Low <10/km²	Population (%)	8.7	11.9	4.4	24.9
	Population density	3	4	3	4
	Land (%)	31.5	35.2	16.7	83.4
	Cropland area (%)	12.0	14.9	13.2	40.1
	Cropland ha per capita	0.41	0.37	0.89	0.48
Total	Population (%)	34.1	45.7	20.2	100.0
	Population density	12	13	12	13
	Land (%)	36.4	43.0	20.6	100.0
	Cropland area (%)	27.3	41.5	31.1	100.0
	Cropland ha per capita	0.24	0.27	0.46	0.30

Source: Authors' calculation based on 2001 LandScan and 2009 FAO Land Cover.

Table 2. Cropland, population, and population density according to agricultural potential

3. Estimating agricultural potential

Realized agriculture potential

Because of the country's diverse agro-ecological conditions, crops produced and consumed often differ spatially. With the absence of official agricultural production statistics in South Sudan and given that the agriculture system in the country is presently dominated by subsistence farming, we use the household food consumption data from the 2009 NBHS to estimate the current spatially disaggregated agricultural production.[4] To be able to understand the country's agricultural potential, it is first necessary to derive a consistent measure of the current agricultural value for different locations, which we herein refer to as the "realized agriculture potential". The calculation considers both quantity of consumption

---

[4] With the exception of cereals, we assume that all agricultural products consumed in South Sudan are produced domestically. For these products, total consumption is assumed to equal domestic production; for cereals, we used a multi-step process because the country imports significant amounts of maize from Uganda and sorghum from Sudan. First, we convert cereal flour consumption into grain by assuming that 1 kg of flour is produced from 1.25 kg of raw grain. Second, following the assumption used by FAO/WFP, we approximate that post-harvest losses at 20 percent. Third, it is assumed that 55 percent of grain purchased by rural households is produced locally, while the rest is supplied by imports; for urban households, we assumed that purchases are mainly supported by imports. Finally, domestic grain production is defined as consumption met by households' own production, stocks, and 55 percent of total rural household's purchases.

and production for 34 individual crops and the corresponding prices for them. The prices[5] used in the calculation are averaged from individual households' self-reported information in the NBHS 2009. Limited by the lack of geo-referenced household identification in 2009 NBHS, we only calculate realized crop values at the state level.

The current value of crop production, which represents the "realized agriculture potential" in South Sudan, is only about US$600 million. Crops, together with livestock and fishery products, make up about US$800 million worth of total agricultural value, but still remains relatively low compared with that of neighbors. Given that the current cropland area is about 2.7 million ha, the average crop value per ha is US$227.

Measured at household level, the total value is about $630 per household, of which $470 is from crops. The difference in per household agriculture value across states is large. For example, Western Equatoria, which has more lands allocated to high value crops, is the richest state with per household agricultural value close to US$1,300. On the other hand, with less than US$300, Unity, Northern Bahr Al Ghazal, and Western Bahr Al Ghazal have the lowest household agricultural values.

Cropland expansion

The low agricultural value can be mainly associated with South Sudan's undeveloped cultivated lands for agricultural production. As previously discussed, the country has abundant land with favorable climatic and soil conditions suitable for crop production; hence there is considerable scope that unutilized land can be converted into crop land under certain necessary conditions. Based on LGP, population density, and type of current land use, we project the potential cropland expansion under a moderate (Scenario 1) and a high expansion scenario (Scenario 2) in five and ten year horizons. In the previous section, we distinguished three types of crop-related land use: the areas identified as "cropland", areas as "grass with crops", and areas as "trees with crops" (Table 1). To start with a benchmark of current cropland area, we assume that 10 percent of areas defined as "grass with crops" and "trees with crops" have been cultivated and thus contributed to the current agricultural production. Hence, the benchmark cropland is the sum of land use under original crop land area (24,779 thousand ha; see Table 1) plus 10 percent of land use each coming from "grass with crops" and "trees with crops". Based on this computation, it is estimated that cropland area is 2.7 million ha or 4.1 percent of total land area in the country (Table 3). However, areas under "grass with crops" are unlikely to become cropland due to unfavorable climatic and soil conditions. Thus, in the moderate expansion scenario, we adopt a hierarchical expansion model in which all land currently identified as "trees with crops" (2.6 percent of national land) is the first to be converted into cropland. Once this potential for expansion is exhausted, further expansion will occur in "tree land" areas (which currently accounts for

---

[5] If the state's average price for particular crop is extremely low or high relative to other states, the national average price is used. If the price is either not available from the survey or extremely low compared with that in neighboring countries, then the lowest relevant price from Kenya or Ethiopia is used.

62.6 percent of national territory). For simplicity, we hereafter refer to both "trees with crops" and "tree land" as just "tree land" and use the following rules for expansion:

a. If a pixel C (current cropland) belongs to Type HH area and is surrounded by pixels under "tree land" then the 8 immediate adjoin pixels (1s in Figure 5), 16 pixels (2s) immediately surrounding the pixels identified with 1s, and the 24 pixels (3s) immediately adjacent to the 2s are assumed to become cropland in the next five to ten years (i.e. all the 1s, 2s, and 3s are candidates);

b. For HL and MH areas, cropland expansion is more modest. It only assumes the 8 pixels (1s) immediately adjoining pixel C and the 16 pixels (identified as 2s) to become cropland in the future if they currently classified as "tree land"; and

c. The expansion is even lower in ML and LH areas as it only considers the 8 pixels immediately adjoining pixel C in the projected cropland conversion. Finally, we assume that any "tree land" of the Type LL area will not become cropland in the future.

Hence, in the moderate expansion scenario and given that each pixel is roughly about 1 km², the maximum possible conversion to cropland is 48 km² in HH areas, 24 km² in HL and MH areas, and 8 km² in ML and LH areas. However, as current cropland areas are often connected, i.e., many pixels (C) are already adjacent each other, only those C pixels at the boundary areas are considered when their surrounded pixels under "tree land" become candidates for cropland expansion in the scenario.

5	5	5	5	5	5	5	5	5	5	5
5	4	4	4	4	4	4	4	4	4	5
5	4	3	3	3	3	3	3	3	4	5
5	4	3	2	2	2	2	2	3	4	5
5	4	3	2	1	1	1	2	3	4	5
5	4	3	2	1	C	1	2	3	4	5
5	4	3	2	1	1	1	2	3	4	5
5	4	3	2	2	2	2	2	3	4	5
5	4	3	3	3	3	3	3	3	4	5
5	4	4	4	4	4	4	4	4	4	5
5	5	5	5	5	5	5	5	5	5	5

Source: Authors' illustration.

Figure 1. Illustration of cropland expansion at pixel level

The high expansion scenario (Scenario 2) doubles the cropland expansion in the moderate scenarios in the next five to ten years and is based on the following assumptions:

a. In HH area, pixels 1, 2, 3, 4, and 5 surrounding pixel C are assumed to be converted to cropland if their current land use is characterized under "tree land";

b. Pixels 1,2,3, and 4 surrounding C in HL and MH areas that are currently covered with "tree land" are assumed to be converted to cropland; and

c.   In ML and LH areas, only the pixels 1,2, and 3 are assumed to become cropland if currently part of "tree land" area.

The resulting cropland expansion of both scenarios is presented in Map 4. It should be noted that the precision and accuracy of the potential cropland expansion are hindered by the lack of additional location-specific information and inability to verify the estimates at the ground level. Moreover, realizing the agricultural potential of new cropland depends on many other important factors such as public investments and policies, which can complicate the process and hence are not considered in the projections. Also, additional factors such as access to markets, land and forest policy regulations, as well as access to resources (tools and labor) required for land clearing and tree cutting, will determine the extent the actual extent of expansion.

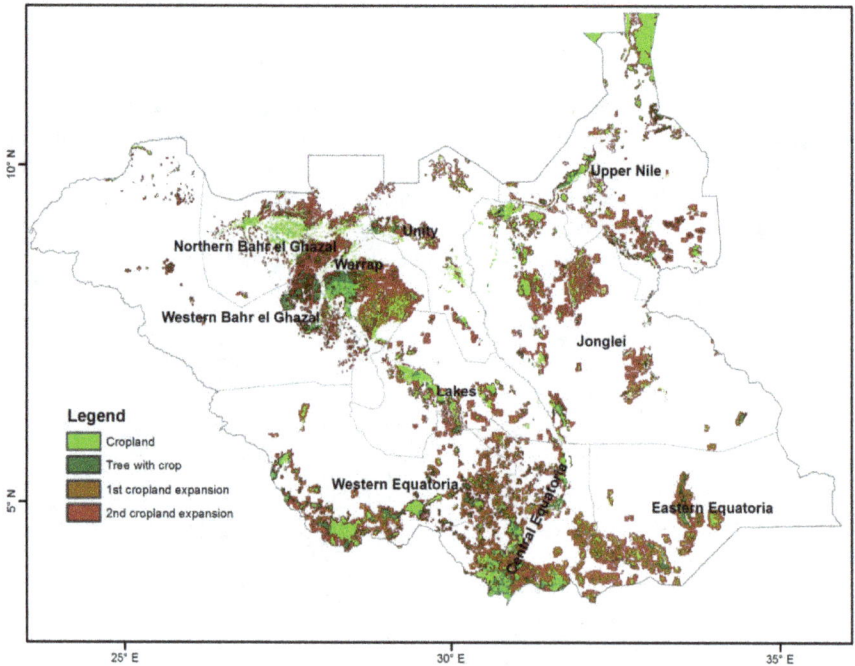

Source: Authors' estimates

Map 4. Cropland expansion under the two scenarios

We focus on the moderate expansion scenario first. Holding other factors constant, cropland area will increase by 2.3 times, from the current 2.7 million ha to 6.3 million ha (Table 3). Cropland becomes 9.7 percent of national total land, up from the current (base) of 4.1 percent. While this increase is significant, it is still far below the agricultural potential assessed by the GOSS, which assesses that 50 percent of South Sudan's land surface is prime agricultural land (GOSS 2010). The share of "tree land" in total area will only slightly decline from the current 63 percent to 60 percent (Table 3).

As expected, most cropland expansion occurred in areas with high agricultural potential. The areas under Type HH, HL, and MH would collectively expand from the current 53 percent to 65 percent of total cropland area. At the state level, the largest expansion into new crop land is expected in Western Bahr el Ghazal and the three Equatorial states (Map 4). Among the livelihood zones, there is huge potential for crop area expansion in Greenbelt and Western Flood Plains.

We also calculate the change in per capita cropland size under the moderate expansion scenario, assuming a 2.5 percent annual population growth rate. If the expansion occurs in five years, the per capita cropland size will increase from the current national average of 0.32 ha to 0.66 ha. If the expansion would take ten years, the land size will increase to 0.59 ha. In either 5- or 10-year simulation, only Western Bahr el Gazal and Western Equatoria will reach cropland size of at least 1.0 ha per capita.

While the rate of cropland expansion is already rapid in Scenario 1, the per capita cropland would still be lower than in neighboring countries. Hence, we consider Scenario 2 that doubles the rate of expansion under the first scenario. Under this more aggressive scenario, there would be a 3.5-fold increase in cropland area of 9.2 million ha (accounting for 14.3 percent of national land). The share of tree land in total land will decline to 55 percent from the current 63 percent. The per capita cropland area under the high expansion scenario would correspondingly increase to 1.0 ha/pc if the expansion is achieved in the next five years and 0.87 ha/pc if expansion takes place in the next 10 years.

Land use categories	Area (in 1000 ha)			Share of total land (%)		
	Current	Scenario 1	Scenario 2	Current	Scenario 1	Scenario 2
Cropland	2,680.9	6,267.4	9,237.4	4.1	9.7	14.3
Trees with crops	1,536.6	0.0	0.0	2.4	0.0	0.0
Tree land	40,526.9	38,477.1	35,507.1	62.6	59.5	54.9
Grass with crops	292.6	292.6	292.6	0.5	0.5	0.5
Grass	9,633.8	9.633.8	9,633.8	14.9	14.9	14.9
Other land use	10,017.3	10,017.3	10,017.3	15.5	15.5	15.5
Total	64,688.3	64,688.3	64,688.3	100.0	100.0	100.0

Source: Authors' calculations.
Note:
(1) Other land use includes Flood land, Water and rock, and Urban as categorized originally in Table 1.
(2) Cropland under "Current" is the sum of land use under original crop land area (24,779 million ha; see Table 1) plus 10 percent of land use under "grass with crops" and 10 percent of land use under "trees with crops".
Table 3. Land expansions in the two scenarios

The increase in cultivated areas through cropland expansion in both scenarios lead to higher agricultural output, and consequently to higher value of agricultural production. Even under the modest cropland expansion (Scenario 1), the value of total agricultural output (including crops, livestock, and fisheries) becomes 2.4 times higher (about US$ 2 billion) than the current US$ 800 million. It is expected that the largest increase will come from the three Equatorial states, Western Bahr el Ghazal, and Warrap. In the high expansion scenario, the potential agricultural production value reaches US$2.8 billion but is still far below the level of output produced in neighboring countries.

Yield improvement

Land expansion is only one of many ways to explore agricultural potential; another avenue is to increase land productivity which also happens to be low in South Sudan. In order to be at par with its neighbors' production levels, yield improvement is necessary. There is a huge gap between the county's actual farm yield and the biophysically achievable yield according to IIASA/FAO Agro-ecological Zone (AEZ) framework (Fischer et al. 2002). The average cereal yield is only about 0.95 ton/ha (FAO/WFP, 2011), but can actually be lower since the cropland area used in the FAO/WFP (2011) is much lower than the areas observed in Land Cover (FAO 2009). This average cereal yield is lower than Uganda where there is minimal use of tradeable inputs (1.6 tons/ha), as well as lower in places with disadvantageous agroecological conditions like Ethiopia (3 tons/ha) and Kenya (2 tons/ha). Such wide yield gap in South Sudan points to a large opportunity to increase average cereal yields.

We design four yield increase scenarios in which the average yield will increase by 50, 100, 200, and 300 percent in a period of 5 or 10 years. An increase by 50 percent is simulated to achieve the average level in Uganda, by 100 percent to attain Kenya's level, and by 200 percent to reach that of Ethiopia. While there is no neighboring country with a cereal yield of 6.0 ton/ha national wide (300 percent increase), such level is observed in certain parts of Ethiopia and Kenya.

Under Scenario 1 of land expansion, a 50 percent yield increase would increase the agricultural production value 3.5 times from the current value. This increase in agricultural value is also 45 percent higher than the increase achieved from Scenario 1 without yield improvement. Accordingly, the value of crop production per ha will grow from the current US$227 to US$340. If yields can increase by 100 percent to mirror the average levels in Kenya, the value of agricultural production in South Sudan (about US$3.7 billion) will overtake the current value in Uganda and crop value per ha will be US$453. Under the most aggressive scenario, with average yield increasing by 300 percent, the total agricultural value will reach US$ 7.9 billion and US$ 1,903/ha.

There are two caveats in our estimation of agricultural potential. First, we do not consider the price effect. At the present, food production of South Sudan is not enough for domestic demand. Urban consumption is primarily met by imports, and food aid is an important food source both for rural and urban households. Thus, we do not expect that a modest increase in crop production to cause an oversupply issue for the country in general. However, it is still possible that significant increases in crop yields, in the absence of opportunities to export surplus can create glut in certain areas during harvest season. When this happens, the prices for many crop products are expected to fall, which indicates that we may overestimate the agricultural potential. The second caveat is related to the livestock sector which we did not consider in the supply increase simulation although this sector also has a huge potential in the country. Without considering productivity increase in livestock production, we may significantly underestimate the agricultural potential.

4. Conclusion

South Sudan, the world's newest nation, has a huge agricultural potential that can be leveraged to improve the national economy and household living standards. The country's endowment of favorable land, water, and weather conditions makes 70 percent of land suitable for agriculture. Yet, less than four percent of total land (about 2.7 million ha) is currently cultivated while more than 80 percent is still under natural vegetation (e.g. trees, shrubs, grass). The production system remains primarily subsistence in nature and crop yield is low. Our analysis shows that the current total value of agriculture production (i.e. "realized potential") only amounts to about US$800 million (US$ 600 million from crops) or less than US$300 per hectare, which is much lower than that of its neighbouring countries. Even with an extremely low population density (13 persons per km²), per capita crop area is only at 0.3 hectare.

In this context, the newly independent country faces challenges in providing enough food for her population that is expected to increase in the short run due to the re-integration of displaced people. Obstacles in developing the country's competitiveness in regional and global markets in the longer term also need to be overcome. In order to have a more realistic agricultural development strategy and investment priorities, it is necessary to understand the country's current agricultural situation and potential for improvement in the near future. We employ a GIS-based analysis and come up with six agricultural potential typologies. HH, HL, and MH are best positioned to be developed, and more than half of current cropland areas fall under these categories. There is possibility of promoting intensive farming systems since areas with "high" and "medium" agricultural potential have population density greater than the 50/km² threshold. However, there are also "high" agro-ecological potential areas with very low population density indicating the difficulty to develop them with a smallholder farming system.

Incorporating these elements together, we then spatially estimate the agricultural potential value in the next five to ten years by simulating: (1) an increase in cultivated area though cropland expansion, and (2) crop yield improvement. If cropland areas expand to 6.3 million or 9.2 million hectares, size of per capita land holding will significantly increase, and consequently results in higher value of agricultural production relative to the current "realized potential". However, the potential agricultural value even in the high expansion scenario is still far below the level of output produced in neighboring countries.

Catching up with crop yield levels achieved by its neighboring countries will be the most important approach to realize agricultural potential. Doubling the current average cereal yield of 0.95 ton/ha, along with moderate cropland expansion, will shoot up the value of agricultural production to US$3.7 billion, a level that can overtake the current agricultural value in Uganda. Given that many challenges in cropland expansion, including high upfront costs of land clearing and low rural connectivity, yield improvement maybe a more effective way to realize agricultural potential in South Sudan over the next years.

Author details

Xinshen Diao, Liangzhi You, Vida Alpuerto and Renato Folledo
International Food Policy Research Institute (IFPRI), Washington DC, USA

Acknowledgement

The chapter is the primary research output of a project funded by Africa Region of the World Bank. Tremendous support has been received from the government of South Sudan, WFP and FAO Sudan offices, researchers in many institutions/organizations in the country, and the World Bank South Sudan office. The principal authors accept responsibility for any errors.

5. References

FAO, 2009. FAO Livestock Population Estimates, Oct 2009. FAOStat, Aug 2010

FAO, 1981. *Report of the Agro-Ecological Zones Project*, World Soil Resources Report No 48, Vol.1-4, Rome, FAO

FAO/WFP. 2009. FAO/WFP crop and food security assessment mission to Southern Sudan, Special Report, 2008/2009, http://www.fao.org/giews.

FAO/WFP. 2010. FAO/WFP crop and food security assessment mission to Southern Sudan, Special Report, 2009/2010, http://www.fao.org/giews.

FAO/WFP. 2011. FAO/WFP crop and food security assessment mission to Southern Sudan, Special Report, 2010/2011, http://www.fao.org/giews.

Fischer, G., H.T. van Velthuizen, and F.O. Nachtergaele. 2002. Global agroecological assessment for agriculture in the 21st century: Methodology and results. RR-02-02. Laxenburg, Austria: International Institute for Applied System Analysis.

GOSS. 2010. The Southern Sudan Food and Agriculture Policy Framework.

Guvele, Cesar. 2009. Agricultural situation in Southern Sudan and the potential for development: a review. A report submitted to USAID.

LandScan, 2008. http://www.ornl.gov/sci/landscan/index.shtml (accessed November 2010).

Musinga, M., J.M. Gathuma, O. Engorok, and T. H. Dargie. 2010 The Livestock Sector in Southern Sudan: Results of a Value Chain Study of the Livestock Sector in Five States of Southern Sudan Covered by MDTF with a Focus on Red Meat. The Netherlands Development Organization.

Southern Sudan Centre for Census, Statistics and Evaluation (SSCCSE). 2010. Poverty in Southern Sudan: Estimates from the NBHS 2009. Juba: SSCCSE.

Southern Sudan Centre for Census, Statistics and Evaluation (SSCCSE), Save the Children UK (SC UK), USAID Famine Early Warning Systems Network (FEWS NET). 2006. Southern Sudan Livelihood Profiles. January 2006.

Uchida, Hirotsugu, and Andrew Nelson. 2008. Agglomeration index: towards a new measure of urban concentration. Background paper for the WDR 2009.

World Food Program (WFP) and The Ministry of Transport and Roads, Government of Southern Sudan (GOSS). 2005. WFP Southern Sudan Emergency Road Rehabilitation Program - Socio-economic impact assessment: A report of 2004 and 2005 road rehabilitation activities

Chapter 9

GIS and *ex situ* Plant Conservation

Nikos Krigas, Kimon Papadimitriou and Antonios D. Mazaris

Additional information is available at the end of the chapter

1. Introduction

In the frame of the global efforts to halt the biodiversity loss by 2010 and with the aim to develop effective conservation strategies extending beyond 2010, stakeholders have recognized as a priority the *in situ* conservation (on site conservation) of target plant species.

Still, the rapid environmental changes including climate change, habitat loss and alteration, could pose some limitations on our ability to conserve target species effectively *in situ* (Sharrock & Jones, 2009). As a result, conservation biologists, policy makers and managers acknowledge the importance of *ex situ* conservation of target plants in botanic gardens and seed banks as an essential back-up solution (Convention on Biological Diversity [CBD], 1992; Glawka et al 1994; Global Strategy for Plant Conservation [GSPC], 2002; European Strategy for Plant Conservation [ESPC], 2009; Sharrock & Jones, 2009).

For the *ex situ* plant conservation, target species mainly refer to plant taxa (species and subspecies) presenting a narrow distribution in the wild (see Krigas & Maloupa, 2008). This category of plants usually includes:

i. Local endemics (endemics of a single mountaintop e.g. *Viola cephalonica* (Katsouni et al., 2009), or endemics of a single island e.g. *Allium samothracicum* (Krigas, 2009), or endemics of a group of nearby areas or islands e.g. *Thymus holosericeus* (Krigas et al. 2010),

ii. Regional endemics (endemics to small parts of a single country, e.g. endemics of southern Spain, endemics of Peloponnese, southern Greece),

iii. National endemics or single-country endemics (e.g. Greek endemics, Italian endemics, Spanish endemics),

iv. Endemics to specified small geographical areas e.g. local Balkan endemics transcending the borders of neighbouring Balkan countries such as *Ranunculus cacuminis* (Krigas & Karamplianis, 2009).

Other target species significant for the *ex situ* plant conservation may include plants which are rare in a certain area (e.g. Europe, Sharrock & Jones, 2009) or plants that are currently threatened with extinction at local, regional or global level i.e. taxa characterized as "Near Threatened", "Vulnerable", "Endangered" or "Critically Endangered" according to the IUCN (2001) criteria. However, given that almost 90% of the Europe's threatened plants are single-country endemics, it should be noted that most of the endemic plants are also threatened with extinction (Sharrock & Jones, 2009). Last but not least, other groups of socioeconomically valuable plants (e.g. saffron) and their progenitors (e.g. plants in the genus *Crocus*) may also be considered as target plants for the *ex situ* conservation (Fernández et al., 2011).

GIS has be given a role in analyzing potential and current spatial distribution of target species, locating and assessing the populations of target plant species and assemblages, measuring biodiversity, monitoring biodiversity patterns and identifying priorities for conservation and management (Iverson & Prasad, 1998; Salem, 2003; Powel et al., 2005; Pedersen et al., 2004).

Habitat evaluation or habitat modeling with the use of GIS has the potential to make a substantial contribution to conservation management of target species within an integrated approach and is suitable for setting conservation priorities at multiple spatial scales (Store & Jokimäki, 2003; Powel et al., 2005). GIS have also been used as a tool for specific conservation programmes including comparisons of ecological patterns between local and regional scales, selection of protected areas according to habitat suitability, analysis of the impact of alien species on endemic plants and selection of sites for representative seed collections of target species (Draper et al., 2003). Flexible GIS-based tools have also been developed to exploit static information of botanical collections in an attempt to evaluate species distributional ranges (Schulman et al., 2007; Loiselle et al., 2008), accounting for potential effect of climate change to predicted models (Loiselle et al., 2008). Other GIS-tools and distribution modeling methods have been applied to examine the hypothesis that wild and cultivated plants of certain species may occur in the same types of habitats (Allison et al., 2006).

Currently, the GIS technology has increasingly been used for predictive purposes in species re-introductions. Combining historical information from specimen labels with up-to-date environmental data, GIS can be used to identify the range of environmental conditions in which plants grow, offering an understanding of the ecological requirements of different species. The relevant environmental conditions can then be used to delimit areas of high, moderate or low survival probability (Sawkins, 1999 as cited in Moat & Smith, 2003; Powel et al., 2005). Recent GIS applications further include studies on the assessment of the importance of landscape connectivity, structure and configuration for native and non-native plant communities, dispersal and invasiveness abilities (Minor et al., 2009). Nevertheless, the most frequently encountered obstacle to the use of GIS technology in conservation planning is lack of data, especially distribution data of target species and digital vegetation maps of areas with conservation interest (Sawkins, 1999 as cited in Moat & Smith, 2003).

The GIS technology has also an enormous potential in seed conservation, particularly in targeting collecting needs for botanic gardens and conservation institutes and in identifying what, where and when to collect (Moat & Smith, 2003). Although concerns are raised regarding limitations and biases (Store & Kangas, 2001), some of the GIS applications have been used for their power to pinpoint by spatial niche modeling, new probable locations where rare endemic species may be found in the wild, thus permitting the search for new populations (Jarvis et al., 2005). The GIS could also be used for assessing the sensitivity of target species to climate change identifying potential distributions as well as vulnerable habitats and species (Vanderpoorten et al., 2006). In this way GIS may deliver important information that can drive seed banking priorities and design (Godefroid & Vanderborght, 2010).

Lately, a new application of GIS has been launched for the *ex situ* plant conservation; GIS was used to describe quantitatively and qualitatively the natural habitats of target plants in order to facilitate their propagation and transfer from the wild habitats to man-made habitats like botanic gardens (Krigas et al., 2010).

2. GIS used for the description of the natural habitats of target species

The geographical data associated with plant collections may be considered as a set of facts about the places in which the plants thrive. In the wild, target plants for *ex situ* conservation could thrive literally anywhere. Given the spatial and temporal heterogeneity of the environmental factors and the dissected topography of the landscape, the target plants may well originate in a variety of quite different habitats in which they are adapted to grow naturally. Hence, it seems quite difficult to be able to emulate their preferable conditions when trying to cultivate them in restricted man-made habitats such as botanic gardens (Krigas et al., 2010). A deeper understanding of the ecology and life cycle of such target species has been considered as a key issue towards a species-specific successful propagation and cultivation method (Baskin & Baskin, 1988).

In this framework, the GIS can be used to offer a reliable, quantitative and qualitative description of the *in situ* habitat conditions preferred and/or tolerated by different target plant species in the wild. This novel GIS application is both ecologically meaningful and useful in horticulture and *ex situ* plant conservation (Schulman & Lehvävirta, 2010).

To demonstrate this application a dataset was chosen including target plants originating from sites of varied landscapes that are also subjected to different climatic conditions (Fig. 1): the Aegean Archipelago, Crete, Ionian Islands, and Peloponnese (Greece, Southern Europe). The target plants included in this dataset fall into the conservation priorities of the Balkan Botanic Garden of Kroussia (BBGK, Krigas & Maloupa, 2008, Krigas et al., 2010). To date, the BBGK has organized several botanic expeditions all over Greece in order to collect appropriate plant material for propagation and *ex situ* conservation. All this plant material is currently maintained under *ex situ* conservation (ca. 25% of the known Greek flora which includes at least 6.300 species and subspecies, Krigas et al., 2010).

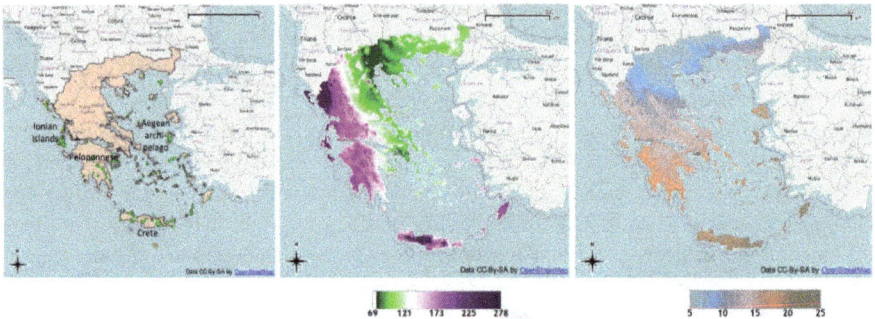

Figure 1. Different target plants (n=256, each dot represents at least one botanic expedition and at least one target plant collected) from various sites originating from the selected area of Greece including the Aegean Archipelago, the Ionian Islands, Crete and Peloponnese (left). All plant material is currently maintained under *ex situ* cultivation at the Balkan Botanic Garden of Kroussia (BBGK). Raster maps displaying the variations of climatic conditions in Greece as expressed by the Emberger Pluviothermic Quotient (middle) and Precipitation Seasonality (right). All values have resulted from map algebra calculations after point sampling for the Greek territory regarding the collection sites of the selected target plants (n=256).

The positions of the collection sites of target plant species were captured in the wild using handheld GPS trackers. The obtained geographic coordinates were consequently imported as point layer into a Geographical Information System (GIS) and point sampling was performed for each one on multiple raster and polygon layers. In order to probe values for topography, terrain, soil, climatic, land cover and habitat attributes at the captured sites, the following datasets were selected (Fig. 2):

a. Raster elevation data from SRTM (USGS 2004), with resolution of 1 km^2 (30 arc second) which were used for terrain and topography attributes.

b. Soil data from the European Soil Database (ESDB) v2.0 (EC, 2004), which is composed of the Soil Geographical Database of Eurasia at a scale of 1: 1,000,000 (version 4 beta) and the Pedotransfer Rules Database (v2.0), with a raster resolution of 1 km^2, presenting 72 soil parameters.

c. Temperature and precipitation data from the WorldClim database (Guarino et al., 2002; Hijmans et al., 2005), with a raster resolution of 1km^2, which has been used for the description of climatic conditions (average values of 50 years).

d. Land cover data from CORINE (Coordination of Information on the Environment) comprehensive hierarchical vector geodatabase at a scale of 1: 100,000 (with a minimum mapping unit of 25ha) which present the spatial distribution of different landcover/land-use types (CORINE land cover classification, EC & ETC/LC, 1999) and soil types (CORINE soil classification EC, 1985).

Additionally, part of the initial data was processed to produce more specific attributes for the collection sites of the target plants, as follows:

a.   The elevation data (DEM) was used for raster based terrain analysis, which resulted in slope and aspect maps of the Greek territory.

b.   The DEM was further used as the base layer for the digitization of the vegetation zones of Greece according to Mavromatis (1980).

c.   The rasters describing the spatial variations of precipitation, maximum and minimum temperatures from the WorldClim geodatabase (Guarino et al., 2002), were used in map algebra calculations for the production of a raster map which displays the spatial variation of the Emberger Pluviothermic (Ombrothermic) Quotient as: 2000*P/(Tmax+Tmin)*(Tmax-Tmin), where P: the layer of annual precipitation (mm), T: the layers of mean maximum (Tmax) and minimum (Tmin) monthly temperature of the warmest and coldest month in Kelvin degrees. Precipitation (P) Seasonality was calculated also, using the following map algebra expression: 100* (Pw -Pd) / Py, where Pw and Pd are the raster layers for the spatial variations of precipitation for the wettest and driest quarter of the year respectively, and Py the total annual precipitation (Fig. 1).

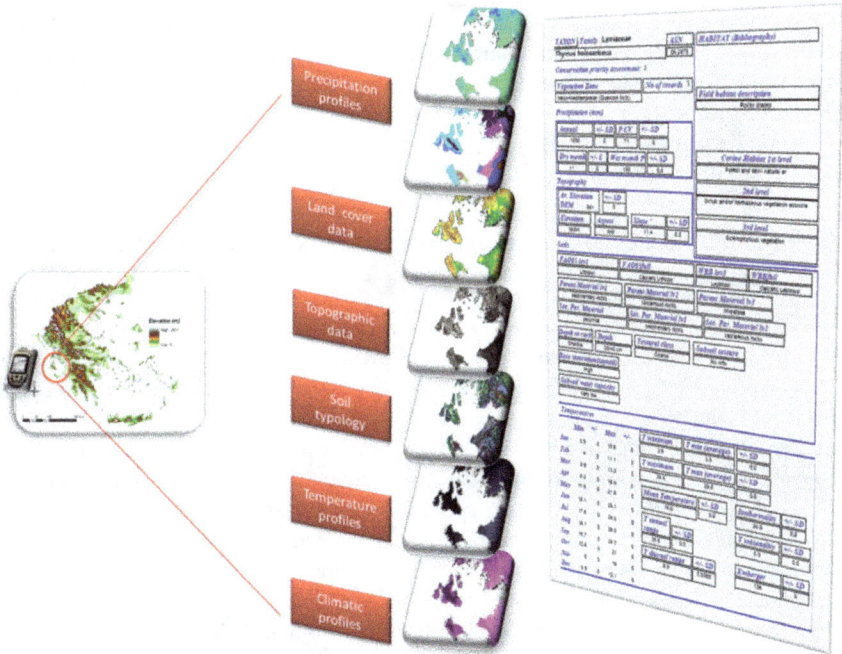

Figure 2. Methodology concept: The GPS data can be used for point sampling in a GIS environment regarding the collection sites of a target species over multiple polygon and raster layers extracted from selected geo-databases. This approach leads to the attribution of selected values regarding precipitation, land cover, terrain, topography, soil typology, temperature and climate to the wild habitats of a target species, resulting in a summarized fact sheet which reflects the ecological preferences of a target species (e.g. *Thymus holosericeus* originating from the Ionian Islands, SW Greece).

Based on this GIS application, simple or advanced ecological fact sheets may be constructed for different target species which actually reveal their preferences in the wild (Fig. 2, 3, 4). An ecological fact sheet illustrates the ecological profile of a wild growing target plant and can be designed in a way to include different kind of information such as: **Vegetation zone** (different types), **Climatic data** which is a combination of **Precipitation and Temperatures** (for the different sites: total annual and annual range of precipitation, driest and wettest month, minimum, mean and maximum temperatures, annual and diurnal temperature range, isothermality and seasonality, Emberger pluviothermic quotient), **Topographic** (for the different sites: elevation, aspect, slope), **Habitat** (bibliographic, field notes), **Land Cover Types** (for the different sites: CORINE classification in three levels) and **Geological-Pedological data** (for the different sites: Food and Agriculture Organization's [FAO] soil classification, World Reference Base Soil classification, dominant parent material in three levels, depth to rock, textural class, topsoil base saturation, subsoil water capacity) (see Fig. 2, 3, 4).

Additionally, other information may also be associated with the ecological description of the wild habitat of target species such as **Taxonomic data** i.e. taxon's name, family, accession number (ASN) given, **Collection data** (e.g. geographical coordinates, description of state, province, exact locality etc), **Conservation assessment** (e.g. ranking according to priorities assigned), and **Mother Plantations** (number of total plant accessions or number of records). It should be mentioned that when the number of records introduced in GIS for a specific information field are >1, then any quantitative information may be described by the mean value and its standard deviation (± SD) (see Fig. 2 and 3).

3. GIS-facilitated development of protocols for *ex situ* conservation of target species

Considering the recent challenges in regional conservation planning, it becomes apparent that there is an urgent need for increased applied research in order to develop propagation and cultivation protocols for target plants threatened with extinction, towards species recovery and populations' reinforcements (Bunn et al. 2011, Maunder et al. 2001 Sarasan et al. 2006). Moreover, the GSPC (2002) and the ESPC (2009) have included this urgent need under their conservation targets at European and global level (Target 8). The GIS may serve such a need and may help at the development of species-specific propagation protocols and *ex situ* cultivation guidelines regarding target species. Such methodologies could also provide basic information and criteria for prioritizing collections of threaten or rare species based on their distributional patterns, population status and the genetic and/or geographic representation of current seed bank collections (Draper et al., 2003; Farnsworth et al., 2006) or even prioritizing sites for seed collections (Jarvis et al., 2005; Ramírez-Villegas et al., 2010).

3.1. GIS and seed germination

The GIS may be used to facilitate the germination of seeds collected from the wild in plant propagation studies as shown by Krigas et al. (2010). This study included plants originating at the Ionian Islands, south-western Greece and provided some example-cases.

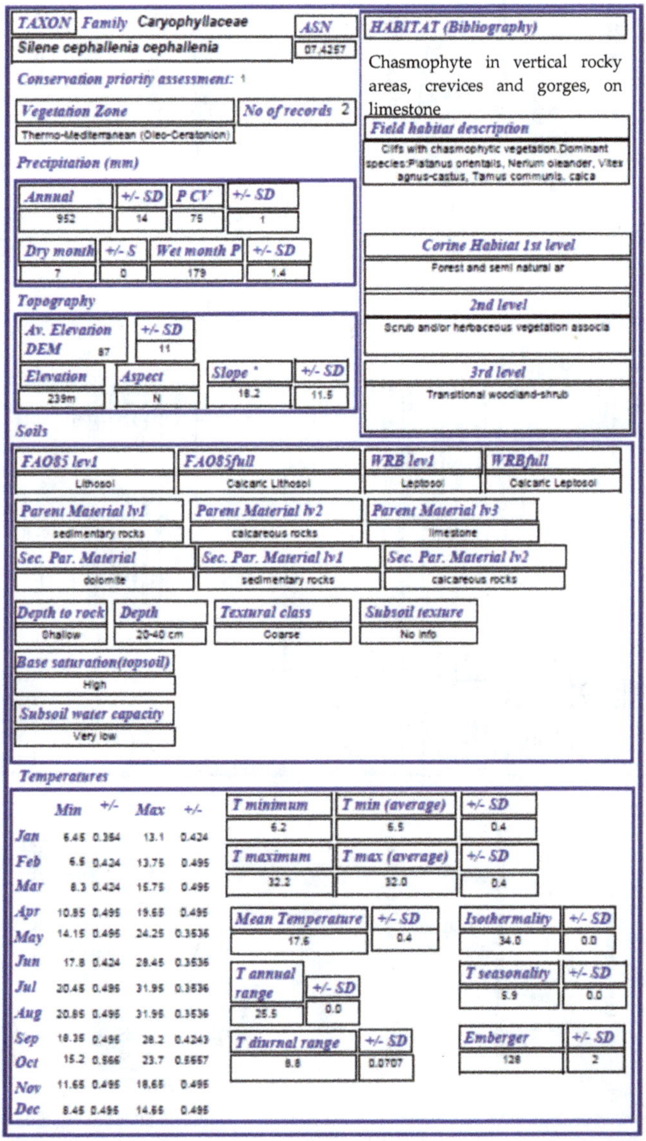

TAXON Family Caryophyllaceae

Silene cephallenia cephallenia

ASN 07.4257

Conservation priority assessment: 1

Vegetation Zone: Thermo-Mediterranean (Oleo-Ceratonion)

No of records 2

Precipitation (mm)

Annual	+/- SD	P CV	+/- SD
952	14	75	1

Dry month	+/- S	Wet month P	+/- SD
7	0	179	1.4

Topography

Av. Elevation DEM	+/- SD
87	11

Elevation	Aspect	Slope °	+/- SD
239m	N	18.2	11.5

HABITAT (Bibliography)

Chasmophyte in vertical rocky areas, crevices and gorges, on limestone

Field habitat description

Cliffs with chasmophytic vegetation.Dominant species:Platanus orientalis, Nerium oleander, Vitex agnus-castus, Tamus communis. calca

Corine Habitat 1st level: Forest and semi natural ar

2nd level: Scrub and/or herbaceous vegetation associa

3rd level: Transitional woodland-shrub

Soils

FAO85 levl	FAO85full	WRB levl	WRBfull
Lithosol	Calcaric Lithosol	Leptosol	Calcaric Leptosol

Parent Material lvl	Parent Material lv2	Parent Material lv3
sedimentary rocks	calcareous rocks	limestone

Sec. Par. Material	Sec. Par. Material lv1	Sec. Par. Material lv2
dolomite	sedimentary rocks	calcareous rocks

Depth to rock	Depth	Textural class	Subsoil texture
Shallow	20-40 cm	Coarse	No info

Base saturation(topsoil): High

Subsoil water capacity: Very low

Temperatures

	Min	+/-	Max	+/-
Jan	6.45	0.354	13.1	0.424
Feb	6.5	0.424	13.75	0.495
Mar	8.3	0.424	15.75	0.495
Apr	10.85	0.495	19.65	0.495
May	14.15	0.495	24.25	0.3536
Jun	17.8	0.424	28.45	0.3536
Jul	20.45	0.495	31.95	0.3536
Aug	20.85	0.495	31.95	0.3536
Sep	18.35	0.495	28.2	0.4243
Oct	15.2	0.566	23.7	0.5657
Nov	11.65	0.495	18.65	0.495
Dec	8.45	0.495	14.65	0.495

T minimum	T min (average)	+/- SD
6.2	6.5	0.4

T maximum	T max (average)	+/- SD
32.2	32.0	0.4

Mean Temperature	+/- SD	Isothermality	+/- SD
17.6	0.4	34.0	0.0

T annual range	+/- SD	T seasonality	+/- SD
25.5	0.0	5.9	0.0

T diurnal range	+/- SD	Emberger	+/- SD
8.8	0.0707	128	2

Figure 3. Simple ecological fact sheet for a single-area endemic (*Silene cephallenia* subsp. *cephallenia)* produced after linking its collection data with those of geodatabases. *S. cephallenia* subsp. *cephallenia* is found exclusively at Poros gorge in Cephalonia (Ionian Islands, SW Greece), is protected by the Greek Presidential Decree 67/1981 and recently it has been included in the Red Data Book of Rare and Threatened Plants of Greece as "Critically Endangered" (Karagianni et al., 2009) and in Annex 2 of the European Threatened Species (Sharrock & Jones, 2009).

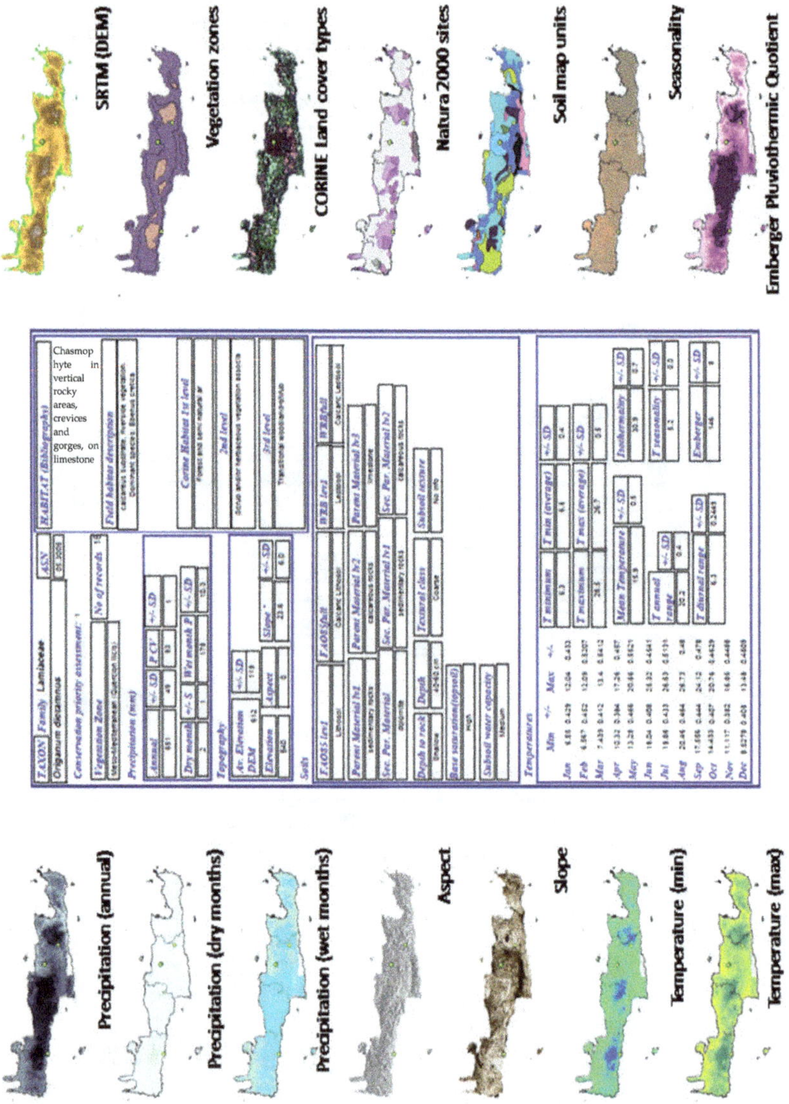

Figure 4. Advanced ecological fact sheet of *Origanum dictamnus* produced after linking data for 18 collection sites with those of geodatabases, enriched with thematic maps displaying the spatial variations of multiple ecological attributes over the area of Crete (the 18 original collection sites are distributed in three main areas shown with yellow circles). *O. dictamnus* is restricted to Crete (endemic), is protected by the Greek Presidential Decree 67/1981, Bern Convention and the European Directive 92/43/EEC, it has been included in the Red Data Book of Rare and Threatened Plants of Greece as "Vulnerable" (Turland, 1995) and in Annex 2 of the European Threatened Species (Sharrock & Jones, 2009).

Generally, the appropriate season for germination trials inside the greenhouse was selected by comparing the seasonal temperatures profiles of the local greenhouse used with those of the wild habitat of the target species (Ionian Islands). The season deemed as more appropriate was characterized by temperature ranges that could easily mimic those prevailing in the natural environment of the target species.

To illustrate an example, the GIS-derived temperature profiles for the wild habitat of *Silene cephallenia* subsp. *cephallenia* (Figs. 3, 5) have dictated the selection of temperatures to test for seed germination, leading to a germination success of 64% (Krigas et al., 2010). Additionally, its ecological profile has explain the fact that seed germination was inhibited at 10°C, whereas seed germination was considerably increased in only 7 days when higher temperature (21°C±1) was applied, in an attempt to emulate natural conditions.

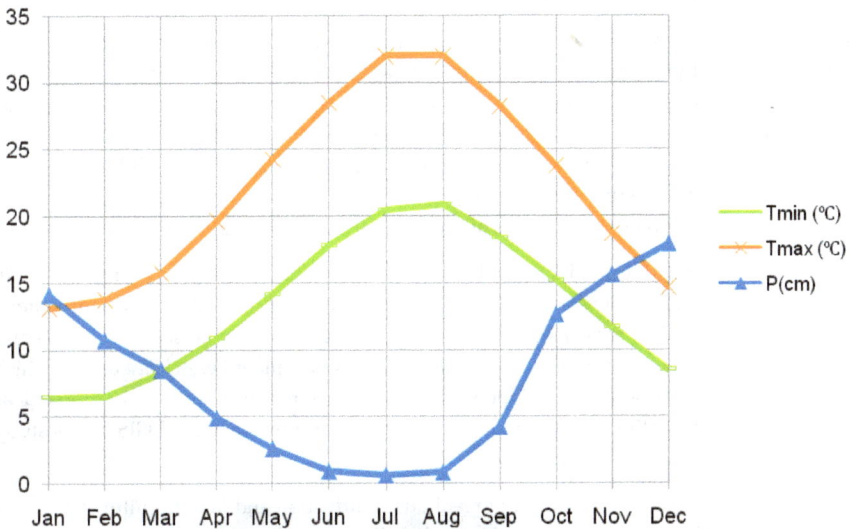

Figure 5. Monthly variations of the climatic conditions (average values of 50 years for precipitation, minimum and maximum temperature) at the original collection sites of *Silene cephallenia* subsp. *cephallenia* derived from its ecological fact sheet.

3.2. GIS and asexual plant propagation by cuttings

Basically, the asexual propagation and the rooting trials of softwood cuttings from plants are performed in greenhouse conditions using a few individuals to produce large amounts of genetically identical plants. Lately, it has been shown that the GIS may also be used to facilitate the asexual plant propagation by cuttings (Krigas et al., 2010). For example, the GIS-derived data for the wild habitat of *Thymus holosericeus* demonstrated that early spring was the most appropriate season for the rooting trials of this target species (Krigas et al., 2010). During early spring, the temperatures of the selected greenhouse in their case have emulated in the best way the temperatures prevailing in the natural habitat of the target plant. In this study, during propagation the rooting substrate's temperature was kept at 18-21°C in order to accelerate rooting as indicated from previous experience (Maloupa et al., 2008). Nevertheless, the air temperature was kept between 18-25°C, in an attempt to emulate as best as possible the original conditions of the natural habitat which were indicated by the GIS. Furthermore, the relative humidity was kept at 80% for the first 7 days and was reduced gradually to 50% during the second week of rooting. This was followed due to the fact that *Th. holosericeus* is a xerophytic species that grows in areas of very low subsoil water capacity, especially during dry months (Fig. 2). In the case of *Th. holosericeus*, the combination of GIS data with previous experience on rooting of other species of the same genus raised propagation success nearly by 90% (from 45% to 80% rooting).

3.3. GIS and *in vitro* plant propagation

The attempts of *in vitro* propagation of rare and threatened plant species are often associated with two inherent problems which may hold back conservation efforts. First of all, there is frequently a lack of published methods regarding the propagation of a certain plant species and secondly, there is often a limited amount of experimental plant material which can be initially available due to scarce populations of limited size found in the wild (Bunn et al., 2011; Krogstrup et al., 2005). Taken these limitations into account, almost every attempt to propagate rare and threatened plants is of great importance but seems difficult to achieve.

Although sophisticated technologies and modern methods have been used to date for the *in vitro* plant propagation (Benson, 1999), the GIS has not been exploited as a tool. Lately, Grigoriadou et al. (2011) have shown how the GIS may also be used to serve the needs and procedures of *in vitro* plant propagation. Using a case-study plant i.e. *Achillea occulta* which is a local endemic of southern Peloponnese (Southern Greece) recently characterized as "Vulnerable" (Constantinidis & Kalpoutzakis 2009), the authors have used GIS in this study for the:

a. Selection of temperatures for both greenhouse cultivation and *in vitro* cultures; the GIS-derived ecological profile for *A. occulta* dictated the selection of 22±2 °C day temperature and 15±2 °C night temperature at 16-h photoperiod to be used as most suitable for the development of the *in vitro* cultures.

b. Selection of appropriate period for acclimatization and transplanting of plantlets produced *in vitro*; after balancing out the temperature profiles of the wild habitat with those prevailing in the man-made site of the botanic garden (BBGK), the GIS revealed that spring temperatures were more suitable for the active growth and acclimatization of plantlets.

c. Selection of growing media and substrates for plantlets produced *in vitro*; the type of commercial peat used with the addition of vermiculite provided a very good imitation of the natural soil conditions as indicated by the GIS-derived ecological profile.

d. Selection of appropriate *ex situ* conservation sites for plants raised *in vitro*; after balancing out the temperature profiles of the wild habitat with those prevailing in the available man-made *ex situ* conservation sites of BBGK, the GIS revealed that specific sites seemed more favorable than others for the accommodation of *A. occulta* plants (a site with temperature range as close as possible to the natural temperature range of the wild habitat was chosen).

4. GIS-derived ecological profiles and guidelines for the *ex situ* cultivation of target plants

From the above mentioned and when taking into account the associated technical information (see Krigas et al., 2010, Grigoriadou et al., 2011) it becomes apparent that GIS may be used for the development of effective protocols concerning the propagation and initial cultivation of plant material derived from target species.

Regardless the method used for plant propagation (cuttings, seeds or *in vitro* techniques), the young individuals produced (plantlets or seedlings) require hardening and acclimatization before their *ex situ* conservation. In this sense, the GIS-derived ecological profiles of the target plants could be used to provide specific guidelines for their effective *ex situ* cultivation (Table 1).

4.1. Cultivation guidelines regarding soil media, potting volume, texture, pH and drainage

In general, it is known that for the successful *ex situ* cultivation of target plants, the plant medium to be used must be similar to that of the substrate in the plant's natural habitat, providing similar root aeration and drainage conditions. This is quite important since improved drainage conditions during cultivation can actually inhibit fungal disease risks, induce greater rooting depth and enhance general plant health and vigor (Brady & Weil, 2002).

The GIS-derived ecological profile for the wild habitat of *Silene cephallenia* subsp. *cephallenia* (Fig. 3) showed that the plants originally grow on "Calcaric Lithosol" according to the FAO (1985) classification or according to the European Soil Database, on "Calcaric Leptosol (mountainous), shallow (<25 cm) or extremely gravelly soils directly over continuous rock or soils having <20% (by volume) fine earth material" (Krigas et al., 2010). In an attempt to emulate the natural habitat during hardening and acclimatization of plantlets raised from seeds (seedlings), the type of commercial peat used in BBGK for transplantation contained lime and by adding vermiculite, a good approximation of its originally drained soil type was achieved. This fact has actually allowed the plantlets of *S. cephallenia* subsp. *cephallenia* to continue growing without problems observed.

For the further *ex situ* cultivation of *Silene cephallenia* subsp. *cephallenia* and *Thymus holosericeous*, the GIS-derived ecological profiles (Fig. 2, 3) indicated that the basic soil group in which their wild habitats originate from is classified as "Calcaric Lithosol" (stony calcareous with a high concentration of $MgCO_3$ and >15% free $CaCO_3$, developed on dolomitic limestone). Such a substratum is characterized by a good drainage capacity, neutral to alkaline soil pH, low depth and a high ratio of stones to fine earth. The above indicated that for the *ex situ* cultivation of these target plants, a potting medium of relative high pH would be required and a fair degree of drainage was deemed essential. To accomplish this, supplementary perlite, sand and/or fragmented stone were added in their growing media. For these plants, medium sized pots (4.5 l) were suggested as suitable and additional Mg dressing was added in order to match the original chemical composition of their natural habitats.

By adopting these guidelines (Table 1), the natural habitat of the target plants was technically emulated with common commercial materials and the plants have adapted well and continued to grow without apparent problems (Krigas et al., 2010).

4.2. Cultivation guidelines regarding temperature range

When cultivating plants from the wild to man-made habitats, the temperature conditions at the plant's growing site (nursery or outdoors) may often be quite different from those prevailing in the plant's original wild habitat. Although it is known that plants usually have a degree of tolerance for a wide range of temperatures, it is important to know the temperature range that a plant species may tolerate, in order to facilitate its growth by maintaining temperatures within these limits during cultivation.

The GIS-derived ecological profile indicated that *Silene cephallenia* subsp. *cephallenia* (Fig. 3, 5) may experience and/or tolerate in its wild habitat mean temperatures ranging from 6.7 °C to 32.2 °C throughout the year, while for *Thymus holosericeus* the mean yearly temperatures may range from 3.9 °C to 29.5 °C (Krigas et al., 2010 and Fig. 2).

For the acclimatization procedure of the seedlings of *S. cephallenia* subsp. *cephallenia* and of the plantlets of *Th. holosericeous* produced in BBGK, the selection of appropriate sites was achieved after balancing out the temperature profiles of the available areas for *ex situ* conservation in northern Greece with the natural range of temperatures of their wild habitats, as revealed with the use of GIS (Table 1). It was suggested that *Th. holosericeus* may be marginally stressed during hot summer days and short term measures could possibly be taken to avoid heat shock (e.g. periodical shading or translocation of mother plants to cooler conditions). However, high summer temperatures did not seem to be a constraint for *S. cephallenia* subsp. *cephallenia*. Moreover, it was suggested that accommodation in a protected greenhouse would be crucial for the winter survival of both plants in northern Greece (Krigas et al., 2010). Indeed, the propagated plants that were transferred indoors (greenhouse at sea level, with controlled winter indoor temperatures ranging from 5 to 25 °C), have actually shown increased height, leaf area and number of flowers during cultivation in comparison to those transplanted outdoors (Krigas et al., 2010).

4.3. Cultivation guidelines regarding watering regimes

When transferring wild plants to man-made sites for their *ex situ* cultivation, the amount of water to be offered to them remains largely unknown. The watering regime followed is based merely on observations regarding the natural habitat of the species in question or exploits previous general horticultural experience associated with the cultivation of other species with presumably similar needs. However, it is generally accepted that the design of an irrigation system for the *ex situ* cultivation of target plants should take into account the natural preferences of the species, the medium type and the irrigation frequency requirements. The latter not only depends on the specific plants in question, but it is also influenced by the growing season and the indoor or outdoor temperatures prevailing in the cultivation sites.

The precipitation profiles derived from GIS for the target plants may indicate differences regarding the water requirements preferred and/or tolerated by each one of them in the wild. For example, the GIS-derived ecological profile of *A. occulta* published by Grigoriadou et al. (2011) for its *ex situ* conservation, indicated a watering regime equal to a mean total annual precipitation of 744 mm (7–136 mm per month) which was followed for the proper cultivation of the *in vitro* propagated plants.

Seasonal variation of the watering regime is equally important as the total amount of water offered to plants. To illustrate an example, the GIS-derived ecological profiles of *Thymus holosericeous* and *Silene cephallenia* subsp. *cephallenia* may be taken into account (Fig. 2, 3, 5, 6). It becomes evident that during the wettest and driest quarters of the year, *Thymus holosericeus*'s habitat receives from 40 to 504 mm, while that of *Silene cephallenia* subsp. *cephallenia* from 26 to 476 mm. During autumn, no noticeable differences exist regarding mean monthly precipitation in the natural habitats of these plants. However, from January to August mean monthly precipitation is consistently higher in the intermediate altitude habitats of *Th. holosericeus*, while until autumn *S. cephallenia* subsp. *cephallenia* (which is naturally found close to sea level) receive comparatively lower precipitation (Fig. 5, 6).

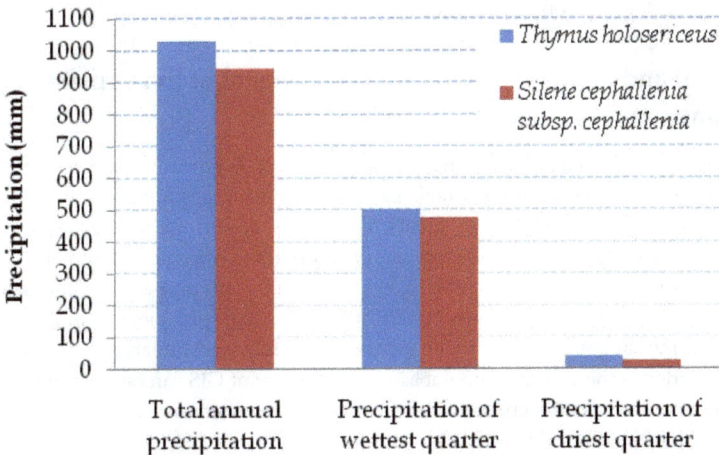

Figure 6. Precipitation profiles for the original collection sites of *Thymus holosericeus* and *Silene cephallenia* subsp. *cephallenia* derived from their ecological fact sheets.

Hence, the GIS-derived precipitation profiles suggest that the watering need (both in terms of water amount received and frequency of watering) is not the same for these plants (Table 1), although both should be under a rather low water regime (regardless of whether it is winter or summer). *S. cephallenia* subsp. *cephallenia* has comparatively lower water demands both annually and seasonally which should result in a more restricted watering schedule, while *Th. holosericeus* has somewhat intermediate watering needs, although a relatively low water regime seems to be equally suitable since it is considered as a xerophytic species (Krigas et al., 2010).

4.4. Guidelines regarding positioning of target plants in displays

The qualitative and quantitative description of the natural habitat of a target species may offer information for appropriate spatial and temporal positioning in plant displays of botanic gardens. Different vegetation types where a target plant is naturally found (deciduous or evergreen i.e. depending on the canopy cover of the vegetation) or land cover types (e.g. rock formations, woodland, pastures etc) from which the target species originate from, are attributes that can be exploited as important ecological information. Such attributes (a) may indicate appropriate or unsuitable sites for the *ex situ* cultivation of target plants, (b) could be useful for the selection of the amount of light or shading needed for the plants or to be avoided by them, and (c) could define the species assemblages with common requirements for specific indoors or outdoors plant displays (Table 1).

To illustrate an example, the GIS-derived ecological profile of *Achillea occulta* published by Grigoriadou et al. (2011) for its *ex situ* conservation indicated that (semi-) shady limestone cracks and rock bases with calcaric lithosols should be used as habitats in order to host the propagated plants and a natural positioning at south, south-eastern and south-western exposures of rock formations was chosen as more favorable for plant growth.

5. Potential and implications for the management of living plant collections

The novel GIS-facilitated application presented here is a powerful tool able to extract ecologically meaningful environmental information from geodatabases regarding the collection sites of different target plants which are useful in applied research and horticulture. This application is able to identify important ecological differences that can contribute to the development of species-specific baseline plant propagation and cultivation protocols. Given the impracticality and lack of on-the-spot field temperature and precipitation measurements and proper soil sampling followed by laboratory analysis, the nothing (just delete the words) geodatabases with the use of GIS can be used to extract at a fraction of time, information crucial for the success of *ex situ* conservation of target species. This application can be used to (Table 1):

- Understand the amplitude of the *in situ* ecological conditions of different target plants both quantitatively and qualitatively (Fig. 1, 2, 3, 4),

Variable / Source used	Selected attributes used	Conservation guidelines produced	
Soil moisture / ESDB v.2 (EC, 2004)	Available topsoil and subsoil water capacity	Calculation of different watering regimes for different groups or target species	
Soil classes and types / ESDB v.2 (EC, 2004), CORINE Soil Classification (EC, 1985)	Textural class Different soil classes Dominant parent material	Selection of different growing media (regarding texture, pH, drainage) for different groups or target species	
Soil nutrient / ESDB v.2 (EC, 2004)	Topsoil and subsoil base saturation Cation exchange capacity	Selection of different growing media and development of fertilization regimes for different groups	
Soil limitations / ESDB v.2 (EC, 2004)	Depth to a gleyed horizon Depth to rock Depth of an obstacle to roots Volume of stones	Selection of different growing media (regarding texture, drainage) and potting volume for different groups or target species	Potential to rank or filter plants in terms of different quantitative or qualitative variables or to group target plants sharing common requirements
Climate / WorldClim Database (Guarino et al., 2002, Hijmans et al., 2005)	**Temperatures** Mean minimum or maximum temperatures of the coldest or the warmest month Annual mean temperature range Temperature seasonality Mean diurnal temperature range **Precipitation** Precipitation of the driest month or the wettest month Mean monthly precipitation Annual precipitation Mean precipitation of driest, wettest, coldest or warmest quarter	Selection of appropriate sites and conditions both in greenhouse and in *ex situ* cultivation sites for different groups or target species Selection of temperatures for seed germination and asexual propagation for different groups or target species Shading and ventilation for temperature regulation Calculation or watering regimes, calibration of different watering regimes in different seasons and/or in different months for different groups or target species	
Topography / Digital Terrain Model created	Aspect, Slope, Altitude (elevation from sea level)	Customization of microclimate in *ex situ* cultivation sites and appropriate positioning in plant displays	
Vegetation zones / Mavromatis (1980)	11 different vegetation types identified for the Greek territory	Selection of the *ex situ* conservation sites for different groups or target species	
Land cover classes and types / CORINE Land cover (EC & ETC/LC, 1999)	45 different land-use classes and types	Selection of shading regime for different groups (seasonal, if originating from deciduous vegetation or all year round, if originating from evergreen vegetation) or for target species Creation of species assemblages Selection of sites for specific plant assemblages and displays	

Table 1. Guidelines for the *ex situ* conservation of target plants based on links of their collection data with geodatabases (from Krigas et al., 2010, with modifications).

- Indicate the preferable *ex situ* growing conditions for each target plant (linking the *in situ* natural conditions with the *ex situ* cultivation regimes) and provide guidelines regarding species-specific treatments (Fig. 2, 3, 5, 6),
- Rank or filter included target species in terms of different quantitative variables or group target plants sharing common preferences or requirements (Fig. 7, 8),

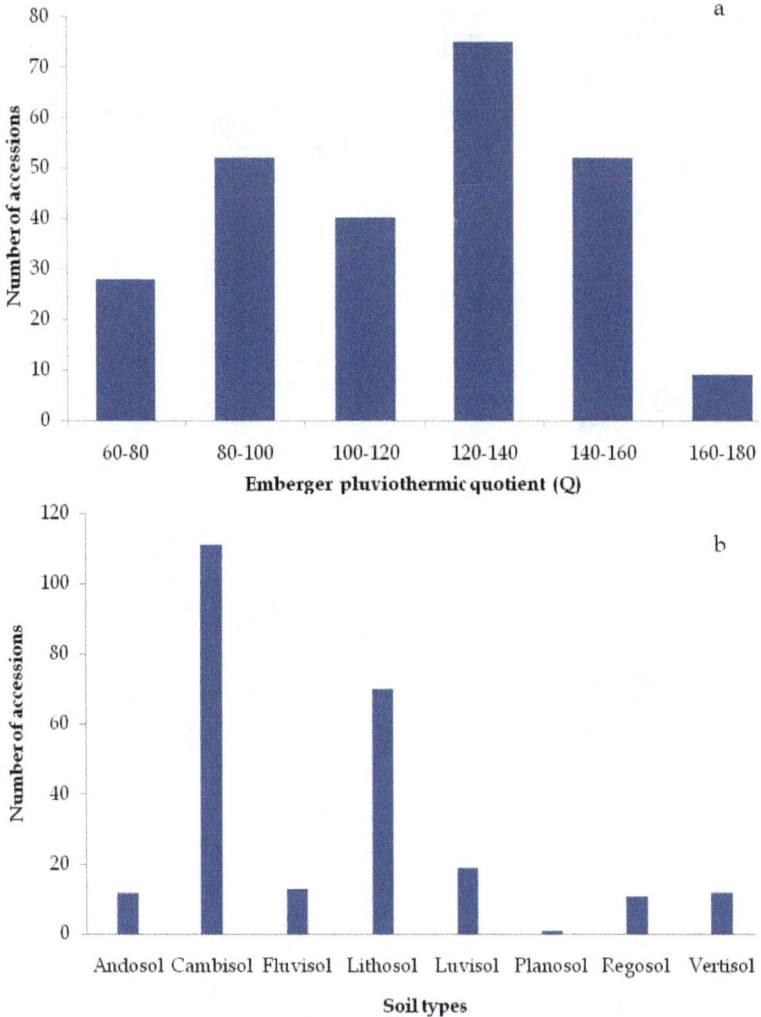

Figure 7. Groupings of target plants based on GIS-derived ecological criteria: Number of accessions of target plants (n=256) originating from the Aegean Archipelago, the Ionian Islands, Crete and Peloponnese, southern Greece grouped into different Emberger's Pluviothermic Quotient classes (a) and according to basic soil types (b). For each class or type specific plant lists can also be generated.

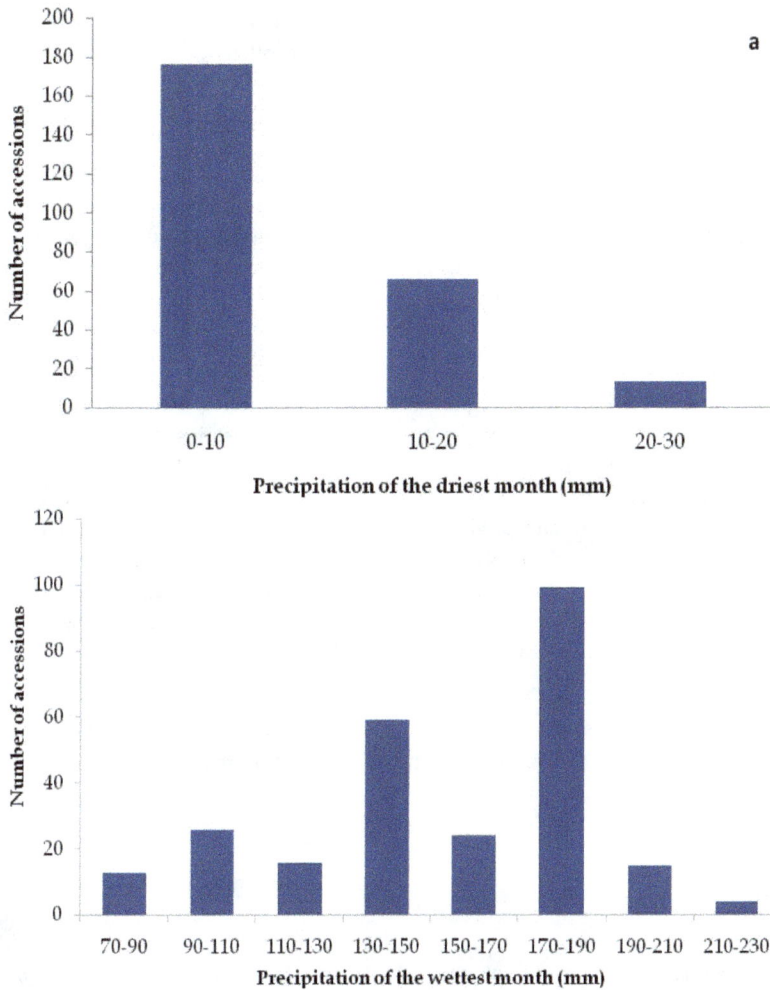

Figure 8. Groupings of target plants based on GIS-derived ecological criteria: Number of accession numbers of target plants (n=256) originating from the Aegean Archipelago, the Ionian Islands, Crete and Peloponnese (Greece) grouped according to precipitation classes during the driest (a) and wettest (b) months of the year. For each class specific plant lists can also be generated.

- Produce lists of species or frequency graphs within specified ranges of environmental variables or combinations thereof (Fig. 7, 8),
- Organize groupings of target plants based on ecological criteria (groups of plants with similar preferences or requirements), thus possibly improving their growing conditions as well as saving human-hours and space in *ex situ* conservation areas (Fig. 7, 8),
- Pinpoint target plants that may or cannot be accommodated in specific areas,

- Reduce trial-and-error losses during the *ex situ* cultivation of target species which seems to be common in the community of botanic gardens, often due to absence of previous experience (Krigas et al. 2010),
- Formulate and establish species-specific guidelines for the *ex situ* cultivation of target plants in botanic gardens,
- Facilitate the gap analysis of the botanic expeditions organized for the collection of plant material (which may also permit better planning of future ones),
- Reveal gaps in the representation of target plants from different altitudes, vegetation zones, habitat types, phytogeographic and climatic regions of specified geographical areas,
- Pinpoint by spatial niche modelling new probable locations where target species may be encountered, permitting the search for new populations in the wild (Jarvis et al., 2005),
- Assess conservation strategies and actions of institutions related to plant conservation.

By exploiting the GIS-derived ecological information for target plant species, the Balkan Botanic Garden of Kroussia (Greece) was the first to initiate this pilot GIS application dedicated to the *ex situ* plant conservation (Krigas et al., 2010). To date the majority of the target plants collected from the wild and propagated at the grounds of the BBGK are able to grow and flower regularly and produce fruits with no problems reported so far (BBGK, pers. comm.). Additionally, their seeds are regularly collected and deposited in a seed bank for future studies. This valuable material of conservation important target plants -in case of catastrophic events and if deemed necessary- could serve as a means to enrich the wild plant populations with individuals raised *ex situ,* actually reducing the risk of their extinction (Bowes, 1999; Bunn et al., 2011).

This novel GIS application described here presents an invaluable (time and money saving) tool with a broad-scale potential in enhancing the prospects of the *ex situ* plant conservation of target species collected from diverse environmental conditions and transferred to man-made sites such as botanic gardens, nurseries and private gardens.

Author details

Nikos Krigas
Laboratory of Systematic Botany & Phytogeography, Department of Botany, School of Biology, Aristotle University of Thessaloniki, Greece

Kimon Papadimitriou
Office for Sustainability, Aristotle University of Thessaloniki, Greece

Nikos Krigas and Antonios D. Mazaris
Department of Ecology, School of Biology, Aristotle University of Thessaloniki, Greece

Acknowledgement

The authors would like to thank Dr E. Maloupa and the Balkan Botanic Garden of Kroussia (Greece) for funding earlier stages in this research and appreciate access to the plant material conserved *ex situ* in its grounds. The field work of N. Krigas in the Ionian Islands, SW Greece was partially supported by the Stanley Smith Horticultural Trust (UK) in the frame of the Ionian Island Project (Collections of rare, threatened and endemic plants of the Ionian Islands for their ex situ conservation in Greek and British botanic gardens). The work of A. D. Mazaris was partially supported by the EU FP7 SCALES project ('Securing the conservation of biodiversity across administrative levels and spatial, temporal and ecological Scales'; project #226852).

6. References

Allison, J., Miller, A.J. & Knouft, J.H. (2006). GIS-based characterization of the geographic distributions of wild and cultivated populations of the Mesoamerican fruit tree *Spondias purpurea* (Anacardiaceae). *American Journal of Botany*, 93, pp. 1757-1767, ISSN 0002-9122

Baskin, C.C., & Baskin, J.M. (1988). Germination ecophysiology of herbaceous plant species in a temperate region. *American Journal of Botany*, 75(2), pp. 286–305, ISSN 0002-9122

Benson, E.E. (1999). An introduction to plant conservation biotechnology, In: E.E. Benson (Ed), *Plant conservation biotechnology*, pp. 3-10, Taylor & Francis, ISBN 0203484193, London, UK & Philadelphia, USA

Bowes, B.G. (1999). *A colour atlas of plant propagation and conservation*, Manson Publishing Ltd, ISBN 9781874545927, London, UK

Brady, N.C. & Weil, R.R. (2002). *The nature and properties of Soils*, Prentice Hall, Upper Saddle River, ISBN 130167630, New Jersey, USA

Bunn, E., Turner, S.R. & Dixon, K.W. (2011). Biotechnology for saving rare and threatened flora in a biodiversity hotspot. *In vitro Cellular & Developmental Biology - Plant*. 47(1), pp. 188-200, ISSN 1054-5476

CBD-*Convention on Biological Diversity* (1992). United Nations Environment Programme, Rio de Janeiro, Retrieved from http://www.cbd.int/doc/legal/cbd-en.pdf

Constantinidis, T. & Kalpoutzakis, E. (2009). *Achillea occulta* Constantin. & Kalpoutz., Vulnerable (VU), In: *The red data book of rare and threatened plants of Greece, vol 1 (A–D)*, D. Phitos, T. Constantinidis, & G. Kamari (Eds), pp. 40–42, Hellenic Botanical Society, ISBN 9789609407120, Patras, Greece (in Greek)

Draper, D., Rosselló-Graella, A., Garcia, C., Tauleigne Gomes, C. & Sérgio, C. (2003). Application of GIS in plant conservation programmes in Portugal. *Biological Conservation* 113, pp. 337–349, ISSN 0960-3115

EC & ETC/LC (1999). *CORINE land cover—technical guide*, European Community, European Environment Agency, Retrieved from http://www.eea.europa.eu/publications/tech40add

EC (1985). *Soil Map of the European Communities (CORINE soil geodatabase) at 1:1000000*, The Commission of the European Communities, Directorate General for Agriculture, Coordination of Agricultural Research, Luxembourg, Retrieved from http://www.eea.europa.eu/data-and-maps/data/soil-type

EC (2004). *The European Soil Database distribution version 2.0*, European Commission and the European Soil Bureau Network (CD-ROM) EUR 19945 EN

ESPC (2009). *A sustainable future for Europe: The European Strategy for Plant Conservation 2008–2014*, Planta Europa Network, Council of Europe & CBD, Retrieved from http://www.plantaeuropa.org/assets/New%20European%20Strategy%20for%20Plant%20 0Conservation%20(2008-2014).pdf

Farnsworth, E.J., Klionsky, S., Brumback, W.E. & Havens, K. (2006). A set of simple decision matrices for prioritizing collection of rare plant species for *ex situ* conservation. *Biological Conservation* 128, pp. 1-12, ISSN 0960-3115

Fernández, J.A., Santana, O., Guardiola, J.L., Molina, R.V., Heslop-Harrison, P., Borbely, G., Branca, F., Argento, S., Maloupa, E., Talou, T., Thiercelin, J.M., Gasimov, K., Vurdu, H., Roldán, M., Santaella, M., Sanchís, E., García-Luis, A., Suranyi, G., Molnár, A., Sramko, G., Gulyas, G., Balazs, L., Horvat, O., Rodríguez, M.F., Sánchez-Vioque, R., Escolano, M.A., Reina, J.V., Krigas, N., Pastor, T., Renau-Morata, B., Raynaud, C., Ibadli, O., Polissiou, M., Tsimidou, M.Z., Tsaftaris, A., Sharaf-Eldin, M., Medina, J., Constantinidis, T., Karamplianis, T., De-Los-Mozos-Pascual, M. (2011). The World Saffron and Crocus collection: strategies for establishment, management, characterization and utilization. *Genetic Resources & Crop Evolution* 58(1), pp. 125-137, ISSN 0925-9864

Glowka, L., Burhenne-Guilman, F., Synge, H., McNeely, J. & Güdling, L. (1994). *Guide to the convention on biological diversity*, Environmental and law paper no 30, International Union for the Conservation of Nature and Natural Resources, ISBN 2831702224, Gland, Switzerland & Cambridge, UK

Godefroid, S. & Vanderborght, T. (2010) Seed banking of endangered plants: are we conserving the right species to address climate change? *Biodiversity and Conservation* 19, pp. 3049-3058, ISSN 0960-3115

Grigoriadou, K., Krigas, N. & Maloupa, E. (2011). GIS-facilitated *in vitro* propagation and *ex situ* conservation of *Achillea occulta*. *Plant Cell Tissue & Organ Culture* 107(3), pp. 531-540, ISSN 01676857

GSPC-*Global Strategy for Plant Conservation* (2002). Secretariat of the convention of biological diversity in association with Botanic Gardens Conservation International, Montreal, Quebec, Retrieved from http://www.cbd.int/gspc/

Guarino, L., Jarvis, A., Hijmans, R.J. & Maxted, N. (2002). Geographic information systems (GIS) and the conservation and use of plant genetic resources. In: *Managing plant genetic diversity*, J.M.M. Engels, V. Ramanatha Rao, A.H.D. Brown & M.T. Jackson (Eds), pp. 387-404, ISBN 0851995225, CABI Publishing, International Plant Genetic Resources Institute, Rome, Italy

Hijmans, R.J., Cameron, S.E., Parra, J.L., Jones, P.G. & Jarvis, A. (2005). Very high resolution interpolated climate surfaces for global land areas. *International Journal of Climatology* 25, pp. 1965-1978, ISSN 1097-0088

IUCN (2001). *IUCN red list categories and criteria, Version 3.1*, IUCN Species Survival Commission, Gland, Switzerland & Cambridge, UK, Retrieved from http://www.iucn.org

Iverson, L.R. & Prasad, A. (1998). Estimating regional plant biodiversity with GIS modeling. *Diversity and Distributions* 4, pp. 49-61, ISSN 1472-4642

Jarvis, A., Williams, K., Williams, D., Guarino, L., Caballero, P.J. & Mottram, G. (2005). Use of GIS for optimizing a collecting mission for a rare wild pepper (*Capsicum flexuosum* Sendtn.) in Paraguay. *Genetic Resources & Crop Evolution* 52, pp. 671–682, ISSN 0925-9864

Karagianni, V., Kamari, G. & Phitos, D. (2009). *Silene cephallenia* subsp. *cephallenia* Heldr., Critically Endangered (CR), In: *The red data book of rare and threatened plants of Greece, vol 2 (E-Z)*, D. Phitos, T. Constantinidis & G. Kamari (Eds), pp. 319–320, Hellenic Botanical Society, ISBN 9789609407120, Patras, Greece (in Greek)

Katsouni, N., Karagianni, V. & Kamari, G. (2009). *Viola cephalonica* Bornm., Critically Endangered (CR), In: *The red data book of rare and threatened plants of Greece, vol 2 (E-Z)*, D. Phitos, T. Constantinidis & G. Kamari (Eds), pp. 385-387, Hellenic Botanical Society, ISBN 9789609407120, Patras, Greece (in Greek)

Krigas, N. & Karamplianis, Th. (2009). *Ranunculus cacuminis* Strid & Papan., Vulnerable (VU), In: *The red data book of rare and threatened plants of Greece, vol 2 (E-Z)*, D. Phitos, T. Constantinidis & G. Kamari (Eds), pp. 269-271, Hellenic Botanical Society, ISBN 9789609407120, Patras, Greece (in Greek)

Krigas, N. & Maloupa, E. (2008). The Balkan Botanic Garden of Kroussia, Northern Greece: a garden dedicated to the conservation of native plants of Greece and the Balkans. *Sibbaldia* 6, pp. 9–27, ISBN 1872291449

Krigas, N. (2009). *Allium samothracicum* Tzanoud., Strid & Kit Tan, Endangered (EN), In: *The red data book of rare and threatened plants of Greece, vol 1 (A-D)*, D. Phitos, T. Constantinidis & G. Kamari (Eds), pp. 68-69, Hellenic Botanical Society, ISBN 9789609407120, Patras, Greece (in Greek)

Krigas, N., Mouflis, G., Grigoriadou, K. & Maloupa, E. (2010). Conservation of important plants from the Ionian Islands at the Balkan Botanic Garden of Kroussia, N Greece: using GIS to link the *in situ* collection data with plant propagation and *ex situ* cultivation. *Biodiversity & Conservation* 19, pp. 3583–3603, ISSN 0960-3115

Krogstrup, P., Find J.I., Gurkov, D.J. & Kristensen M.M.H. (2005). Micropropagation of Socotran fig, *Dorstenia gigas* Schweinf. ex Balf. f.—A threatened species, endemic to the island of Socotra, Yemen. *In vitro Cellular & Developmental Biology - Plant* 41, pp. 81–86, ISSN 1054-5476

Loiselle, B. A., Jørgensen, P.M., Consiglio, T., Jiménez, I., Blake, J.G., Lohmann, L.G. & Montiel, O.M. (2008). Predicting species distributions from herbarium collections: does climate bias in collection sampling influence model outcomes? *Journal of Biogeography* 35, pp. 105–111, ISSN 1365-2699

Maloupa, E., Grigoriadou, K., Papanastassi, K. & Krigas, N. (2008). Conservation, propagation, development and utilization of xerophytic species of the native Greek flora towards commercial floriculture. *Acta Horticulturae* 766, pp. 205–213, ISSN 0567-7572

Maunder, M., Higgens, S. & Culham, A. (2001). The effectiveness of botanic garden collections in supporting plant conservation: a European case study. *Biodiversity & Conservation* 10, pp. 383–401, ISSN 0960-3115

Mavromatis, G. (1980). The Bioclimate of Greece, relationship of climate and natural vegetation, bioclimatic maps, Institute of Forest Research of Athens, Athens, Greece

Minor, E.S., Tessel, S.M., Engelhardt,K. A.D. & Lookingbill, T.R. (2009). The role of landscape connectivity in assembling exotic plant communities: a network analysis. *Ecology* 90, pp. 1802-1809, ISSN 0012-9658

Moat, J. & Smith, P.P. (2003). Applications of Geographical Information Systems in Seed Conservation, In: *Seed Conservation: Turning Science into practice*, R.D. Smith, J.B. Dickie,

S.H. Linigton, H.W. Pritchard & R. J. Probert (Eds), Chapter IV, pp. 81-87, Royal Botanic Gardens, ISBN 1842460528, Kew, UK, Retrieved from http://www.kew.org/science-research-data/kew-in-depth/msbp/publications-data-resources/technical-resources/seed-conservation-science-practice/index.htm

Pedersen, Å. Ø., Nyhuus, S., Blindheim, T., & Wergeland Krog, O. M. (2004). Implementation of a GIS-based management tool for conservation of biodiversity within the municipality of Oslo, Norway. *Landscape & Urban Planning* 68, pp. 429–438, ISSN 0169-2046

Powel, M, Accad, A. & Shapcott, A. (2005). Geographic information system (GIS) predictions of past, present habitat distribution and areas for re-introduction of the endangered subtropical rainforest shrub *Triunia robusta* (Proteaceae) from south-east Queensland Australia. *Biological Conservation* 123, pp. 165–175, ISSN 0006-3207

Ramirez-Villegas, J., Khoury, C., Jarvis, A., Debouck, D.G. & Guadrino, L. (2010). A gap analysis methodology for collecting crop genepools: a case study with *Phaseolus* beans. *PloS ONE* 5(10), pp. 1–18, ISSN 1932-6203

Salem, BB. (2003) Application of GIS to biodiversity monitoring. *Journal of Arid Environments* 54, pp. 91–114, ISSN 0140-1963

Sarasan, V., Cripps, R., Ramsay, M.M., Atherton, C., McMichen, M., Prendergast, G., Rowntree, J.K. (2006). Conservation *in vitro* of threatened plants—progress in the past decade. *In vitro Cellular & Developmental Biology - Plant* 42, pp. 206–214, ISSN 1054-5476

Schulman, L. & Lehvävirta, S. (2010) Botanic gardens in the age of climate change. *Biodiversity & Conservation* 20, pp. 217–220 (Editorial note), ISSN 0960-3115

Schulman, L., Toivonen, T. & Ruokolainen, K. (2007). Analysing botanical collecting effort in Amazonia and correcting for it in species range estimation. *Journal of Biogeography*, 34, pp. 1388-139, , ISSN 1365-2699

Sharrock, S. & Jones, M. (2009). *Conserving Europe's threatened plants: progress towards Target 8 of the Global Strategy for Plant Conservation*, Botanic Garden Conservation International, Richmond, UK, ISBN 9781905164301, Retrieved from http://www.bgci.org/files/Worldwide/Publications/euro_report.pdf

Store, R. & Jokimäki, J. (2003). A GIS-based multi-scale approach to habitat suitability modeling. *Ecological Modeling* 169, pp. 1–15, ISSN 0304-3800

Store, R. & Kangas, J. (2001). Integrating spatial multi-criteria evaluation and expert knowledge for GIS-based habitat suitability modeling. *Landscape & Urban Planning* 55, pp. 79–93, ISSN 0169-2046

Turland, N. (1995). *Origanum dictamnus* L., Vulnerable (VU), In: *The red data book of rare and threatened Plants of Greece*, D., Phitos, A., Strid, S., Snogerup & W., Greuter (Eds), pp. 394-395, WWF, ISBN 9607506049, Athens, Greece

USGS-United States Geological Survey (2004). Shuttle Radar Topography Mission, 30 Arc Second scene SRTM_GTOPO_u30_mosaic, Unfilled Unfinished 2.0, Global Land Cover Facility, University of Maryland, College Park, Maryland, February 2000, Retrieved from http://glcf.umiacs.umd.edu/data/srtm/

Vanderpoorten, A., Sotiaux A. & and Engels, P. (2006). A GIS-based model of the Distribution of the rare liverwort *Aneura maxima* at the landscape scale for an improved assessment of its conservation status. *Biodiversity and Conservation* 15, pp. 829-838, ISSN 0960-3115

Use of GIS to Estimate Productivity of Eucalyptus Plantations: A Case in the Biobio Chile's Region

Rolando Rodríguez and Pedro Real

Additional information is available at the end of the chapter

1. Introduction

In the past, today and the future, forest productivity has become an important issue to ensure the sustainability of forest resources. In a modern sense of concept, forest sustainability may be defined as the combination of biological (biotic), environmental (abiotic), and cultural factors that determine the rate at which the forest overcomes "environmental resistance" and achieves the potential productivity of a site (Medlyn et al., 2011).

Governments and managers need to ensure that society is provided with forest products in both ecologically and economically correct ways, considering social aspects. Therefore, forest productivity is an important criterion of sustainability because of its strong relationship with economics and profitability.

Managers are insistently demanding precise estimates of forest biomass productivity and potential growth rates at global and local scales. This situation creates the necessity of research in growth and yield simulation, since accurate prediction is decisive for decision-making. The quality of decision is strongly influenced by the types of models. Innovative simulation strategies are essential to predict potential impacts of future changes in the global environment (Medlyn et al., 2011).

On the one side, classical growth and yield forest models have been criticized as being empirical, with models revealing little about physiological mechanisms that control the adaptation to environmental conditions. On the other side, complex mechanistic models of growth have been criticized as being cumbersome, requiring too many hard to measure inputs variables and relying heavily on untested assumptions (Pinjuv, 2006). The above considerations lead to the conclusion that the best option is to combine both types: empirical and process-based models in a joint and calibrated hybrid model (Almeida, et al., 2004).

Rather than using empirical measurements or complex mechanistic models of growth, process-based modeling attempts to simulate the general ecological mechanism of a given ecosystem. Within the advances in computing technologies and understanding of ecological process, process-based simulations models are providing means to address scientific and management questions at all spatial scales, from individual trees to the entire globe (Kirk & Burk, 2004).

Forest managers are interested not only in the characteristics of specific trees or forest stands, but also in trends that extend across large areas such as watershed, landscapes or ecoregions (Shindler, 1998). While scientifically based management at these coarse scales is desired, collecting appropriate data is a major challenge. Process-based models provide a viable alternative to large-scale field sampling for several reasons (Kirk & Burk, 2004).

This paper discusses a generalized framework for developing a regional-scale process-based forest productivity model implemented for the study of the environmental limitations on the growth of *Eucalyptus nitens* plantations. Maps of potential productivity are developed and regulations of productivity by environmental factors are discussed. Additionally, comments are added on the usefulness of GIS in the study of spatial patterns that govern forest productivity.

2. Modeling forest system

2.1. Empirical model

Forests are very complex ecosystems, and historically many approaches to predict yield and productivity of forests sites have been developed. Growth and yield of forest sites have been modeled in different ways, and different productivity measures exist as site index, site quality and potential site (Daniel et al., 1979).

Vanclay (1994) has defined stand growth model as "an abstraction of the natural dynamics of a forest stand, which may encompass growth, mortality, and other changes in stand composition and structure". Common usage of the term "growth model" generally refers to a system of equations which can predict the growth and yield of a forest stand under a wide variety of conditions. Thus, a growth model may comprise a series of mathematical equations, the numerical values embedded in those equations, the logic necessary to link these equations in a meaningful way, and the computer code required to implement the model on a computer. In its broadest sense, the term may also embrace yield tables and curves, which are analogous to equations, but which have been stated in a tabular or graphical form, rather than a mathematical form (Vanclay, 1994).

An important but simple model in forestry is the plantation yield table, which may comprise only of two columns of figures: age and the expected standing volume at that age. The yield table may also be expressed graphically as a series of curves, with the horizontal axis indicating age and the vertical axis indicating volume produced. It may also be expressed more concisely as a mathematical equation (Vanclay, 1994).

In Chile, empirical growth and yield simulators have been developed to be used in intensively managed plantations of *Pinus radiata, Eucalyptus globulus* and *Eucalyptus nitens*. To improve its predictive ability, the Chilean model has been adapted to growth zones and, stratification is used to improve the productivity measure (Site Index) of the simulator. In Chile, the need to incorporate physiographic, soil and climate data as continuous co-variables instead of zones is relevant. However, the weakness of the discrete territory handling that is implicit in the modeling strategy has become evident over time.

In general, limitations of the approach taken are: 1) predictions are based on historical data records, 2) lack of necessary flexibility to estimate changes in the stand growth in response to environmental changes, 3) limited capabilities of extrapolation when new treatments are not considered in the fitting data base, and finally, 4) the development and maintenance of plot networks are expensive to establish and re-measured (Bernier et al., 2003).

Empirical models are used in most forest companies and remain valuable tools for management, but they do not provide answers to some critical questions that arise in the planning process, especially those related to factors that are affected by silviculture and climate (Constable & Friend, 2000). However, recent approaches tend to improve the applicability of empirical models under changing conditions. Examples of these approaches are 1) the dynamic state-space approach, or 2) the development of productivity-environments relationship (Fontes et al., 2010).

2.2. Process-based model

Limitations described for empirical models, have led to the exploration of new and more consistent productivity definitions, such as concepts and methods used by ecologists, of which the most common productivity measurements are gross primary productivity (GPP), net primary productivity (NPP) and net ecosystem productivity (NEP). Growth can be defined as the net accumulation of carbon and other organic materials in plants. An indirect measurement of growth is used for the photosynthetic and respiration rates (Hari et al., 1991).

The processes governing forest growth are well-recognized and understood (Landsberg, 1986; Landsberg & Gower, 1997). Photosynthesis, driven by radiant energy intercepted by the foliage, produces carbohydrates, which are respired to provide the energy needed for protein synthesis and formation of new tissues, and partitioned to branches and stems, coarse and fine roots and new foliage (Landsberg & Waring, 1997). Photosynthetic efficiency is affected by the nutrient status of the foliage, and also depends on the uptake of CO_2 through stomata, which may be affected by the water status of the leaves. Growth processes are influenced by temperatures which affect process rates: extreme temperatures may cause damage and disrupt growth temporarily or permanently (Landsberg, 2003). These processes and their interactions produce carbohydrates, some of which become wood – the product of interest to the commercial forester.

However, these processes can be affected by competition. Under these approaches, productivity is determined by the amount of light intercepted by the canopy, efficiency of light use proportion of assimilates allocated to wood and mortality losses (Cannell, 1989). Consequently, process-based modeling can be defined as a procedure by which the behavior of a system is derived from a set of functional components and their interactions with each other and the system environment, through physical and mechanistic processes occurring over time (Bossel, 1996). Even though, process-based models were originally designed for research purpose more recently they have been developed towards use in practical forest management (Fontes et al., 2010). Therefore, process-based models attempt to predict the products by describing the processes that lead to them, their responses to external driving variables and the interactions between them (Landsberg, 2003). This is observed on the schematic representation of the 3-PG process-based model (Fig.1).

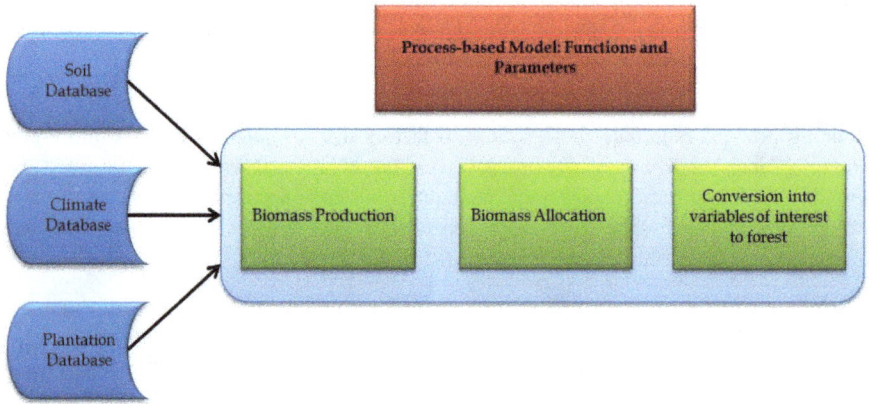

Figure 1. Schematic representation of 3-PG process-based model.

The 3-PG model is a stand-level model that uses monthly or annual time steps. It requires as inputs initial stand data, soil and weather data. The 3-PG model consists of five simple sub models: biomass production obtained from the assimilation of carbohydrates, the allocation of biomass between foliage, roots and stems, the determination of stem number, soil water balance, and the conversion of biomass values into variables of interest to forest managers.

In recent years, forest managers have expressed interest in the application of process-based models in management decision making (Mäkelä, 2000, Korzukhin et al., 1996, Johsen et al., 2001). Battaglia & Sands (1997) have identified five potential uses of process-based forest productivity models as management tools: prediction of growth and yield, selection of new plantations sites, identification of site limitations on productivity, assessment of risk associated with locations or management options and use of models as surrogates for field experiments.

However, all these models share the problems of parameter estimation caused by lack of precise data and an incomplete understanding of some important processes.

2.3. Linking process-based and empirical forest model

The term "hybrid modelling" refers to approaches that are grounded in both empirical and process-based concepts of forest dynamics, thus trying to capitalize on the advantages of each approach (Fontes et al., 2010). Specifically, the underlying idea is to benefit from the predictive ability and parsimony in the calibration data needs of empirical approaches as well as the explicit environment-dependence of process-based formulations (Fontes et al., 2010). This approach offers potentially the best prospects for developing models to support forest management (Battaglia et al., 1998).

The development of process-based models to predict forest growth has been developing rapidly in the last few years. However, operational applications in forest plantations are still at an early stage (Almeida et al., 2003). All these models cited in scientific literature share in common the need to assess and to evaluate various parameters and processes, modeled with sub-models, based on field measurements and independently calibrated (Sharpe & Rykiel, 1991). Some sub-models, or a subset of sub-model parameters, will fail calibration because of the lack of adequate data, problems of scaling up, or poor understanding of processes (Mäkelä et al., 2000). In such case, and in order to predict system-level behavior (Mäkelä et al., 2000), these sub-models or sub-model parameters are best estimated by using empirical equations relating to the whole system (Sievänen & Burk, 1993). Korzukhin et al. (1996) presented a detailed analysis of the relative merits of process-based models and empirical models, which highlighted the value of both classes of models and indicated how they can be applied in forest ecosystem management. The above considerations lead to the conclusion that the best option is to combine both types: empirical and process-based models in a joint and calibrated hybrid model.

Hybrid models were developed from complementary merging of well understood processes and reliable tree/stand empiricism, which aim to achieve a process model for the manager in which the shortcomings of both approaches can be overcome to some extent. There is a combination of causal (at the level of the process such as: carbon balance, water balance, soil carbon cycling, soil carbon cycling) and empirical (at the higher stand level the model is empirical) elements (Almeida et al., 2003).

Shifting views of the forest from primarily one as a production system for many products to an ecosystem with spatially and temporally complex interrelationships is changing the demand for information about the forest (Korzukhin et al., 1996). These new information needs are characterized by greater complexity, limited availability of mechanistic hypotheses, and paucity of data. In contrast to empirical models, process-based model seek primarily to describe data using key processes that determine an object's internal structure, rules, and behavior (Korzukhin et al., 1996). Both types of solutions can be developed across the full range of spatial scales. In Figure 2 we present the hybridized 3-PG model.

A hybrid approach combining the main advantages of process-based models and empirical models has been adopted in some cases. Baldwin et al. (1993) combined a single-tree empirical model called PTAEDA2 (Burkhart et al., 1987) with process-based model

MAESTRO (Wang & Jarvis, 1990). Using PTAEDA2 they projected to a certain age the stand variables used by MAESTRO to calculate biomass production, which was fed back to TAEDA2 to adjust its predictions. These steps were repeated to the end of rotation. In turn, Battaglia et al. (1999) used the process-based model PROMOD and the empirical model NITGRO developed for *Eucalyptus nitens* plantations. The model PROMOD predicted the mean annual increment (MAI; m^3 ha^{-1} $year^{-1}$) and estimated site index (SI) applying relationship MAI and SI. However, according to Johnsen et al. (2000) major constraints in the implementation of hybrid models for practical forestry are those related to soil and nutrient dynamics. The improvement of both process-based and empirical models will lead to better hybrid models. This architecture has been demonstrated to be successful as a practical tool in operational forestry (Almeida, et al., 2003).

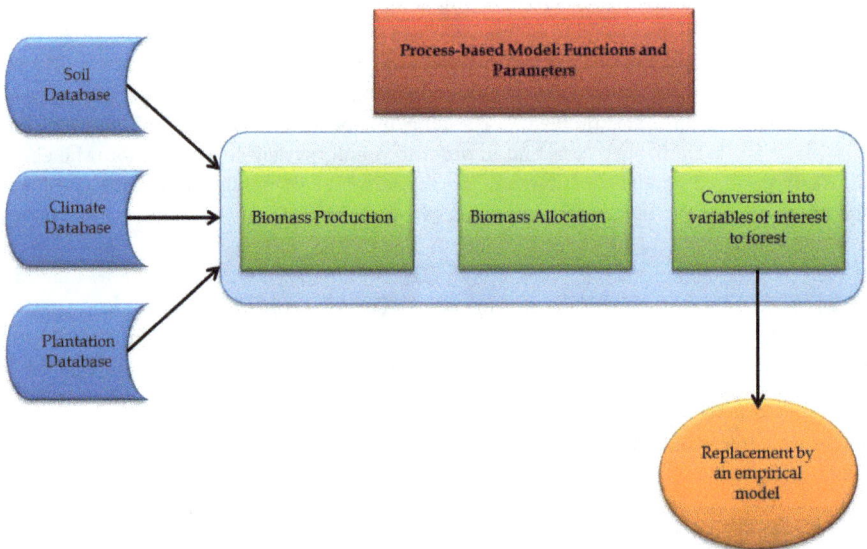

Figure 2. Schematic representation of 3-PG process-based mode hybridized (adapted from Fontes (2007)). In addition to Fig. 1, functions of replacement by an empirical model are added to the process-based model to get variables of interest to the forester.

2.4. Overview of the 3-PG model hybridized

The model called 3-PG (Physiological Principles Predicting Growth) is a process-based model that requires few parameter values and only readily available data inputs, and was developed by Landsberg & Waring (1997) with the intent of being both simple and applicable to management.

The 3-PG model is based on physiological principles and parameterized with empirical data. It uses the concept of radiation use efficiency, where potential carbon gain is then constrained by the biology of the species and the physical characteristics of the site, and includes allometric relationship for carbon allocation. The 3-PG model is generalized, with monthly time-step, which means that biomass is the output. The important features of the model are that the output variables it produces are those of interest and utility to forest managers.

The 3-PG model lies between traditional growth and yield and detailed physiological representations, physiological process predicting growth, and the model is a "big leaf" carbon balance. A "big leaf" model is one that represents a forest canopy as a homogeneous structure, usually leaf area index (LAI), that is the area of leaves per unit area of ground surface. This model is based on two fundamentals ideas (Mason, 2010): net primary productivity is a linear function of absorbed photosynthetically active radiation (φp.a.), with the slope of the function varying with environmental states and, allocation of carbon can be estimated from allometry and that allocation to roots will vary with fertility of soils. In addition, the 3-PG model uses Beer's law to estimate absorbed photosynthetically active radiation given any amount of radiation and LAI.

As described by Landsberg & Waring (1997), the basic structure 3-PG model:

i.   Estimates Gross Primary Production (PG) based on the utilizable photosynthetically active radiation (φp.a.u) and the canopy quantum efficiency (αC). The value of φp.a.u is obtained by reducing the photosynthetically active radiation (φp.a.) through non-dimensional modifying factors whose values vary between 0 and 1. The modifying coefficients reflect constraints imposed on the utilization of φp.a. by vapor pressure deficit (VPD), drought, and stand age. Soil fertility determines allocation of carbon to above and belowground components. Drought is defined for soils of different textures by the ratio of water in the root zone to the maximum available (θ). The model also includes suboptimal temperature and frost modifiers.

ii. Estimates Net Primary Production (P_N) from PG. The model uses the proportion $PN/PG = c_{pp}$, which has been found to be 0.47 ± 0.04 for diverse types of forests and geographic locations (Waring, 2000).

iii. Estimates carbon in root biomass using two basic relations: the inverse relation between stem growth and the P_N fraction allocated to soil, and the effects of drought and nutrition on the annual allocation of carbon to root biomass.

iv. Calculates changes in stand density over time using a sub-model derived from the self-thinning law coupled with stem growth rates.

v.   Apportions carbon among aboveground tree components using allometric relations:

$$\hat{w}i = a_i \hat{w}^{Ni}$$

Where $\hat{w}$ is the tree's total biomass and i is the tree component. The parameter N_i reflects the species' genetic characteristics.

vi. The growth rate of trees declines with age. To this respect, the hydraulic limitations theory proposes that the decline of forest productivity with age is a consequence of whole-plant and leaf-specific hydraulic conductance with tree height caused by increased friction. Based on this theory, the 3-PG model estimates the effect of age on tree growth using the linear relation between hydraulic conductivity and PN, in which stem conductance declines with age and thereby induces a lower stomatal conductance (g_c).

3. Potential use of GIS coupled with process-based model for asses' forest productivity

A review of forestry applications in GIS reveals an extensive range of activities. GIS for forest management may be characterized by two broad and related categories: resource inventory including monitoring and analysis, modeling or prediction to support decision making (McKendry & Eastman, 1991).

The massive data management and its transformation into information requested by users with different goals of information, has been promoted heavily from the 80's with the rapid development of so-called Geographic Information Systems (GIS). System design involves the existence of a set of elements or parts that are interrelated, in the case of Geographic Information Systems the following elements can be distinguish: 1) User information, 2) Spatial and / or space, 3) Software for managing spatial data and 4) Hardware.

Unlike traditional information systems, GIS is a specialized system, designed to handle geographic data with spatial reference. Since human activity and the action of nature and their interactions occur in a geo-referenced environment (the earth), GIS provides a platform capable of structuring, integrating and connecting data of different nature, and enhancing the capabilities of analysis compared to Information Systems with no spatial reference.

A process-based model to estimate productivity in plantations, involves handling a wide range of spatially referenced data that needs to be collected from different sources, input to the GIS, standardized, organized, integrated and analyzed, generating new data for use as information in the process-based model that generates the results of growth and yield.

The data to be integrated corresponds to climate, soil, topography and plantations. Data should be organized into the GIS as a spatial data base model, creating subsystems that provide information to the process simulator.

The organization of a GIS system for this purpose is shown schematically in Figure 3.

The use of Geographic Information Systems, in our experience, demonstrated the high flexibility of analysis, data integration capabilities provided by GIS and functionality required for forest productivity studies. As shown in Figure 3, the concept of GIS used is modular and contains four subsystems described below.

3.1. Data acquisition

Corresponds to the use of input and storage facilities provided for the GIS software designed to input, store and edit basic data collected. This stage includes the standardization of information and the development of thematic maps and their associated databases.

The climate data include rainfall, temperatures (maximum, minimum and optimum for photosynthesis of the species), solar radiation, and vapor pressure deficit and frost days. In the case of soil chemical properties, N, P, K and trace elements were considered to construct a ranking of soil fertility. Soil physical properties were also used, such as depth, texture, structure, bulk density, field capacity and permanent wilting point allowed us to estimate available soil water holding capacity. Along with soil data details of geography and terrain such as latitude, longitude, slope, aspect, altitude and topographic position were also considered.

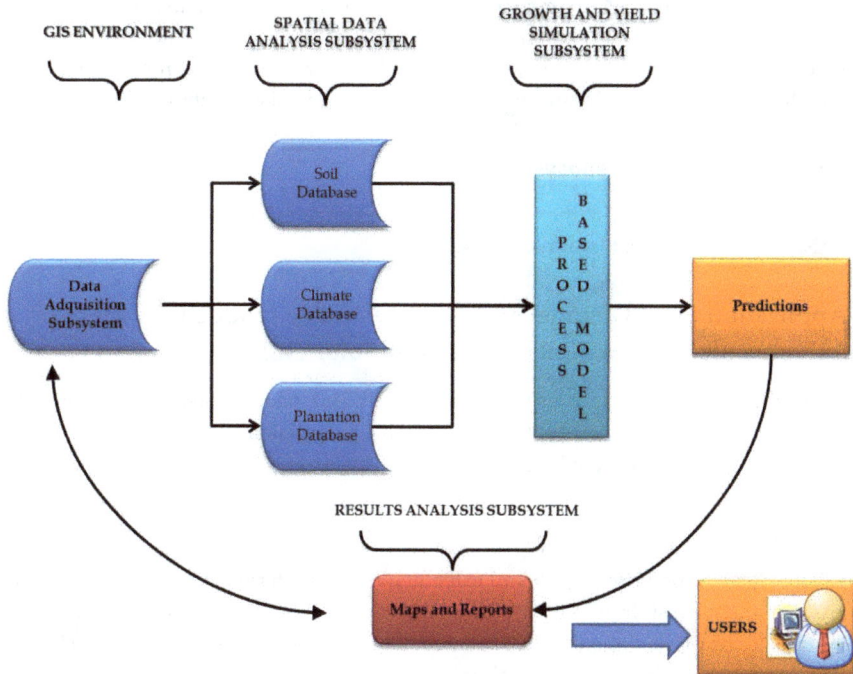

Figure 3. Diagram of the integration process and data analysis of GIS to quantify forest plantation productivity.

3.2. Data integration and analysis

This subsystem uses spatial analysis capabilities data to perform geospatial interpolation from discrete spatial data. As an example, the interpolation of climate can be mentioned where data from weather stations are interpolated to create surfaces with appropriate pixel size (pixel: unit of spatial resolution in the map). Soil and topography information were produced and integrated, using the GIS overlay capabilities that allow the integration of graphical data and their associated databases. Orographic description was made by the creation of Digital Terrain Models and their derived products: slope and aspect. Data from the forest and its spatial distribution were integrated to define areas with the specie and validate the response of the system.

3.3. Growth and yield simulator

This subsystem corresponds to the software platform used to simulate forest development under different climate, soil and environmental conditions. This subsystem includes functions and species-specific physiological parameters that explain growth, such as specific leaf area, light conversion efficiency of photosynthesis, light conversion efficiency constraints on associated with temperature, fraction of absorbed radiation by the canopy, allometric equations, foliage turn-over, maximum leaf stomatal conductance, stomatal conductance canopy maximum and wood density. This subsystem produces for each geographic area productivity results for the species and forest site-specific conditions.

3.4. Results and analysis subsystem

The results may be returned to the GIS platform to be converted into thematic maps, editing and distribution to different information users.

4. Application of 3-PG model and GIS to estimate productivity of Eucalyptus plantations in the region of Biobío, Chile

To assess the appropriateness of a simplified process-based model to predict potential forest growth over the study area, we applied a geographic information system (GIS) with spatial layers of climate and soils to the 3-PG model.

4.1. Study area

The Biobío Region comprises 36,929 km^2 extending north-south from 36° 00' to 38° 30' south latitude and east-west from 71° 00' west longitude to the Pacific Ocean. The Andes Mountains to the east and the Coastal Mountain Range to the west define four agro-climatic areas located longitudinally: the coastal dry sector, the inner dry sector, the central valley, and the Andean foothills. The climate and soils of each agro-climatic area determine potential land-uses and productive capacity (Del Pozo & Del Canto, 1999).

4.2. Functions and parameters of the model

Considering the basic structure 3-PG model, Table 1 shows functions and parameters used in this study.

These functions and parameters differ from the original model (Landsberg & Waring, 1997), following Sands (2000) for 10-year old *Eucalyptus nitens*. αC, canopy quantum efficiency, MJ^{-1}; φp.a.u, photosynthetically active solar radiation utilized, MJ m^{-2} per month^{-1}; φ T$_{op}$, optimum temperature for photosynthesis, $^\circ$C; Tmi, minimum temperature for growth, $^\circ$C; T$_{ma}$, maximum temperature for growth, $^\circ$C; L, LAI, m^{2}m^{-2}; dia, average stem diameter (mm); m, soil fertility rank (m is 1 fertilized soil and decease to 0,1 in non-fertilized soils). To obtain maximum potential MAI, nutritional status was not limiting photosynthetic production and m is 1.

Variable	Values of functions and parameters	References
Light conversion efficiency of photosynthesis	Maximum αC ranges from 1.8 to 4.2 g C MJ^{-1} φp.a.u	Landsberg, 1986; Waring, 2000
Constraints on light conversion efficiency associated with temperature	Top was set at 20°C, Tmi 2°C and Tma 32°C	Battaglia et al. (1998)
Specific leaf area	4.0 m^2 kg^{-1}	Landsberg & Waring (1997)
Fraction of radiation absorbed by the canopy	1-(1.38 exp(-0,5*L)	Landsberg & Waring (1997)
Allocation equation for stem mass	Stem mass, kg=0.00007*dia. (mm$^{2.65}$)	Tickle et al. (2001)
Allocation equation for foliage mass	Foliage mass, kg=0,00005*dia. (mm$^{2.65}$)	Tickle et al. (2001)
Wood density	500 kg m^{-3}	Sands (2000)
Foliage turn-over	2% per month^{-1}	Tickle et al. (2001)
Maximum leaf stomatal conductance	0.005 m s^{-1}	Sands (2000)
Maximum canopy stomatal conductance	0.02 ms^{-1}	Sands (2000)
Fraction of production allocated to roots	ηr= (0,8*0,23)/(0,23+(0,8-0,23)mφp.a.u	Sands (2000)

Table 1. Functional forms and parameters used in this study.

4.3. Predicting potential productivity

Battaglia et al. (1998) developed a simple analytic model for the relation between LAI and dry matter production in *Eucalyptus nitens*. In their model, "increasing LAI increased light interception and hence dry matter production, but simultaneously increased canopy respiration. Consequently, for a given light utilization coefficient, there was a value of LAI that maximized PN" (Battaglia et al. 1998). We ran the 3-PG model until reaching an optimum value of LAI, which we estimated using the equation of Battaglia et al. (1998) as:

$$LAI = 1/k \ln [(s\varepsilon\Omega k)/(r_0 NF + \gamma)]$$

Where LAI is the canopy leaf area index (m^2 m^{-2}), k is the canopy light extinction coefficient, s is the specific leaf area (m^2 kg^{-1} DM), ε is the light utilization efficiency (kg C MJ^{-1}), Ω is the annual incident radiation (MJ m^{-2} yr^{-1}), r_0 is maintenance respiration rate per unit canopy N content (kg C kg^{-1} N yr^{-1}), NF is the average leaf N concentration (kg N kg^{-1} DM) and γ is rate of carbon loss as litterfall (kg C kg^{-1} DM yr^{-1}). We used the parameter values of Battaglia et al. (1998) with the exception of Ω, which was measured in the Biobío region, and ε which, as described below, we calculated on a site specific basis.

Light utilization efficiency (ε), therefore the optimal LAI, varies with site conditions, particularly water availability and temperature (Battaglia et al., 1998). To estimate how environmental conditions affect the distribution of optimal LAI throughout the Biobío Region, we used Landsberg's (1986) equation relating ε to mean annual temperature (T) and water stress index (W), in the form:

$$\varepsilon = 0.00195 f_W f_T$$

where $f_T = \max\{0,(1-(T-13.2/9.9)^2)\}$ and $f_T = \max [0,(1+0.9 \ln W)]$

In this modeled relationship, 13.2 °C is a value that maximizes net canopy production and 9.9 is a value of temperature stress. W was calculated as the annual mean of the daily ratio of actual to potential evaporation (Battaglia et al., 1998). Actual evapotranspiration was calculated as the product of potential evaporation and a factor crop as estimated by Worledge et al. (1998). Potential evaporation was estimated from Mean Class A pan evaporation distributed in the Biobío region (Del Pozo & Del Canto, 1999). Note that the LAI values estimated by the in PN-maximizing model should not be interpreted as those necessarily present in the plantations.

4.4. Analysis of plot information and validation of the 3-PG model

To validate the 3-PG model for *Eucalyptus nitens* in the Biobío Region, data from 46 permanent, 1000 m² plots were used. The plots were part of the national eucalypts growth and yield cooperative and were scattered over the main area of *Eucalyptus nitens* plantations in the Biobío region. The plots are measured annually and have between 6 and 12 measurements, at ages that vary between 7 and 17 years old. Data collected on each plot include age, diameter at breast height (Dbh), volume, and density together with climatic and soil variables that describe site conditions.

The accuracy of validation of the 3-PG model was evaluated by the Mean Square Error (MSE), in measured units and percentages. To evaluate bias the Aggregated Difference (AD), in measured units and percent, was used.

4.5. Geographic information system

Equal productivity zones were delineated using GIS Software (Chang, 2004). The following databases were used:

4.5.1. Spatial soil model

The Earth Ordering System (scale 1:250,000) developed by Schlatter et al. (2004) and soil origin data provided by the Instituto de Investigaciones de Recursos Naturales (1964) were used to construct the soil layer. The maximum available soil water was estimated from retention curves provided by Carrasco et al. (1993).

4.5.2. Spatial climate model

The spatial distribution of monthly maximum, mean and minimum temperatures was obtained from Santibáñez & Uribe (1994). Estimates of monthly solar radiation values were derived from maximum and minimum data at direct measurements performed by the Instituto de Investigaciones Agropecuarias using the procedure of Bristow & Campbell (1984). Maximum potential solar radiation on a flat surface was obtained by correcting from latitude and elevation (Running et al, 1987). This potential value was corrected by the monthly variation of the angle of solar declination over the earth's surface (Waring, 2000).

4.5.3. Digital elevation model

The slope and aspect models were derived from a digital terrain elevation model generated with information provided by regular Chilean cartography (Instituto Geográfico Militar, 2001). Slope and exposure classes were derived using a Triangle Irregular Network (TIN), which is an elevational data storage system (Crosier et al., 2004).

4.5.4. Map construction

Each digital coverage layer was created, cleaned and processed using the Arc Info Work-Station System version 7.1 running in a UNIX platform. GIS results were exported to the 3-PG model. Finally, the 3-PG estimations and the basic layers where converted to shape format and manipulated in Arc View 3.2, where the cartography was built. The resulting maximum MAI values were grouped using the legend tool to define potential productivity zones.

The classification method used is the Natural Breaks default classification method in Arc View. This method identifies breakpoints between classes using a statistical formula (Jenk's optimization). The Jenk's method minimizes the sum of the variance within each of the classes to find groupings and patterns inherent in the data (Crosier et al., 2004). We used the program 3PGpjs (Sands, 2000) to predict potential productivity with a 3-PG interface from the CSIRO Forestry and Forest Products & CRC for Sustainable Production Forestry. The calibration of these models for diverse Eucalyptus species has been described by Landsberg et al. (2003).

4.6. Results of the application

To evaluate the effects of spatial variation in climate and soils on forest productivity across broad regions is of main importance to a forester. In our study, we are interested in predicting potential MAI, as measure of productivity, and how it is influenced by environmental factors.

4.6.1. Model validation for MAI prediction

MAI values observed in the growth plots varied between 23 and 50 m^3 ha^{-1} at age 10 and were highly correlated with MAI simulated by the 3-PG model. Additionally, a low bias was recorded with AD of 0.26 in absolute terms and 0.85% in percentage terms. The model also had acceptable precision; MSE had an absolute value of 4.55 and a percentage value of 14.6%.

4.6.2. Potential MAI in the Biobío region

There are three main physiographic landscapes dominating the Biobío Region. The coastal range along to the Pacific Ocean and the Andes Cordillera enclose the Central Valley from north to south and create a great variety of climates, soils, and forest sites.

Predicted MAI classes for *Eucalyptus nitens* plantations ranged from 17 m^3 ha^{-1} yr^{-1} to 62 m^3 ha^{-1} yr^{-1} in the Biobío Region general, predicted MAI increased in mountainous terrain in the foothills of the Andes following gradients in precipitation, annual temperature, and soil fertility (Fig. 4; Table 2). Not surprisingly, highest predicted MAI were in areas with highest precipitation, coolest annual temperatures, and the most fertile soils. Elevation by itself was not a good predictor of MAI, as highly productive sites occurred at both low and high elevations. Soil nitrogen concentrations on the most productive sites varied widely but on average, soils were considerably more fertile and held more water than sites with lower productivity (Table 2).

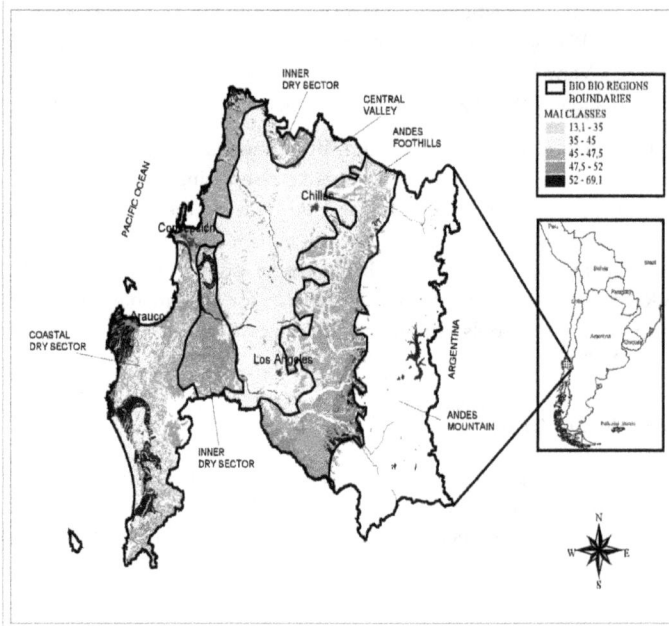

Figure 4. Potential productivity classes of MAI (m^3 ha^{-1} yr^{-1}) for the Biobío Region, as estimated by the 3-PG model. (The predicted values of MAI in each site are not possible to reach under the operational silviculture).

Site Class	MAI ($m^3 ha^{-1} yr^{-1}$)	Ω ($MJ\ m^{-2} yr^{-1}$)	P ($mm\ yr^{-1}$)	T (°C)
Very high	>55	5.5	1250-2000	13.0
Very high	>55	6.2	1500-2550	12.0
Moderate	47.6-52.0	5.8	640-1100	14.7
Very low	>35	6.6	750-1250	14.0

(a)

Site Class	MAI ($m^3 ha^{-1}yr^{-1}$)	Soil order	Altitude (m)	Texture	WHC (mm)	N (%)
Very high	>55	Alfisol	160	Fr-arc	410	0.05-0.25
Very high	>55	Inceptisol	830	Fr-lim	570	0.18-0.35
Moderate	47.6-52.0	Alfisol	130	Fr-arc-are	340	0.04-0.09
Very low	>35	Alfisol	170	Are	52	0.01-0.07

(b)

Note: In table 2a) and 2b), MAI is the potential mean annual increment as estimated by 3-PG model; N is soil nitrogen; WHC is available soil water holding capacity; Ω is the annual incident radiation; P is the annual precipitation, and T is the mean annual temperature. Data were provided by the Instituto de Investigaciones de Recursos Naturales (1964) and Instituto de Investigaciones Agropecuarias (Del Pozo & Del Canto, 1999).

Table 2. a) Climates variables related to the potential productivity of 10-yeard old *Eucalyptus nitens* on the Biobío Region; b) Soil variables related to the potential productivity of 10-yeard old *Eucalyptus nitens* on the Biobío Region.

Combined with spatial information derived from GIS, the 3-PG model can generate valuable information for the development of operational silviculture (Almeida et al., 2004). The AD values (0.85%) and MSE values (14.6%) obtained for MAI confirm that process-based models hybridized have precision levels similar to traditional inventories. Our work corroborates the suggestion of Tickle et al. (2001) that process-based models are a valuable tool for predicting growth at a regional level. Additionally, 3-PG model provide insights into the role of environmental variables in determining MAI that are very difficult to obtain with traditional inventories (Waring, 2000; Almeida et al., 2004).

To better understand the factors that explain potential productivity of *Eucalyptus nitens* plantations in the Biobío region, we used soil and climate information from Del Pozo & Del Canto (1999) and Carrasco (1993). In addition typical values of observed environmental variables are presented for very high, moderate and very low MAI classes (Table 2a and 2b).

Highest productivity sites occur on inceptisols and on those alfisols in areas with relatively high precipitation and regularly subjected to freezing temperatures. In the Biobío Region, inceptisols are derived from volcanic ash, and have high water holding capacity, good drainage, and high nitrogen concentrations (Rodríguez et al., 2009). They also occur in areas of relatively high precipitation. Alfisols are marine terrace soils with good drainage. In addition to high precipitation, those alfisols with high productivity also have high water holding capacity, with relatively high average soil nitrogen concentrations (Table 2). High productivity areas span a wide range of elevation, mean annual temperature, and annual incident radiation. Consequently, the productivity of *Eucalyptus nitens* in the Biobío Region does not respond directly to any of these three factors (Rodríguez et al., 2009). However, the factors considered here may affect productivity through several distinct mechanisms, in which factors can interact or factors are related with seasonal variation (Battaglia & Sands, 1997). Because water is considered to be the main factor limiting productivity, productivity is consistently related to the capacity of the soil to store water (Dye et al., 2004); this is the case with the inceptisols derived from volcanic ash and alfisols derived from marine terraces, that have high water holding capacity and can explain the highest productivity in these sites (Table 2).

The relatively small change in MAI over a large elevational span is consistent with the ecophysiology of *Eucalyptus nitens*. Optimal temperatures for photosynthesis span between a wide range (14 to 20°C) for the species (Battaglia et al. 1996) – this commonly occurs with temperatures in the study area. Our results support the hypothesis of Battaglia et al. (1996) that the broad photosynthetic response of *Eucalyptus nitens* enables it to acclimate to a wider range of environments than many other Eucalyptus species. *Eucalyptus nitens* in the Biobío Region occurs in elevations from 0 and 830 m.a.s.l, with temperatures varying between 8 and 32°C in the summer and -3.0 to 13ºC in winter; the higher elevations experience frequent winter frosts. Similar to the results of Battaglia et al. (1998) for *Eucalyptus nitens* in Australia, frost has little or no effect on this species' MAI in Chile.

Moderate levels of MAI were predicted for the Coastal Range. The soils of Coastal Range are derived from granitic and metamorphic rocks that compose the so-called "Batholitic Coastal", and have been seriously impacted by erosion. MAI values for the Coastal Range were lowest in the north and highest in the south following a gradient of precipitation and plant-available soil water holding capacity (Table 2). Lowest growth rates occurred in Central Valley at middle elevations with decreasing precipitation, high evapotranspiration rates, and a long drought period (approximately six months). The sandy soils of the Central Valley also have low inherent fertility and very low water holding capacity (Fig. 4; Table 2).

The very low productivities occurred in sandy soils derived from metamorphic and granitic rocks. These soils have low water retention capacity. The very low productivity sites were in Central Valley on entisols; these are young soils originating in the Biobío area from various rock types (e.g. granitic, metamorphic, andesitic, and basaltic). They are sandy soils with excessive drainage, low soil water holding capacity (less than 52 m^3 ha^{-1} of available water), and low nitrogen concentrations. Compounding the low water holding capacity of soils, these areas also have low annual precipitation (750 to 1000 mm) and high solar radiation potentials (> 6.5 MJ m² yr¹), which would increase evaporative demand (Table 2). These factors suggest that productivity in these areas is strongly controlled by water holding capacity and perhaps nitrogen limitations. Battaglia et al. (1998) found that LAI of *Eucalyptus nitens* plantations in Australia increased sharply as annual mean temperature increased from 7 to 11 degrees C and reached a plateau between 13 and to 14 degrees (the highest temperature measured for *Eucalyptus nitens*). We found a similar pattern; however, the LAI of *Eucalyptus nitens* plantations in the Biobío Region plateaus reach lower temperatures than those present in Australia. However, although mean annual temperature had a strong influence of LAI in our study, productivity was influenced more strongly by the constraints on leaf efficiency exerted by water availability and soil fertility (for more details see Rodríguez et al., 2009).

This finding is consistent with the hypothesis of Ollinger et al. (1998) and Coops et al. (2001) that forest productivity is strongly correlated with precipitation and suggests that water is an important factor controlling regional pattern of productivity in forest types, as well as in *Eucalyptus nitens*, which was demonstrated in our study. Additionally *Eucalyptus nitens* was very sensitive to water soil availability (Fig. 4) in areas with low water holding capacity of soils. These findings are consistent with the hypothesis of White et al. (2000) that water stress reduces dramatically the growth rate of *Eucalyptus nitens* due to the decrease of the osmotic potential and bulk elastic modulus in response to water stress.

Understanding how productivity varies across complex environmental gradients requires modeling that combines influences of radiation, temperature, humidity deficits, and drought and soil fertility as they change spatially and temporally.

5. Conclusion

In our work, we linked process-based and empirical forest model with GIS to asses' site quality for *Eucalyptus nitens* en the Biobío region, Chile. We defined GIS system as modular and contain four subsystems: Data acquisition, Data integration and analysis, Growth and yield simulator and Results and Analysis subsystem. The 3-PG model with GIS showed the usefulness for managing spatial data and analyzing temporal trends, and provided regional estimates of forest productivity.

Combined with spatial information derived from GIS, the 3-PG hybridized model can generate valuable information for the development of operational silviculture, and measure of forest productivity as MAI can be derived in a GIS from readily-available topographic data, climate data and soils map. A process-based model coupled with GIS showed potential to serve as a useful tool to screen areas as prospective plantations sites.

Our results in the Biobío region indicate that annual precipitation; available soil water holding capacity and low levels of soil nitrogen are the principal factors influencing

Eucalyptus nitens productivity. Our analysis was able to separate the effects of limiting water from those of nutrients, and it is reasonable to conclude that soil water availability is an important factor controlling regional patterns of productivity. Highest productivity sites occur in areas with relatively high precipitation and regularly subjected to freezing temperatures in *Eucalyptus nitens*.

In our study *Eucalyptus nitens* reduced growth dramatically in areas with low water holding capacity of soils. Additionally, we found that large changes in elevation have only a small influence of the productivity of *Eucalyptus nitens* in the Biobío Region. Consequently, it is reasonable to conclude that soil water availability is an important factor controlling regional patterns of productivity as has been demonstrated in similar studies.

Author details

Rolando Rodríguez and Pedro Real
Forest Science Faculty, Universidad de Concepción, Chile

6. References

Almeida, A.C.; Maestri, R.; Landsberg, J.J. & Scolforo, J.R.S. (2003). Linking process-based models and empirical forest model in Eucalyptus plantations in Brazil. In: Modeling Forest System, Ana Amaro et al. (Eds.), CABI Publishing. ISBN 0-85.199-693-0.

Almeida, A.C.; Landsberg, J.J.; Sands, P.J.; Ambrogi, M.S.; Fonseca, S.; Barddal, S.M. & Bertolucci, F.L. (2004). Needs and opportunities for using a process-based productivity model as a practical tool in Eucalyptus plantations, Forest Ecology and Management, Vol. 193, pp 67-177.

Baldwin, V.C.; Burkhart, H.E.; Dougherty, P.M. & Teskey, R.O. (1993). Using a growth and yield model (PTAEDA2) as a driver for a biological model (MAESTRO). US Department of Agricultural, Forest Service, Southern Forest Experimental Station, New Orleans, Research Paper SO-276, 9 pp.

Battaglia, M.; Beadle, C. &. Loughead, S. (1996). Photosynthetic temperature responses of *Eucalyptus globulus* and *Eucalyptus nitens*, Tree Physiology, Vol. 16, pp 81-89.

Battaglia, M. & Sands, P. (1997). Modeling site productivity of *Eucalyptus globulus* in response to climate and site factors, Australian Journal of Plant Physiology, Vol. 24, pp 831-850.

Battaglia, M.; Cherry, M.L.; Beadle, C.L.; Sands, P.J. & Hingston, A. (1998). Prediction of leaf area index in eucalypt plantations: Effects of water stress and temperature, Tree Physiology, Vol. 18, pp 521-528.

Battaglia, M.; Sands, P.J. & Candy, S.G. (1999). Hybrid growth model to predict height and volume growth in young Eucalyptus, Forest Ecology and Management, Vol. 120, 193-201.

Bernier, P., Landsberg, J.; Raulier, F; Almeida, A.; Coops, N.; Dye, P.; Espinosa, M.; Waring, R.& Whitehead, D. (2003). Using process-based models to estimate forest productivity for management purposes. In: XII World Forestry Congress. Available at: http://www.fao.org/DICREP/ARTICLE/WFC/01515-B4.htm.

Bossel, H. (1996). TREEDYN3 Forest Simulation Model, Ecological Modelling, Vol. 90, pp 187-227.

Bristow, K.L. & Campbell, G.S. (1984). On the relationship between incoming solar radiation and daily maximum and minimum temperature. Agricultural and Forest Meteorology, Vol. 32, pp 159-166.

Burkhart, H.E.; Farrar, L.R.; Amateis, R.L. & Daniels, R.F. (1987). Simulation of individual tree growth and stand development in loblolly pine plantations on Cutover, site prepared areas, Virginia Polytechnic Institute and State University, School of Forestry and Wild Life Resources, Blacksburg, FWS 1-87, 47 pp.

Cannel, M.G.R. (1989). Physiological basis of wood production: a review. Scandinavian Journal of Forest Research, Vol. 4, pp 459-490.

Carrasco, P.; Millán, J. & Peña, L. (1993). Suelos de la cuenca del río Bío Bío, características y problemas de uso. In: Gestión de los Recursos Hídricos de la Cuenca del Río Bío Bío y del Area Marina Costera Adyacente, F. Faranda & O. Parra (Eds), Universidad de Concepción, 12-108 pp (in Spanish).

Chang, K. (2004). Introduction to geographic information system, McGraw Hill, New York.

Constable, J.V. & Friend, A.L. (2000). Suitability of process-based tree growth models for addressing tree response to climate change, Environmental Pollution, Vol. 110, pp 47-59.

Coops, N.C. & Waring, R.H. (2001). Estimating forest productivity in the eastern Siskiyou Mountain of southern Oregon using satellite-derived process-based model, 3 PGS, Canadian Journal of Forest Research, Vol. 311, pp 143-154.

Crosier, S.; Booth, B.; Dalton, K.; Mitchell, A. & Clark, K. (2004). ARGIS 9 Getting started with ARCGIS, ESRI Pr, ISBN: 9781589480919.

Daniel, T.W.; Helms, J.A. & F.S. Baker. (1979). Principles of silviculture, McGraw-Hill Company, New York.

Del Pozo, A. & Del Canto, S. (1999). Áreas agroclimáticas y sistemas productivos en la VII y VIII Regiones, Ministerio de Agricultura, Chillán, 115 pp. (in Spanish).

Dye, P.J.; Jacobs, S. & Drew, D. 2004. Verification of 3-PG growth and water-use predictions in twelve Eucalyptus plantations stands in Zululand, South Africa, Forest Ecology and Management, Vol. 193, pp 197-218.

Fontes, L. (2007). Process-based and hybrid forest models. Available at http://www.efi.int/attachments/luis_pontes-presentation/PDF.

Fontes, L.; Bontemps, J-D.; Bugmann, H.; Van Oijen, M.; Gracia, C.; Kramer, K.; Lindner, M.; Rötzer, T. & Skovsgaard, J.P. (2010). Models for supporting forest management in a changing environments, Forest System, Vol. 19, pp 8-29. ISSN: 1131-7965.

Hari, P.; Nikinmaa, E. & Korpilathti, E. (1991). Modelling: canopy, photosynthesis, and growth. In: Physiology of Trees, Ed. A.S. Raghavondra, John Wiley and Sons, New York.

Instituto de Investigaciones de Recursos Naturales. (1964). Suelos. Descripciones Proyecto Aerofotogramétrico Chile/O.E.A./B.I.D. Publicación N°2. Santiago, Chile. (in Spanish).

Instituto Geográfico Militar. (2001). Colección Geografía de Chile. IGM. Santiago, Chile. (in Spanish).

Johnsen, K.; Samuelson, L.; Teskey, R.; Mc Nulty, S. & Fox, T. (2001). Process models as tools in forestry research and management, Forest Science, Vol. 47 No. 1, pp 2-8.

Kirk, R. W. & Burk, T.E. (2004). Regional-scale forest production modeling using process-based models and GIS. In: Proceedings of the 4th Southern Forestry and Natural Resources GIS Conference, 68-88 pp. Available at http://facstaff.elon.edu/rkirk2/files/rkirk_cv.pdf.

Korzukhin, M.D., Ter-Mikaelian, M.T. & Wagner, R.G. (1996). Process versus empirical models: which approach for forest ecosystem management, Canadian Journal of Forest Research, Vol. 26, pp 879-887.

Landsberg, J.J. (1986). Physiological ecology of forest production, Academic Press, London.

Landsberg, J.J. (2003). Modelling forest ecosystems: state-of-the-art, challenges and future directions, Canadian Journal of Forest Research, Vol. 33, pp 385-397.

Landsberg, J. J.; Waring, R.H. & Coops, N.C. (2003). Performance of the forest productivity model 3-PG applied to a wide range of forest types, Forest Ecology and Management, Vol. 172, pp 199-214.

Landsberg, J.J. & Gower, S.T. (1997). Applications of physiological ecology to forest management, Academic Press, San Diego.

Landsberg, J.J. & Waring, R.H. (1997). A generalized model of forest productivity using simplified concepts of radiation-use efficiency, carbon balance, and partitioning, Forest Ecology and Management, Vol. 95, 209-228.

Mäkelä, A., Landsberg, J.J.; Ek, A.R.; Burk., T.E.; Ter-Mikaelian, M.; Agren, G.I.; Oliver, C.D. & Puttonen, P. (2000). Process-based model for forest ecosystem management: current state of the art and challenges for practical implementations, Tree Physiology, Vol. 20, No 5-6, pp 289-298.

McKendry, J.E. & Eastman, J.R. (1991). Applications of GIS in Forestry: A review. Available at www.nrac.wvu.edu/classes/for326/GISInForestryReviewPaper.pdf.

Mason, E.G. (2010). Growth and yield modeling in New Zealand. In: Forest Growth & Yield Hybrid Modeling, Cristian Higueras (Ed.), University of Concepción Press.

Medlyn, B.; Duursma, R.A. & Zeppel, M.J.B. (2011). Forest productivity under climate change: a checklist for evaluating model studies, Climate Change, Vol. 2, pp 332–355.

Ollinger, S.V.; Aber, J.D. & Federer, C.A. (1998). Estimating regional productivity and water yield using model linking to a GIS, Landscape Ecology, Vol. 13, pp 323-334.

Pinjuv, G.L. (2006). Hybrid forest modelling of *Pinua radiata* D. Don in Canterbury, New Zealand. Thesis submitted fulfillment of the requirement for the Degree of Doctor of Philosophy in Forestry, University of Canterbury, New Zealand.

Rodríguez, R; Real, P.; Espinosa, M.A. & Perry, D.A. (2009). A process-based model to evaluate site quality for *Eucalyptus nitens* in the Bio-Bio Region of Chile, Journal of Forestry, Vol. 82, No 2, pp 149-162.

Runing, S.W., Nemani, R.R. & Hungenford, R.D. (1987). Extrapolation of synoptic meteorological data in mountain terrain and its use for simulating forest evapotranspiration and photosynthesis. Canadian Journal of Forest Research,Vol. 17, pp 472-483.

Sands, P. 2000. 3PGpjs –a user-friendly interface to 3-PG, the Landsberg and Waring model of forest productivity, Cooperative Research Centre for Sustainable Production Forestry and CSIRO Forestry and Forest Products, Technical Report 29, Tasmania, 16 pp.

Santibáñez, F. & Uribe, C.J. (1993). Cartas del mapa agroclimático de Chile, Universidad de Chile. Facultad de Ciencias Agrarias y Forestales, 99 pp. (in Spanish).

Sharpe, P.J.H. & Rykiel, E.J. (1991). Modelling integrated response of plants to multiple stresses. In: Response of Plants to Multiple Stresses, H.A. Mooney et al. (Eds.), Academic Press, New York.

Schlatter, J.E. (2004). Sistema de ordenamiento de la tierra: Herramienta para la planificación forestal aplicada a las regiones VII, VIII y IX, Universidad Austral de Chile, Serie Técnica, 114 pp. (in Spanish).

Shindler, B. (1998). Landscape-Level Management: It's All About Context, Journal of Forestry, Vol. 98, pp 10-14.

Sievänen, R. & Burk, T.E. (1993). Adjusting a process-based model for the dimensional growth model for varying site conditions through parameter estimation, Canadian Journal of Forest Research, Vol. 23, pp 1837-1851.

Tickle, P.K., Coops, N.C. & Hafnes, S.D. (2001). Comparison of a forest model (3-PG) with growth and yield models to predict productivity at Bago State Forest, NSW, Australian Forestry, Vol. 64, pp 111-122.

Vanclay, J. (1994). Modelling forest growth and yield: Applications to mixed tropical forests, Cab International, ISBN 0 85198 913 6.

Wang Y. P., & Jarvis P. G. 1990. Description and validation of an array model-MAESTRO, Agricultural and Forest Meteorology, Vol. 5, pp 257-280.

Waring, R. H. (2000). A process model analysis of environmental limitations on the growth of sitka spruce plantations in Great Britain, Forestry, Vol. 73, pp 65-79.

White, J.D., Coops, N.C. & Scott, N.A. (2000). Estimates of New Zealand forest and shrub biomass from the 3-PG model, Ecological Modelling, Vol. 131, pp 175-190.

Worledge, D.; Honeysen, J.L.; White, D.A.; Beadle, C.D. & Hetherington, S.J. (1998). Scheduling irrigations in plantations of *Eucalyptus globulus* and *Eucalyptus nitens*: A practical Guide, Tasforest, Vol. 10, pp 91-101.

Chapter 11

GIS Applied to Integrated Coastal Zone and Ocean Management: Mapping, Change Detection and Spatial Modeling for Coastal Management in Southern Brazil

Tatiana S. da Silva, Maria Luiza Rosa and Flávia Farina

Additional information is available at the end of the chapter

1. Introduction

Information is the basis for sustainable development. If a decision is taken without any quality information to back it up, it relies on guesswork. Accurate, comprehensive and periodic environmental information is then crucial for the success of the decision-making and environmental planning processes.

Coastal zones are characterized by fragile, complex and productive environments, typical of the sea-land system. They deserve special attention from the government and society. Besides, most of the world population lives by the sea and there is a permanent trend of demographic concentration in these areas. The health and well-being of the coastal populations, and sometimes even their survival, depend on the status of coastal and marine ecosystems. Managing this complexity implies the cooperation between the levels of the government and the society.

In Brazil, the National Coastal Management Plan (PNGC) was established through the Law 7661 of 1988 in order to plan the use of coastal areas. Today in its second version, the PNGC provides the following instruments: (1) the State Coastal Management Plan (PEGC), clarifies the development of the PNGC at the state level, aiming to implement a State Coastal Management Policy; (2) the Municipal Coastal Management Plan (PMGC), clarifies the development of the PNGC and PEGC at the municipal level, aiming to implement a Municipal Coastal Management Policy; (3) the Coastal Management Information System (SIGERCO) is a component of the National Environmental Information System (SINIMA) and gives support to the state / municipal subsystems; (4) the Environmental Monitoring

System (SMA-ZC) is the operational framework of continuous data collection to monitor the social-environmental indicators and support the management plans in the coastal zone; (5) the Environmental Quality Report (RQA-ZC) is the consolidation of results from the SMA-ZC in periodic reports and, above all, aims to assess the efficiency of the management actions undertaken; (6) the Ecologic-Economic Coastal Zoning (ZEEC) is the spatial regulatory instrument to plan land use in a given territory; and (7) the Coastal Zone Management Plan (PGZC) is a set of coordinated and programmatic actions, built in a participatory manner, and applicable to different levels of government.

Unfortunately, more than 20 years after the implementation of the 7661 Law, the institutionalization of coastal management is still incipient (Jablonski and Filet, 2008), even in the Rio Grande do Sul State, where the environmental control is very restrictive compared to the rest of the country. In some municipalities of the Rio Grande do Sul coastal plain, environmental plans were built and consist in the only environmental regulatory instruments at municipal level, but they are not truly implemented. At the national level, on the other hand, the adoption of the I3Geo as component of SINIMA was a successful initiative. The I3Geo is a platform for publishing spatial data and interactive mapping applications to the web, helping the establishment of cooperative networks, showing the advantages of using a GIS platform as the core of an environmental information system.

In a general sense, all the coastal management instruments, as they are designed in the PNGC, explicitly depend on spatial tools or at least would be benefited through the use of maps. The universities focused on the coastal and marine ecosystems of the Rio Grande do Sul generate a number of GIS-based products as research results. They have been helpful in building the environmental and master plans of many coastal municipalities, but there is still a great potential to include spatial information in other mechanisms of coastal management. A significant knowledge about the natural resources in this region has been gained from GIS and remote sensing in the last 30 years. Geotechnology has been successfully used to understand the spatial structure and dynamics of the coastal landscapes and, more recently, in simulation modeling and change prediction. The acquired knowledge should be included into the existing instruments, and also used to improve them, promoting the adaptation to the current pace and reach of human activities over the coastal zone.

The Rio Grande do Sul coastal zone (figure 1), in Southern Brazil, is characterized by a wide coastal plain generated by the sea level changes during the Quaternary, which resulted in a complex lagoon system all over its extent (Asmus et al., 1988). The Patos Lagoon is the most expressive of such water bodies, comprising almost 10.000 km² and a number of valuable marginal ecosystems and aquatic species, some of them of economic interest. Research has been focused on coastal issues for more than 20 years. And since then, space has been the base of approach.

Thus, this chapter aims to provide a discussion about the applicability of the mapping, change detection and spatial modeling efforts in the Southern Brazil coastal plain to the policy instruments defined by the National Coastal Management Plan. We also recommend strategies to promote an adaptative management concerning coastal zones through the use of GIS. Geotechnology will be presented as a way to enhance the exchange and feedback among academic researchers, stakeholders and community.

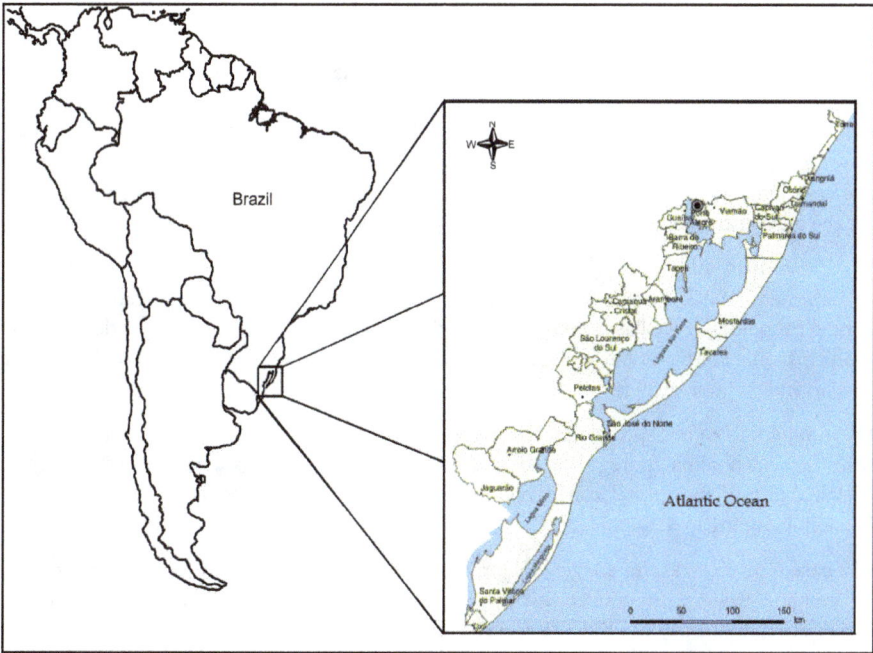

Figure 1. Rio Grande do Sul coastal zone.

2. The Brazilian program on coastal management: Implementation status and spatial nature of policy instruments

The second National Plan of Coastal Management – PNGC II was approved in 1997, in substitution of PNGC I. The National Program on Coastal and Ocean Management (GERCOM) aims to put PNGC into operation, in order to plan and manage the economic activities so to ensure the sustainable use of the coastal environments. GERCO is coordinated by the Ministry of Environment and executed by the 17 coastal states. Rio Grande do Sul is one of the most advanced states regarding the implementation of the coastal management plan, but actions are highly concentrated in the north littoral where urbanization is more spread and severe.

The State Envionmental Protection Agency (FEPAM) is the state-level authority in charge of environmental management in Rio Grande do Sul. FEPAM is legally in charge of implementing the Coastal Management Program. The participation of FEPAM, as the ultimate client for the management plans, is crucial to guarantee long-term sustainability of applied research endeavors. To the extent that FEPAM takes ownership of projects results, and makes the proposed plans its own, this would guarantee that such plans are incorporated in the State's budged and overall environmental policy (Tagliani *et al.*, 2003).

The Rio Grande do Sul coastal zone comprises 27 coastal municipalities: Torres, Arroio do Sal, Três Cachoeiras, Três Forquilhas, Maquiné, Capão da Canoa, Terra da Areia, Xangrilá, Osório, Imbé, Tramandaí, Cidreira, Palmares do Sul, Viamão, Mostardas, Barra do Ribeiro, Tapes, Tavares, Camaquã, Arambé, São José do Norte, São Lourenço do Sul, Rio Grande, Pelotas, Arroio Grande, Jaguarão, and Santa Vitória do Palmar. Only 9 of them have environmental plans approved by FEPAM. However, conservation units and preservation areas are considered in most of the master plans. In some municipalities, they are the only instrument of environmental and territorial control.

The State Plan of Coastal Management (PEGC) and the Municipal Plan of Coastal Management (PMGC) are highly related to territorial ordination planning. Environmental zoning proposals as well as the detection of priority areas for management subside state and municipal plans. This is particularly important for those municipalities where the lack of human and technological resources is more dramatic.

Environmental zoning proposals can also base the Coastal Ecological-Economic Zoning (ZEEC), which aims to regulate the territorial use in order to achieve environmental sustainability of the coastal zone development, respecting the directives of the Ecological-Economic Zoning in national scale.

GIS-based project results in general are potential inputs for the Coastal Management Information System (SIGERCO), a component of the National System of Environmental Information (SINIMA). SIGERCO integrates PNGC information, aiming to give support and capillarity to the subsystems structured and managed by coastal states and municipalities.

Research results that also include a temporal dimension (land use change studies, for example) give GIS procedures the status of a monitoring method, once they can be replicated in other times or locations. This type of spatial data can be used as environmental quality indicators and, thus, can be incorporated by the Coastal Zone Environmental Monitoring System (SMA-ZC). Consequently, they are able to support the Coastal Zone Environmental Quality Report (RQA-ZC), which is the periodic consolidation of the results obtained by the environmental monitoring.

Once the institutions involved in public management have absorbed the available technology and information, the possibilities of use them in the decision making process are diverse. The possibility to visualize any attribute in a map makes easier to understand the coastal space, promoting knowledge-based public participation in the decision making process. The chance of governance to succeed is higher in this way.

In a general sense, all the coastal management instruments, as they are designed in the PNGC, explicitly depend on spatial tools or at least would be benefited through the use of maps. Some of the GIS-based research works have already been helpful in building the environmental and master plans of some coastal municipalities (Pelotas, Rio Grande, São Lourenço do Sul, Turuçu, among others) but there is still a great potential to include this information in other mechanisms of coastal management. Geotechnology has been successfully used to understand the spatial structure and dynamics of medium littoral of the Rio Grande do Sul coastal plain. The acquired knowledge should also be included into the existing instruments to improve them, providing a more prospective focus and promoting the adaptation to the current pace and reach of human activities over the coastal zone.

3. Rio Grande do Sul coastal plain: Mapping, change detection and spatial modeling as decision support resources

3.1. Mapping the coastal zone

PNGC delegates power for states and municipalities to legislate issues related to the use of soil, water and forest resources. Based on that, municipalities are highly recommended to create master plans and legal instruments to manage land use, mantain environmental quality, and use properly the natural resources at municipal level. Several methods are involved in the implementation of a master plan, like statistical analysis, mapping, zoning, registration survey, field research, among others. Thus, GIS represents an extremely useful tool for municipal planning purposes. GIS gathers together a great set of application to collect, storage, restore, change, and represent spatial data, as well as related attributes. GIS implementation, as the basis for planning and management, means a huge step toward a greater efficiency of municipal administration.

Basic GIS questions, such as *when, what, what distance, and which,* sometimes need integrating tools to be answered. In this sense, Multi Criteria Evaluation (MCE) techniques are effective procedures for several purposes. Gómez & Barredo (2005) have analyzed GIS analytical functions integrated to MCE techniques, which allows introducing objective and/or heuristic knowledge to simulate the possible results of different decisions and develop virtual scenarios to evaluate the implementation of policies.

3.1.1. Geological and geomorphologic mapping of coastal plain as the basis for land use planning

The knowledge about the substrate and geological evolution is a very important point for planning the use of coastal environments. When considering the substrate characteristics, we might evaluate the best way and location for development, integrating human interests with the particularities of each environment.

Rio Grande do Sul coastal plain geology has been studied over the years especially through surface geological mapping, supported by remote sensing data, drilling, and more recently, geophysical surveys. This information can be integrated and made available through GIS, allowing its use in several applications.

Rio Grande do Sul coastal plain represents the top and youngest portion of the Pelotas sedimentary basin. It is mainly formed by alluvial fan and barrier-lagoon sedimentary deposits (figure 2). According to previous works, these deposits were formed in response to sea level changes, which were controlled by glacioeustasy during the Quaternary. Mineralogical and geomorphological patterns result in four barrier-lagoon systems. The oldest system is designated as I and the youngest (still active) as IV (Tomazelli & Villwock, 1996). The age of the oldest systems was established mainly by correlation with oxygen isotope curves. Each system represents the maximum of the Postglacial Marine Transgression (PMT): 400 ka (I), 325 ka (II) and 120 ka (III).

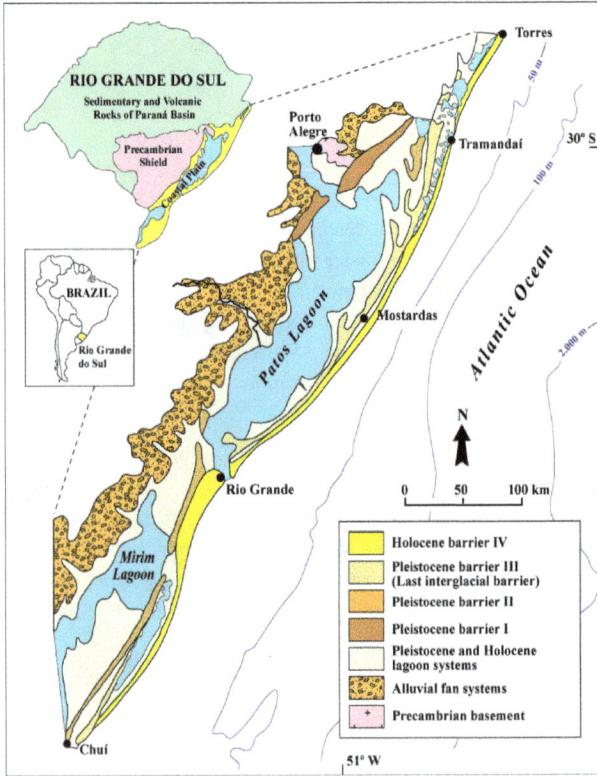

Figure 2. Map of Rio Grande do Sul coastal plain, showing its main geological units (Tomazelli & Villwock, 1996).

The youngest system (IV) is a record of the last glacial cycle, which evolved after the last glacial maximum about 18 ka ago. This is the best kwon geological unit of the Rio Grande do Sul coastal plain, where most of the coastal population lives. Sea level was placed around 120-130 m below the current level about 18 ka ago (Corrêa, 1995). Thereafter, sea level started to rise, exceeding the current level at 7.7-6.9 ka, and reaching its maximum at about 6 ka (Martin et al., 1979; Angulo & Lessa, 1997). By this time, it reached a high 2 to 4 m above the present sea level (Dillenburg et al., 2000). Since then, it has begun lowering to the current position (Angulo & Lessa, 1997).

Sea level changes controlled the evolution of depositional systems found in the Rio Grande do Sul coastal plain. The deposits associated to this cycle reflect such variations. The understanding of elements composing coastal environments and how they evolved over the last thousands years may be applied to several purposes.

Dillenburg *et al.* (2000) have studied the Holocenic coastal barrier, defining sectors with divergent behaviors in a scale of centuries to millennia. Some sectors are under erosion, others are stable or quasi-stable, and we also find sectors where the beach is growing. Toldo Jr. *et al.* (2005) showed that 442 km of 621 km of coastline have been eroded. One of the main indicators of this process is the mud outcrop of lagoonal origin in the present beach (figure 3). [14]C dating procedures revealed the age of these muds: 5760±120 ka according to Travessas (2003), 4,330±60 ka according to Tomazelli *et al.* (1998), and 3.5 ka according to Dillenburg *et al.* (2004), indicating that the coastal barrier is regressing in these sectors.

Figure 3. Pictures showing beach erosion indicators. The left picture shows mud outcropping at the beach and the right one the buildings destruction due to sea level rises during storms.

GIS-based analysis of outcrops along with satellite images allowed to clarify some specific characteristics of these sectors. Given their particular locations along the coast, they are frequently associated to the occurrence of great dune fields (figure 4). Dune fields are generally found in the far northeastern of the sectors under erosion, what is related to longshore drift (Tomazelli & Villwock, 1992; Toldo Jr. *et al.*, 2004).

The human occupation of some sectors under erosion already has negative consequences. When an elevation of sea level occurs, as a result of meteorological tides, we face the destruction of residences and infrastruture (figure 3). Sectors under erosion should be treated differently from those under accretion in the planning process. It is absolutely necessary to concern about safety measures, such as the maintenance of buffer zones free of development near the beach.

After the expansion of the coastal zone occupation, energy resources are increasingly required. An alternative adopted along Brazilian southern coast is the installation of wind generators. For determining the better place to install them, we must consider several criteria, including legal and socio-environmental aspects.

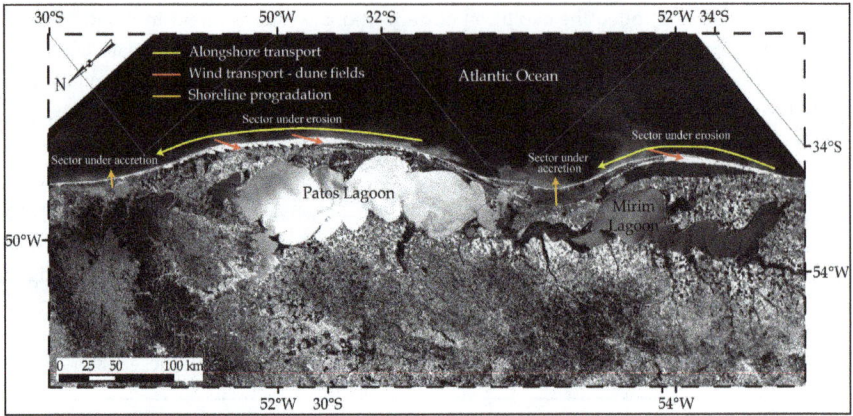

Figure 4. Image of Rio Grande do Sul coastal plain (Landsat 7, ETM+, Band 2 - 130° inclination) showing its embayments and projections. Alongshore transportation is indicated. Extensive transgressive dune fields occur mainly in the NE portion of erosion sectors (Rosa, 2010).

According to the Brazilian environmental law, dunes are preservation areas. However, the understanding of dune fields dynamic allows to predict dune behavior. Many dune fields in southern Brazil have been monitored by historical aerial photography. Many of them are endangered. Cidreira dune fields are an example of that (figure 5). This field used to be maintained by a sand supply transported by NE winds. Urban development has blocked sand transportation in this area. Thus, dunes and sand sheets are about to disappear, once there is no alternative sand supply to keep sand dunes in place and the current dunes tend to migrate toward coastal lagoons, away from the beaches. The environmental agency in charge authorized a wind farm installation in this area, once sustainable dune fields no longer exist.

In a general sense, when we apply coastal dynamic knowledge to land use planning, environmental impacts of development are likely to be reduced, and a better balance between human needs and the coastal systems carrying capacity is reached.

3.1.2. Modeling of urban sprawl based on GIS: The Rio Grande city case, south coastal plain

The human activities planning process imply combining multiple criteria, including law, ecological function of environments, and human needs. In this sense, a GIS-based urban sprawl model was built by Farina (2003), using a multi-criteria evaluation as a tool for decision making. Rio Grande was chosen to develop a case study of urban sprawl. It has been urbanized and industrialized in an unplanned manner, resulting in environmental degradation of many valuable ecosystems. Rio Grande is located in the south littoral of the Rio Grande do Sul coastal plain (Figure 1). The urban area is physically limited by water, once it is surrounded by the Atlantic Ocean on the east and by the Patos Lagoon on the north and west. Given the physiographic characteristics of this municipality, it has a great potential to port and industrial development, which in turn drives urban expansion.

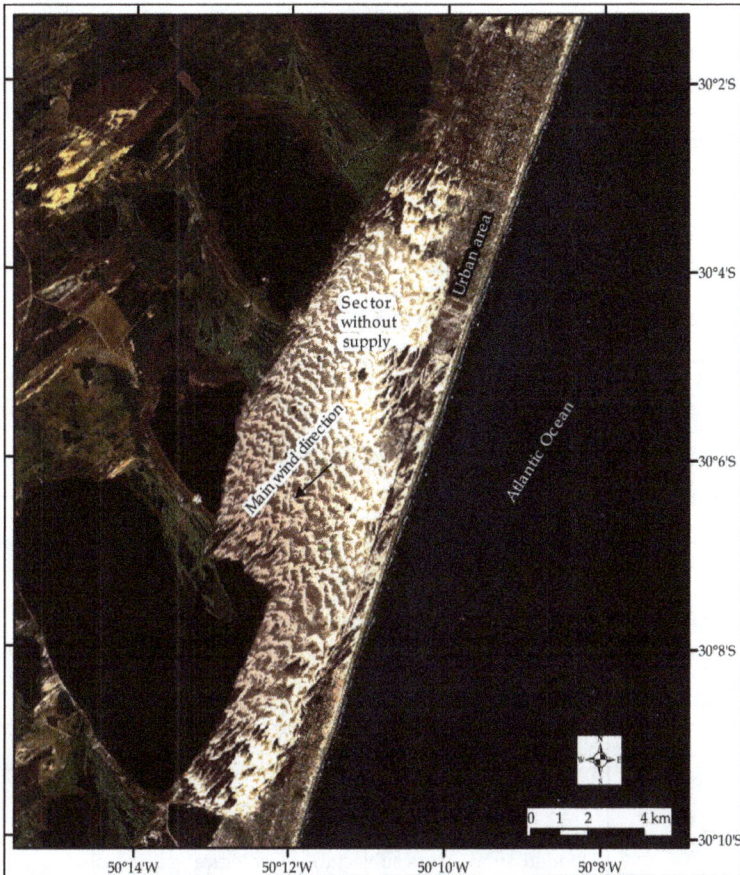

Figure 5. Image of Cidreira dune field (Landsat 7, ETM+, R3G2B1 composition, fused with panchromatic band) showing the northern area where sand supply was canceled by urban occupation.

Decision theory is concerned with the logic by which one arrives at a choice between alternatives. A criterion is some basis for a decision that can be measured and evaluated. Criteria can be of two kinds: factors and constraints, and can pertain either to attributes of the individual or to the entire data set (Eastman *et al*, 1995). A criterion can enhance or detract the suitability of a specific alternative for a given activity. Constraints limits the alternatives under consideration, usually expressed in the form of Boolean maps. Factors, on the other hand, are commonly measured in a continuous scale of suitability. Decision rule is the procedure by which criteria are selected and combined to arrive at a particular evaluation (Gómez & Barredo, 2005, *apud* Eastman *et al*, 1993).

For the Rio Grande urban development model, the following criteria were adopted: proximity to existing urban areas; proximity to the road network; environmental and cultural function of vegetation types; occurrence of flood areas; and occurrence of areas legally constrained to urban development. The geographic data used are presented in the Table 1. Details about the satellite images classification procedures can be found in Farina (2003) and Tagliani (1997). Criteria defined based on the proximity to target areas were obtained through by distance images, where each pixel records the distance to the nearest target feature.

Variable	Layers	Source
Political boundary of the municipality	Study area	Rio Grande master plan
Urban area	Urban area consolidated; distance from the urban network consolidated	RGB543 color composition of Landsat7 ETM+ images
Road network	Main road; distance from the main road	Topographic maps, updated based on RGB543 color composition of Landsat7 ETM+ images
Vegetation	Vegetal cover	Classification of Landsat7 ETM+ images (Maximum Likelihood Method)
Area of historical and ecological interest	Active dunes, steady dunes; lands recently emerged; swamp; archaeological sites	Rio Grande master plan
Hydrography	Lakes, lagoons, canals, streams, coves	Topographic maps updated based on RGB543 color composition of Landsat7 ETM+ images
Legislation	Distance from water resources; vegetation protected by law; master plan; areas with slope above 30%	Topographic maps Rio Grande master plan RGB543 color composition of Landsat7 ETM+ images
Geology Geomorphology	Flooded areas; areas subject to seasonal flooding; stable substrate	Geological/geomorphologic map produced by Tagliani (1997)

Table 1. Variables and layers relevant to urban occupation.

The following criteria were defined as constraints: a) occurrence of urban areas already consolidated; b) water bodies and water courses; and c) occurrence of preservation areas. The following criteria were defined as factors: a) proximity to urban areas and road network; b) environmental and cultural function of vegetation types; c) flood risk; and d) geologic/geomorphologic suitability to urban development.

Constraints were defined by Boolean images, where areas unsuitable to urban development are coded with a 0 and those opened for considering are coded with a 1 (figure 6). Factors, on the other hand, were standardized to a common numeric range (from 0 to 255), and then combined by means of a weighted average. This procedure is known as Weighted Linear Combination (WLC) (Eastman, 1995).

GIS Applied to Integrated Coastal Zone and Ocean Management: Mapping, Change
Detection and Spatial Modeling for Coastal Management in Southern Brazil

207

Urban suitability was considered inversely related to the distance from water courses/bodies based on a sigmoidal basis. We choose a sigmoidal function once the need for water resources conservation is greater closer to water bodies/courses, given the higher vulnerability of marginal ecosystems. Such ecosystems tend to be more attractive to irregular settlements. People without access to sanitation and water supply use nearby waterbodies for consumption and waste disposal. The initial control point (where the suitability index start to grow) correspond to 30m from water bodies/courses, once this buffer consists in a preservation area, excluded from consideration by legal constraints.

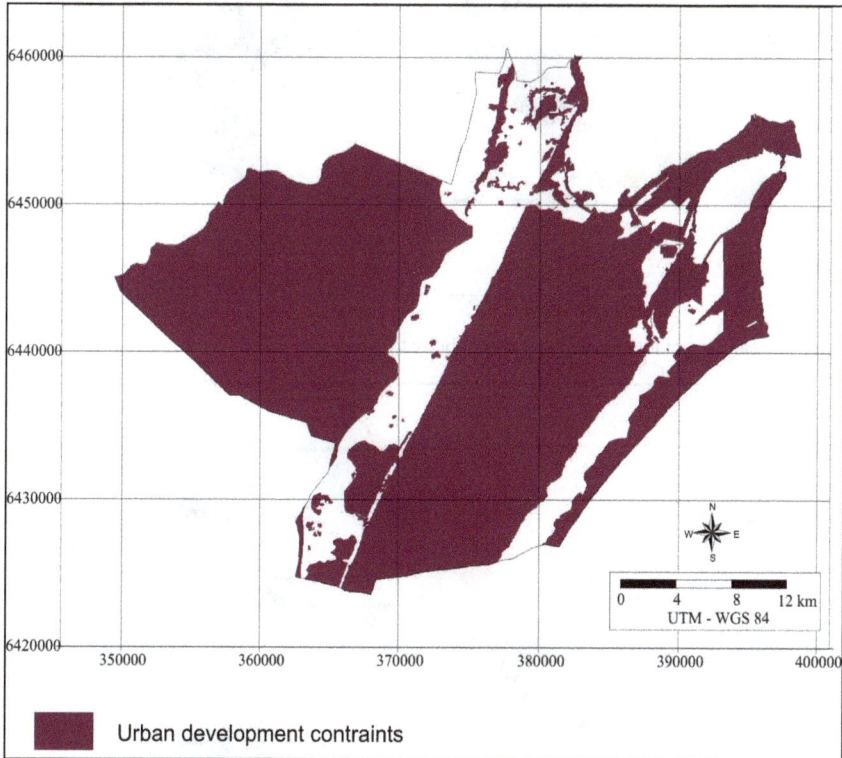

Figure 6. Areas constrained to urban development.

The proximity to the urban areas reduces infrastructure installation costs and travel time. Thus, urban suitability is related to distance from urban areas in a linear function. The same was applied regarding the proximity to road network. Control points were set at 0 and 10 km (figure 7).

The weight assigned to factors ranged from 1 to 5. The final result is an image of suitability to urban development, where a value of 0 is assigned to constrained areas and a value ranging from 0 to 255 (highest suitability) is assigned to the areas under consideration, based on a weighted linear combination of factors. To better visualize the results, the urban suitability image was classified as follows: Class I - Very high suitability (170 –255); Classe II - High suitability (170 – 90); Class III – Medium suitability (90 – 40); and Class IV - Low suitability (40 – 0). The urban suitability map is shown in the Figure 8.

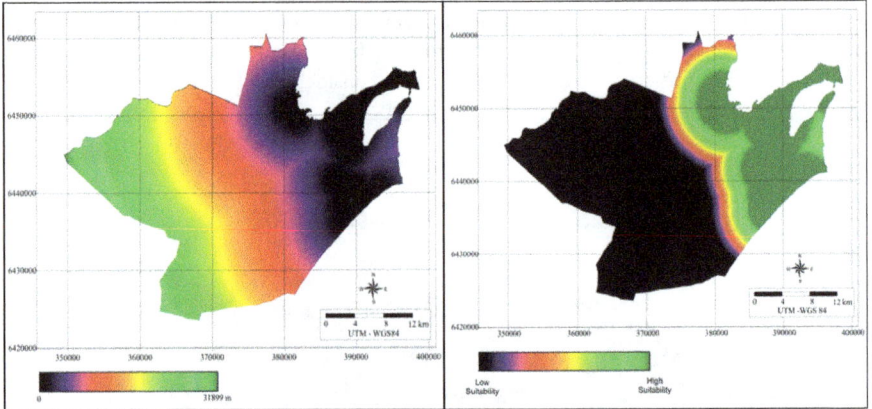

Figure 7. Distance from the urban area and suitability to urban area.

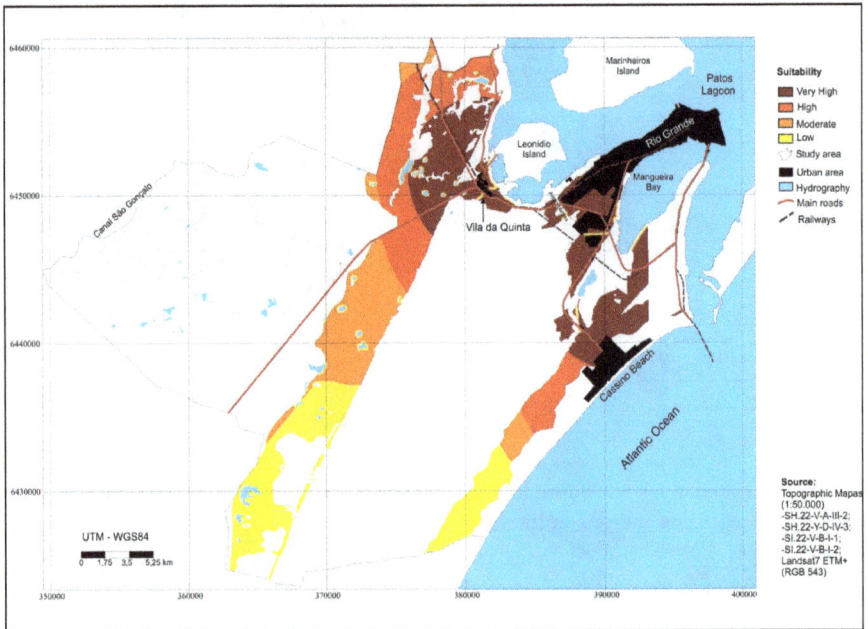

Figure 8. Suitability to urban development.

About 723 km² of the study area, which comprises 926 km², consists in constrained areas to urban development. However, unconstrained areas according to the model are 5.7 times the size of the implemented urban settlements. The Classes I and II together comprise about 110 km². Based on the model, they should be priority areas for urban development. Classes III and IV present lower suitability indexes due to the proximity to water resources. We recommend a specific occupation plan for these areas, including, for example but not limited to, public parks and reserves. Total constraints to urban development represent areas of a high ecological value. Some of them are legally protected, such as, dunes, native woods, wetlands, and aquatic ecosystems.

Spatial modeling consists in a important tool to urban development planning. However, the decision frame adopted for the Rio Grande case must be adapted to be applied in other coastal municipalities, once the relative importance of criteria can change from a place to another, as well as the environmental law at local level.

3.2. Land use change detection and prediction in the medium littoral of the Rio Grande do Sul coastal plain

Land use and cover changes are among the major environmental concerns of humanity today, directly or indirectly contributing to climate change, biodiversity loss, and air, soil and water pollution. They are central in the climate change scenario, once they release greenhouse gases, such as carbon compounds (due to deforestation and soil disturbance), methane (due to hydrological alterations, wetland drainage, and rice crop), and nitrogenates (due to the use of fertilizers, irrigation and burning). Besides, the natural vegetation removal by agriculture not only leaves the soil susceptible to erosion, but releases a large amount of nutrients and sediments to adjacent water bodies, causing many negative impacts on environment.

Silva & Tagliani (2012) mapped the land use changes in the medium littoral of the Rio Grande do Sul coastal plain between 1987 and 2000 (figures 9 and 10). The land use and cover change data were used as an input in a change prediction model. Only the main anthropic changes are used in building the model, that is, changes related to deforestation, urbanization, and afforestation. These three processes synthesizes the major land use and cover changes occurred in the region. The software (GIS) interpret them, respectively, as changes from "forest" to "all" classes, from "all" classes to "urban", and from "all" classes to "silviculture".

The change prediction model results in a susceptibility to change image, varying from zero (low susceptibility) to 1 (high susceptibility).The outputs of the model along with an environmental zoning (figure 11), also proposed by Silva & Tagliani (2012) guide decision making and environmental management toward more sustainable actions. The combination (cross tabulation) between susceptibility to change data and the environmental zoning allows to concentrate efforts in specific target areas. The environmental zoning defines the level of protection of land resources while the change prediction model points priority areas (highly susceptible to change) for each management class. Thus preservation areas highly susceptible to change must be the focus of conservative actions, and become priority areas to preventive management. Development areas highly susceptible to change, on the other hand, should also be the focus of actions but as priority areas to infra-structure installation or regulatory measures.

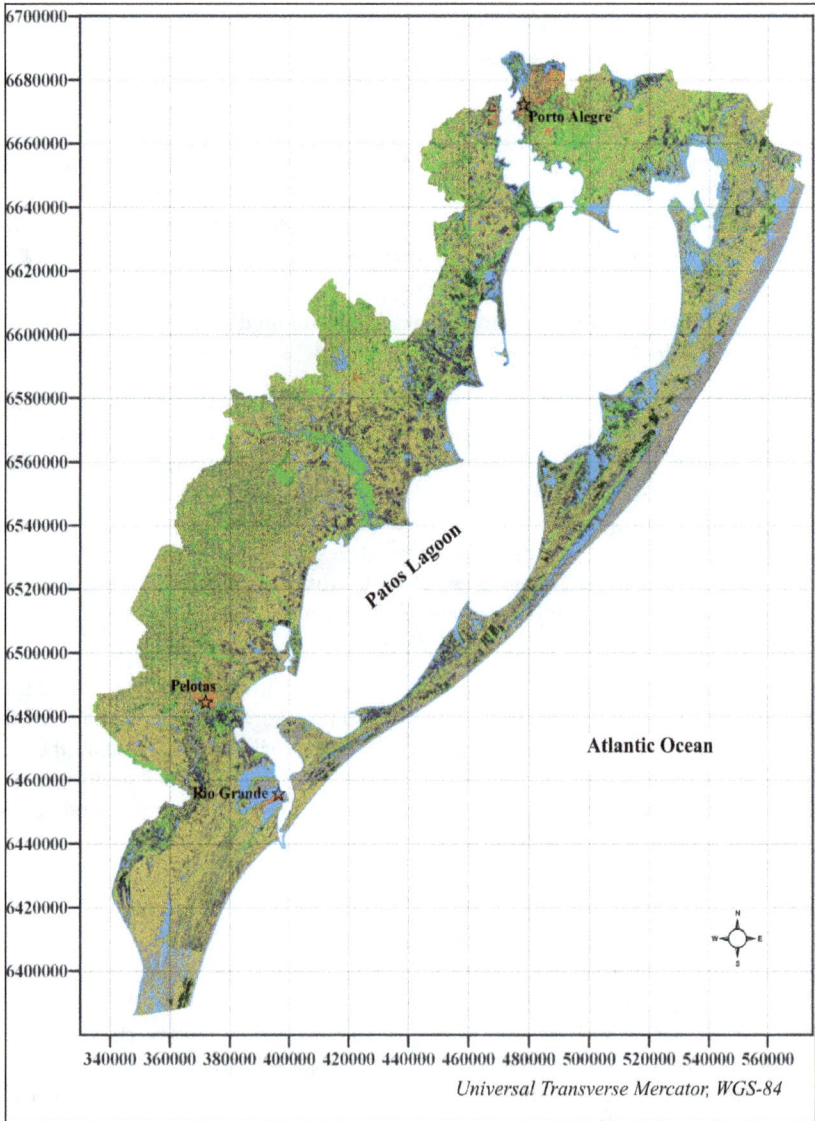

Figure 9. Land use in the medium littoral of the Rio Grande do Sul coastal plain – year 1987.

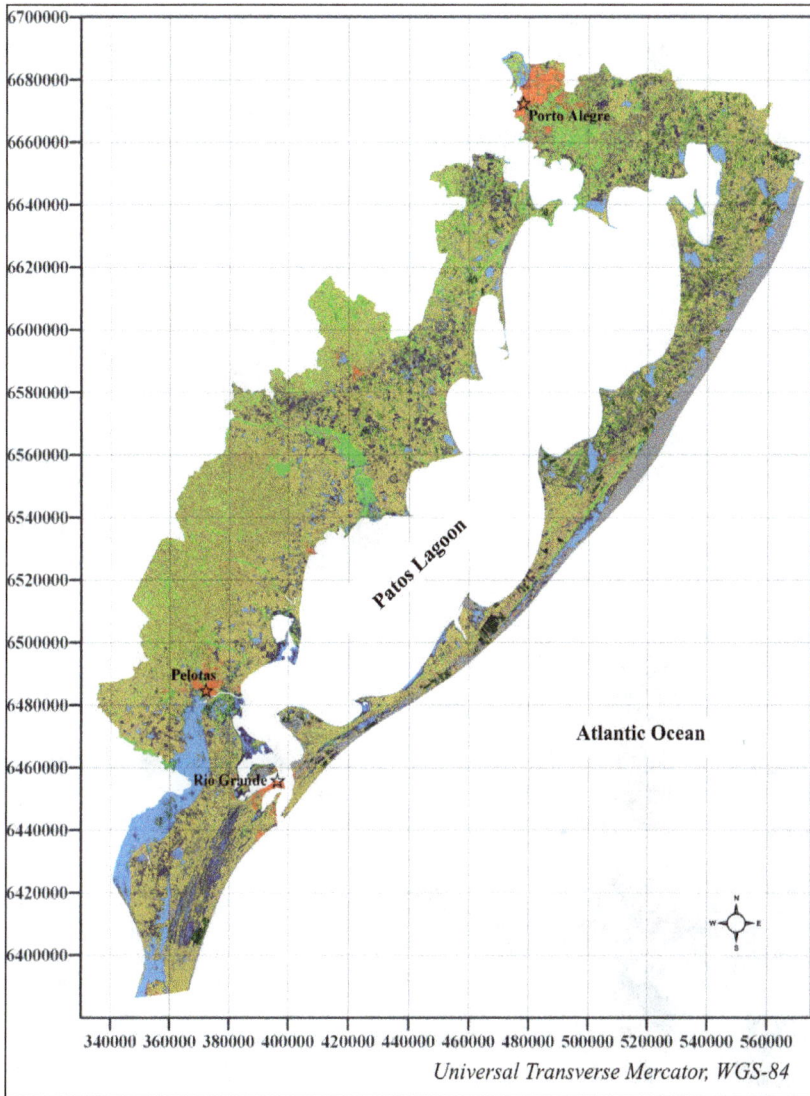

Figure 10. Land use in the medium littoral of the Rio Grande do Sul coastal plain – year 2000.

One of the impediments for the proper development and implementation of the coastal management instruments regards the selection of suitable indicators. Land use (change) seems to be an obvious choice to track several coastal processes, once it can directly or indirectly degrade the health of coastal and marine ecosystems and the services and goods they provide. Besides, it is applicable from local to global scales, acting as a binding element.

Classical land use planning techniques can be applied to coastal zone and protected areas, based on static land use and cover maps. The environmental zoning is an example of that. Land change detection and prediction, on the other hand, can properly be used to monitor the coastal zone and base the environmental planning of these areas. In a general sense, land use based studies provide quality information for virtually all coastal management instruments in Brazil (Silva & Tagliani, *op. cit.*).

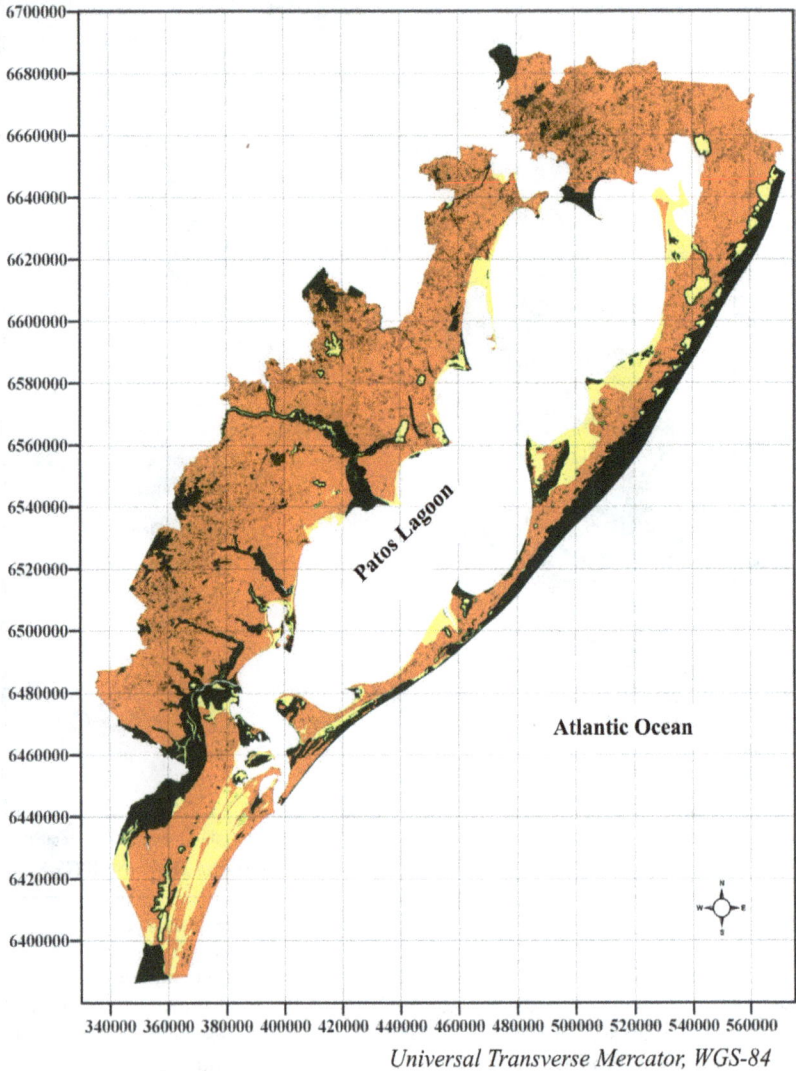

Universal Transverse Mercator, WGS-84

Figure 11. Environmental zoning proposal for the medium littoral of the Rio Grande do Sul coastal plain.

GIS Applied to Integrated Coastal Zone and Ocean Management: Mapping, Change
Detection and Spatial Modeling for Coastal Management in Southern Brazil

213

3.3. Spatial modeling subsides for risk management and coastal adaptation

The flood and landslide control and prediction are one of our biggest challenges given the current scenario of global changes. Once the medium littoral of the Rio Grande do Sul coastal plain is increasingly urbanized and deforestated, the runoff components are altered and flood and landslide hazards worsen. In this context, GIS has excelled in the development of spatial modeling. Beyond the conventional models, GIS is capable of generating vulnerability and suitability indexes, based on map algebra and context operators, pointing areas at high-risk. Whereas Remote Sensing is an unique data source, GIS spatial modeling allows to establish multiple analytical approaches to assess local vulnerability to environmental changes.

The GIS-based models used to spatially define the relative flood and landslide risk in the medium littoral of the Rio Grande do Sul coastal plain (Silva *et al.*, 2011) incorporates a digital elevation model and also rainfall and soil infiltration potential data to calculate flow direction and accumulation. Other criteria are derived from land use and soil data. Absolute constraints for both models comprise the occurrence of water bodies and wetlands. The resulting images shows the flood and landslide risk, in four classes: low, moderate, high, and very high. About 5% of the area had a decrease in the soil infiltration potential due to land use changes between 1987 and 2000. The flood and landslide risk models gives us a hint about what areas should be the focus of concern in the case of extreme rainfall events. It also can be used as a scenario generation tool, prospecting the effect of land use and climate changes on hydrological vulnerability. Additional criteria, if necessary, can be included to the models. The results are intended to support environmental management and development planning of the costal municipalities surrounding the Patos Lagoon, some of them already suffering the socio-economic consequences of hydro-meteorological disasters. Figure 12 shows the flood risk model results for the years 1987 and 2000.

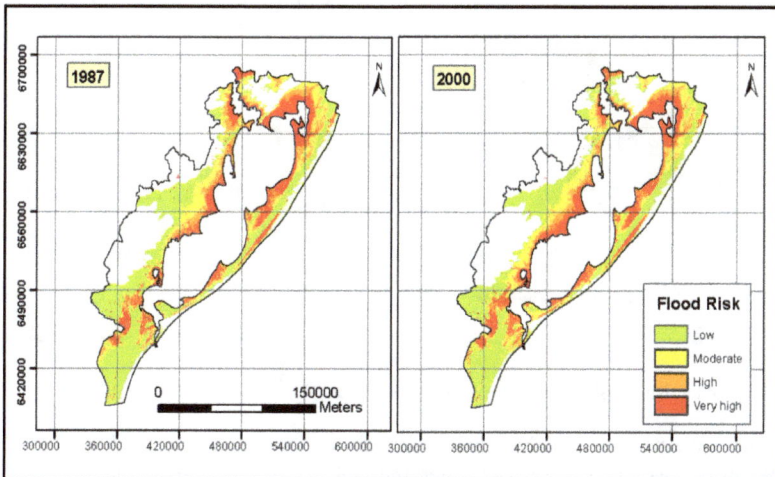

Figure 12. The flood risk images for the years 1987 and 2000. Universal Transverse Mercator, WGS-84.

At least two major hydro-meteorological disasters were recorded in the west coast of Patos Lagoon in the last couple of years. In 2009, 8 people died, 1,200 left homeless, and a bridge was dragged due to an extreme rainfall and flood in Pelotas and Turuçu. In 2011, a similar event resulted in 8 more deaths and 20,000 homeless in São Lourenço do Sul. Half of the urban area was flooded. Two bridges were destroyed and many municipalities isolated. A close view of the model results shows how flood risk responded to land use changes in São Lourenço do Sul, the most impacted city by flood incidents (figure 13).

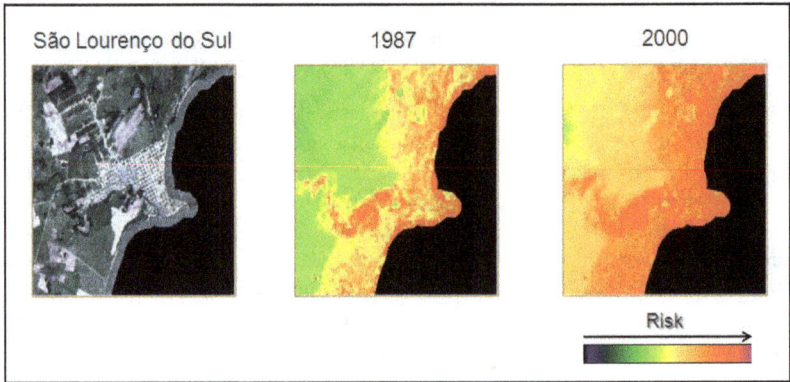

Figure 13. A close view of the flood risk model results for São Lourenço do Sul.

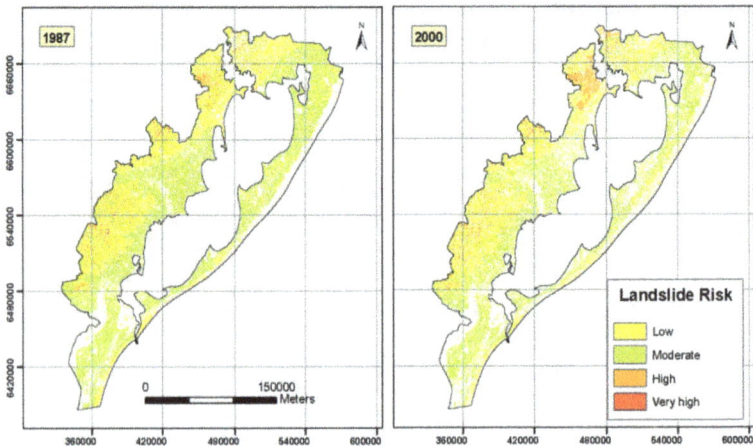

Figure 14. The landslide risk images for the years 1987 and 2000.

Landslides have not affected the region as much as flood, although local incidents have occurred in the last years. However, given the violence of some recent rainfall, combined with land use changes, we may infer about the imminent danger in many parts of the

GIS Applied to Integrated Coastal Zone and Ocean Management: Mapping, Change
Detection and Spatial Modeling for Coastal Management in Southern Brazil

215

medium littoral. Rio de Janeiro already felt the consequences of this process, which resulted in serious disasters. Thus it is not a surprise if we start facing more severe landslides incidents in the Rio Grande do Sul as well. Figure 14 shows the GIS-based landslide risk results for the years 1987 and 2000.

4. Strategies and recommendations to promote GIS as a knowledge translation tool in the realm of integrated coastal zone and ocean management in Southern Brazil

Scientific understanding is crucial for a good ICM decision making (Cicin-Saint et. al., 1998). There is a necessity to establish a relationship between science and management and to incorporate scientific information into the management process. This is especially true in Brazil, where academics play a role beyond scientific production. They are indeed active actors in the coastal management process.

If we take a look at the five phases of the integrated coastal management process, we notice that in most of the cases we only took the first step: issue identification and assessment. It is true regarding integrated coastal management as a hole and also true when we consider the coastal management instruments alone. Even if we consider the I3Geo a successful initiative, we do not have evidences that it has supported the start of new cycles of management, where priorities and policies are adjusted to reflect experience and changing in social/environmental conditions. Besides, the platform works as an one-way tool, giving people access to cartographic information, but it does not promote the communication among sectors of society. The development of webGIS containing public participation tools is a way to overcome this limitation, promoting community empowerment through demand-driven, user-friendly and integrated applications of geoinformation.

The environmental management in Brazil is highly based on licensing. Thus, environmental licensing forces the information exchange between scientists, developers, and environmental agencies. On one hand, licensing is positive, driving research toward real social demands and promoting information interchange. On the other hand, change prediction and scenario generation studies are not required in the licensing process and, consequently, preventive planning is not encouraged. The inclusion of prospective studies in the policy instruments would enhance the efficiency of coastal management, allowing to focus actions on priority areas for development or conservation.

5. Conclusion

Academy is not responsible for defining the planning process, but it plays a major role in supporting it in Brazil, producing relevant information for the decision making process. Research results of some case studies were presented in this sense. Land changes, vital areas suppression, rural impoverishment, urban swelling, they are nothing else but governance failures.

Once the Brazilian National Plan of Coastal Management instruments are spatial in nature, the GIS-based research works have already been helpful in improving the planning process in the Rio Grande do Sul coastal plain. Besides, the acquired knowledge in spatial modeling can be included into the existing instruments to promote their adaptation to the current pace and reach of human activities over the coastal zone.

Geological and geomorphologic mapping of coastal plain is presented here as the basis for land use planning. When coastal dynamic knowledge is applied to land use planning, environmental impacts of development are expected to be reduced.

Geotechnology is also important to urban development planning. The GIS-based model of urban sprawl of the Rio Grande city is an example of that, giving subsidies for management at local level.

Change detection and predictive modeling arise as a vital mean to support the adaption to the current environmental scenario of fast changes. Land use change and environmental risk models seem to bring significant advances in this sense.

A long journey must be undertaken before GIS products and recommendations produced by universities become actions. A new range of opportunities and challenges opens up, whether in the GIS-based environmental plans development and implementation or in further research in spatial modeling as a subside for risk management and coastal adaptation.

Author details

Tatiana S. da Silva, Maria Luiza Rosa and Flávia Farina
Federal University of Rio Grande do Sul, Institute of Geoscience, Brazil

6. References

Angulo, R.J. & Lessa, G.C. 1997. The Brazilian sea-level curves: a critical review with emphasis on the curves from Paranaguá and Cananéia regions. *Marine Geology,* Vol.140, pp. 141-166.

Asmus, H.E., Garreta-Harkot, P.F., Tagliani, P.R.A. 1988. Geologia ambiental da região estuarina da Lagoa dos Patos, Brasil. *Proceedings of VII Congresso Latino-Americano de Geologia.* Belém. November 1988.

Cicin-Sain, B. Knecht, R.W., Kullenberg, G. 1998. *Integrated coastal and ocean management: concepts and practices.* Island Press. ISBN 1559636041. Delaware.

Corrêa, I.C.S. 1995. Les variations du niveau de la mer durant les derniers 17.500 ans BP: l'exemple de la plate-forme continentale du Rio Grande do Sul-Brésil. *Marine Geology,* Vol.130, pp. 163-178.

Dillenburg, S.R.; Roy, P.S.; Cowell, P.J. & Tomazelli, L.J. 2000. Influence of antecedent topography on coastal evolution as tested by the shoreface translation-barrier model (STM). *Journal Coastal Research,* Vol.16, pp. 71-81.

Dillenburg, S.R.; Tomazelli, L.J. & Barboza E.G. 2004. Barrier evolution and placer formation at Bujuru southern Brazil. *Marine Geology*, Amsterdan, Vol.203, pp. 43-56.

Eastman, J. R. (1995). *Idrisi for Windows User's Guide*, Clark University, Worcester, USA.

Eastman, J. R.; Jin, W.; Kyem, P.A.K. & Toledano, J. (1995). Raster procedures for multicriteria/multi-objective decisions. *Photogrammetric Engineering and Remote Sensing*, Vol.61, No.5 (May 1995), pp. 539-547, I SBN 0099111295.

Farina, F. C. (2003). *Utilização de técnicas de geoprocessamento para seleção de áreas adequadas à expansão urbana: caso do município de Rio Grande-RS*. UFRGS, Porto Alegre, Brasil.

Gómez, M. D. & Barredo, J. I. C. (2005). *Sistemas de Informacíon Geográfica y evaluación multicriterio en la ordenación del territorio*, RA-MA, ISBN 84-7897-673-6, Madrid, Spain.

Jablonski, S. & Filet, M. 2008. Coastal management in Brazil – A political riddle. *Ocean & Coastal Management*. 51, pp. 536–543, ISSN 0964-5691.

Martin, L.; Suguio, K. & Flexor, J.M. 1979. Le Quaternaire marin du littoral brésilien entre Cananéia (SP) et Barra de Guaratiba (RJ). *Proceedings of International symposium of coastal evolution in the Quaternary*, São Paulo, Brasil, pp. 296-331.

Rosa, M.L.C.C., 2010. *Estratigrafia de Sequências: aplicação das ferramentas na alta frequência. Um ensaio na Planície Costeira do Rio Grande do Sul*. Ph.D Qualifiyng. Instituto de Geociências. Universidade Federal do Rio Grande do Sul. Porto Alegre, Brasil. 67 p.

Silva, T.S., De Freitas, D., Tagliani, P.R.A., Farina, F.C., Ayup-Zouain, R.N. 2011. Land use change impact on coastal vulnerability: subsidies for risk management and coastal adaptation. *Proceedings of CoastGIS 2011*. Ostend. September 2011.

Silva, T.S. & Tagliani, P.R.T. 2012. Environmental planning in the medium littoral of the Rio Grande do Sul coastal plain - Southern Brazil: elements for coastal management. *Ocean & Coastal Management*, Vol.59, pp. 20-30, ISSN 0964-5691.

Tagliani, C.R.A. 1997. *Proposta para o Manejo Integrado da Exploração de Areia no Município Costeiro de Rio Grande – RS. Um Enfoque Sistêmico*. UNISINOS, São Leopoldo, Brasil.

Tagliani, P.R.A., Landazuri, H., Reis, E.G., Tagliani, C.R., Asmus, M.L., Sánchez-Arcilla, A. 2003. Integrated coastal zone management in Patos Lagoon estuary: perspectives in context of developing country. *Ocean & Coastal Management*. 46, pp. 807-822, ISSN 0964-5691.

Toldo Jr., E.E.; Almeida, L.E.S.B.; Nicolodi, J.L. & Martins, L.R.S. 2005. Retração e Progradação da Zona Costeira do Estado do Rio Grande do Sul. *Gravel*, Vol.3, pp. 31-38.

Tomazelli, L.J. & Villwock, J.A. 1989. Processos erosivos na costa do Rio Grande do Sul, Brasil: evidências de uma provável tendência contemporânea de elevação do nível relativo do mar. *Proceedings of Congresso da Associação Brasileira de Estudos do Quaternário* 2th, p.16, Rio de Janeiro, Brasil.

Tomazelli, L.J. & Villwock, J.A. 1992. Considerações Sobre o Ambiente Praial e a Deriva Litorânea de Sedimentos ao Longo do Litoral Norte do Rio Grande do Sul, Brasil. *Revista Pesquisas*, Porto Alegre, Vol.19, pp. 3-12.

Tomazelli, L.J. & Villwock, J.A. 1996. Quaternary Geological Evolution of Rio Grande do Sul Coastal Plain, Southern Brazil. *Anais da Academia Brasileira de Ciências*, Vol.68, No.3, pp. 373-382.

Travessas, F.A. 2003. *Estratigrafia e evolução no Holoceno Superior da barreira costeira entre Tramandaí e Cidreira (RS).* Instituto de Geociências, Universidade Federal do Rio Grande do Sul. Porto Alegre, Brasil.

Chapter 12

GIS Applied to the Hydrogeologic Characterization – Examples for Mancha Oriental Aquifer (SE Spain)

David Sanz, Santiago Castaño and Juan José Gómez-Alday

Additional information is available at the end of the chapter

1. Introduction

The population on planet Earth, according to FAO forecasts, will increase from 6 billion to 8.1 billion inhabitants in 2030 and will coincide with an increase in water demands to meet human needs. Fresh water has ceased to be an inexhaustible resource to become a rather limited and scarce one.

Earth's hydrosphere has an approximate volume of 1.38×10^{10} km³ of water, which has remained virtually constant since its formation over 3 billion years ago. This volume of water is distributed into four groups:

1. The vast majority is in the oceans, at 97.6% of the total ($1,350 \times 10^6$ km³),
2. In second place is solid water, in glaciers, at 1.9% (26×10^6 km³),
3. Third is groundwater, with 0.5% of the total, which is 7×10^6 km³, and
4. The remainder of water on Earth (0.03% of total) is divided among lakes (0.017), soil (0.01%), the atmosphere (0.001%), Biosphere (0.0005%) and rivers (0.0001%).

Ocean water is salt water and the glaciers are difficult to utilize because they are located far from major populated areas. Therefore, we find that groundwater is the largest volume of freshwater available to man. The volume of groundwater is 4,000 times greater than that of rivers and 30 times higher than the rest of liquid water that is on the surface of the continents. In addition, groundwater has characteristics that make it especially attractive to combating the processes of drought and desertification. Unlike surface water, groundwater does not evaporate, there are no major seasonal variations and flow is very slow, so it is difficult to contaminate (Castaño et al., 2008).

Groundwater is held in a naturally occurring reservoir called an aquifer, a geologic formation capable of storing, receiving and transmitting water so that man can easily take advantage of economically significant quantities to meet needs. The water is contained in any geological formation (ie: river gravel, karstified limestone, porous sandstone and so on).

Like all scarce resources, groundwater management must be approached from a dual approach (Knowledge and Sustainability):

Knowledge: There must be a sound understanding of the hydrogeological aquifer system to be managed. This should include a detailed analysis of the hydraulic aspects (geology, hydraulic parameters and groundwater flow) and should be contrasted with the hydrochemical aspects of water containing (origin of the substances dissolved in groundwater and hydro-chemical changes due to movement through groundwater flow).

Sustainability: Groundwater pumping of a region (an aquifer) shouldn't exceed the water received, that is, the available resources. If users pump a volume of groundwater for short-term needs (ie: drought conditions), beyond the resources available, they use the aquifer reserves. Then, the aquifer must be given time to recover (either by saving water or allowing recharge to increase during periods of more rainfall). Otherwise resources suffer overexploitation, putting the aquifer at risk of becoming depleted. It is obvious that the water volumes involved should be determined as precisely as possible.

In managing groundwater resources, Geographical Information System (GIS) are tools capable of storing and managing spatial hydrogeological data by spatial referencing in digital formats. The correlation of all data with location is the key feature of GIS, which provides the ability to analyze and model hydrologic processes and produce results in maps and in digital formats. Thus, GIS can be considered a support system in decision making and an ideal tool for monitoring certain hydrogeological processes with socio-economic impacts (Goodchild et al., 1996).

Figure 1 shows a diagram of a GIS aquifer system modeling tool (Case study of the Mancha Oriental System). This scheme is integrated into a) a block of hydrogeological data maps of the study area, which supplies data on groundwater from urban and industrial and general hydrological information (surface and groundwater hydrology) necessary to carry out the integration and interpretation of some results, and b) a block of data from remote sensing imagery. Remote sensing allows for classification of crops and their relationship to water supply for irrigation, mapping of wells, and the assessment of recharge by precipitation of rainfall. All this information is transferred to software that simulates groundwater flow in the Mancha Oriental aquifer using intersection tools.

The main goal is to show the methods some GIS applications have in hydrogeological studies. This chapter is divided into two sections which describe some examples of hydrogeological characterization, and secondly, a method for calculating groundwater abstraction. To demonstrate these applications one of the largest aquifers in southern Europe (in terms of area), the Mancha Oriental System has been chosen.

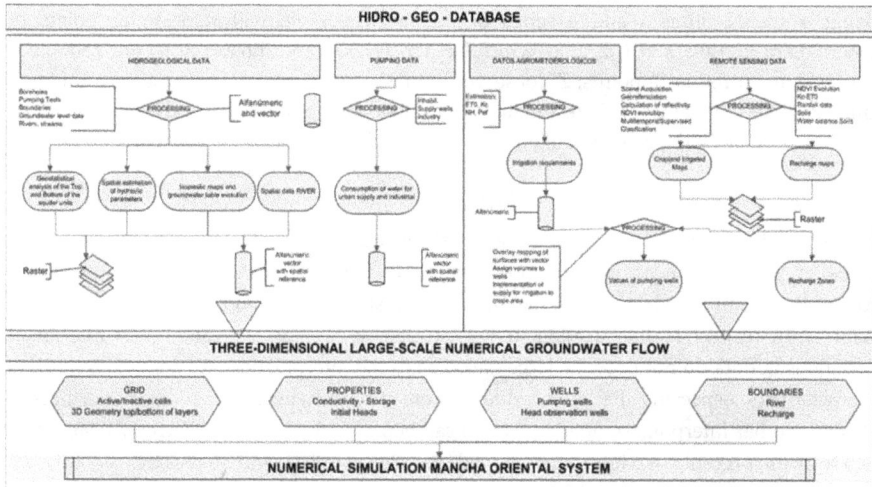

Figure 1. Basic diagram of coupling remote sensing and GIS techniques with the groundwater flow model of the Mancha Oriental System.

2. GIS & hydrogeological characterization

In general, among the Earth Sciences and particularly hydrogeology, sources of data tend to be from points (wells, points of water, lithological columns, etc.) defined by a geographic location (UTM or geographic coordinates) and attributes (topographic top or bottom of a hydrogeologic entity, groundwater level, hydraulic parameters, concentration of a chemical compound). This type of data, usually measured in the field, must be spatially distributed in a continuous manner such that a value is given for any point within the space. To achieve this, interpolation or spatial estimation is used. This method derives an interpolation function that provides estimates for a point in space based on the points measured. GIS tools have incorporated algorithms which perform these operations with discrete entities (vector) and generate spatially continuous entities (raster, line models, etc.). In addition to expanding and the geodatabase and adding values, these techniques create a foundation for spatial modeling (Peña Llopis, 2006).

The most commonly used spatially continuous entities are raster maps, which are characterized by a two-dimensional numerical matrix or digital image. Each element of the matrix, called a picture element or pixel, has an attribute assigned to it in the database. The only requirements are for maps to have attribute values referenced to the same coordinate system and the same number and arrangement of pixels to perform algebra operations with them (ie: isopaches: difference between top and the bottom raster maps of the geologic formation; calculation of storage volumes: difference between raster maps and contour lines or groundwater for different dates, multiplied by the storage coefficient, etc.).

2.1. Theoretical foundations

Using a spatial domain and a series of points (which we will refer to as points of observation) Pi, where i= 1, 2, ...n, which have a series of coordinates xi where variable Z has been measured in Pi points; Zi (observed values), the interpolation or spatial estimation aims to find the value of Z (estimated values) at any point in the known space. An interpolation function must be obtained:

$$Z(x)=\theta(x,xi,Zi) =f(x) \tag{1}$$

The interpolation function should have certain characteristics: a) accuracy: the estimated value in points of measurement should coincide with the measured value, b) spatial continuity, c) ability to be derived: the interpolation should be "smooth" and d) it should be stable with respect to the location of the variable as well as its value such that small variations in data do not provoke large variation in the interpolation. As a function of these characteristics, especially the last condition, there is no universal interpolator and there is always another interpolation method which can be applied (Samper & Carrera, 1996). Most GIS software presents two interpolation methods: Deterministic and Stochastic.

2.1.1. Deterministic methods

This type of method is characterized by associating a mathematical function, such as an interpolation function, to the measured or observed values, in which these points are considered without error. Following the nomenclature followed until now, this mathematical function could be written in the following manner:

$$Z(x) = f'(x) = \sum_{i=1}^{n} c_i f(x)_i \tag{2}$$

where for each x a Z(x) value is measured through a function f(x), which is defined by the sum of all "n" points of observation of a product between a base function f(x)i and coefficients, Ci. For example, in a simple exact interpolation the observed or measured values (Zi) coincide with the Cs values, multiplied by a weighting factor given by the function f(x)i. The deterministic interpolation functions differ from one another in the means of evaluating f(x)i and Ci.

There are various deterministic interpolation techniques. The most commonly used methods are presented here (ESRI, 1997) (Samper & Carrera, 1996):

Nearest neighbor (Thyessen polygons, Polygons of influence).

This method assigns the value of each measured or observed point to each pixel or node of the interpolated area. For each point of observation the Euclidean distance is calculated for all other points and each is given the closest value. The result is a map of polygons with an interpolated value (Fig. 2A). This method is often used for regular grids and/or dense observed data, or to find areas of influence.

Interpolations based on weighting functions.

The estimation or interpolation in this type of method is performed by a weighted average of the observed values. At each point of observation a weight is assigned. The selection criterion is that the weighting function is exclusively dependent upon the distance (d). The weight will decrease with increasing distance between points. The most common strategy for generating this criterion is the Inverse of Distance raised to some exponent (a).

$$f(x) = \frac{1}{d^a} \qquad (3)$$

This exponent shows the "speed" with which the weight of a point of observation decreases with distance from the point of estimation. At times the number of points of influence is restricted, or a radius or maximum distance is assigned for considering points of observation. This interpolation method is exact and is commonly applied, with the only disadvantage being creating the feared "bulls-eyes" (Fig. 2B).

Polynomial interpolation.

In this method the interpolating function is a polynomial function which varies in its exponential order. The choices for polynomial are: a) through exact fit and b) fit by least mean square. The first method aims to resolve the system of equations defined by the n points of observation. If there are many points of observation, the fit of higher order polynomials can become unviable, giving unrealistic interpolations with exaggerated variation among the values (Fig. 2C). In fact, by default these methods limit the polynomial to third order and only use the number of points in a nearby group. One special case of polynomial interpolation is linear interpolation, wherein the interpolation function is a first order polynomial which directly depends on the position of the observed values. It is an exact method and does not take into account the spatial distribution of the variable, with the result of soft surfaces. It is an easy method and is often used, above all in cases when not a lot of data is available and the aim is to study the spatial variation of a certain variable. In general, this interpolation method is not used for spatial estimates on realistic structures (topography, groundwater levels, etc.) but rather to determine the tendency of data (Fig. 2C).

Spline functions.

Within polynomial interpolation, this general method generates a different series of expressions for each subdomain into which the whole interpolation space has been divided, wherein continuity requisites are imposed, especially in the contours common to more than one subdomain. The results of this interpolation tend to be surfaces with small changes in levels (Fig. 2D).

2.1.2. Stochastic methods

This methodology is based on the premise that the variable to be interpolated is a random function associated with probabilistic distribution laws. This type of method gives a

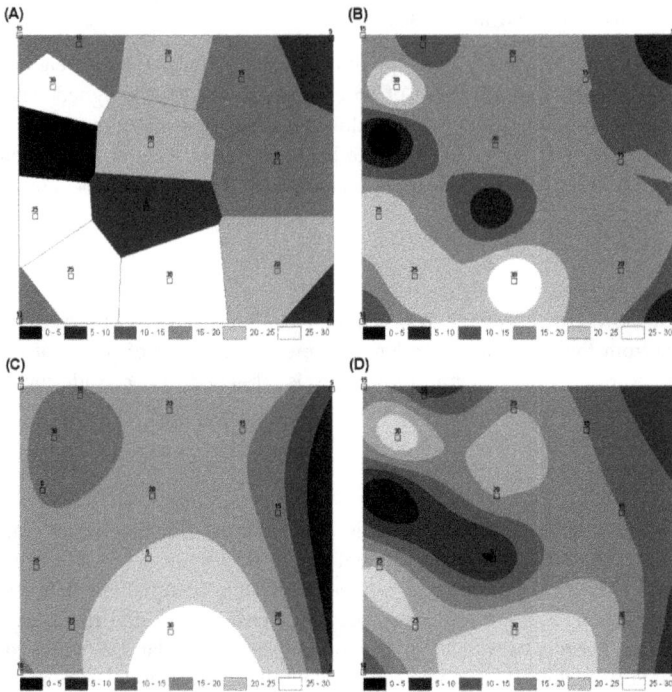

Figure 2. Examples of the different deterministic interpolation methods using GIS tools. A) Polygons of Influence (Thyessen polygons, Nearest neighbor), B) Interpolations based on weighting functions (in this case the function is the inverse distance squared (IDW); C) polynomial interpolation, in this case third order; D) Spline functions. The same data are used for each case and the results are shown as continuous identical spatial entities, with the same size and pixel size.

measure of error of the interpolation based on the data. There are two classes of stochastic interpolation: a) non-parametric, which are not exact because the errors are assumed to be independent and b) parametric, wherein the interpolation function depends on certain parameters calculated as a function of the observed data (IDW or Krigging) (Samper & Carrera, 1996). The most common method, Krigging, which is available in most GIS software packages, is explained below.

Krigging was created under a new discipline, geostatistics, as a result of problems presented by deterministic interpolation in Earth sciences due to the uncertainty and variability of data (Cassiraga, 1999). The starting hypothesis of geostatistics is that the data of study has a correlation spatial structure, as the realization of an infinite amount of possible realizations. For this reason, geostatistics is called the science of regionalized variables. For spatial estimation using Krigging, the steps below should be followed, among others, and the variable to be interpolated should meet the criteria of normality and stationarity (Johnston et al., 2001).

The first step is structural analysis, with the objective of estimating the semivariogram. This relates the Euclidean distance among the points of observation with the variability of the measured values (Samper & Carrera, 1996). First, the variable should be defined as stationary, if there is a tendency among the data, etc. The function of the semivariogram (estimator of spatial variability) is expressed as:

$$\gamma(x) = \frac{1}{2N} \sum_{i=1}^{n} \left[Z(x_i + h) - Z(x_i) \right]^2 \tag{4}$$

where:

Z(xi): experimental data,
h: distance between points of observation (Variogram step),
N: number of pair separated by vector h found in a group of data,
Xi, xi+h: experimental points in an n-dimensional space.

At first, from the observed data an experimental semivariogram will appear. A theoretical function with similar behaviourcan be fit to this in order to calculate a weighting matrix for each point, and statistical error affecting the interpolation can be calculated. The semivariogram is composed of a series of elements (Fig. 3A):

Range: the distance from which the spatial correlation is practically null (Area of influence),
Sill: value that the semivarogram takes in the Range,
Nugget: value of the semivariogram when it intersects at the coordinate axis.

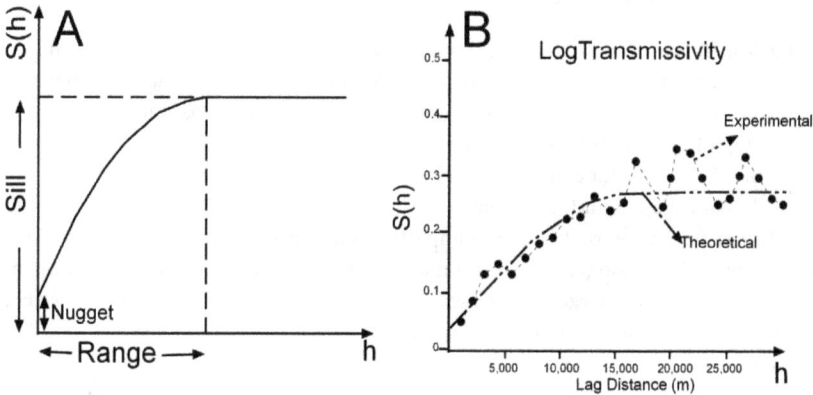

Figure 3. A) Elements of a spherical semivariogram. B) Experimental and theoretical semivariogram sets.

The experimental variogram cannot be used for the geostatistical application. It must be fit to a theoretic model (Fig. 3B). There are different technical variogram models available, with the most popular being the stationary or spherical semivariogram. Once the theoretical semivariogram has been chosen, the Krigging technique performs the spatial estimation of the data. There are diverse Krigging techniques as a function of diverse methodological hypotheses:

Simple: Hypothesis of stationary variable with a known mean and covariance,
Ordinary: Hypothesis of stationary variable with an unknown mean and known covariance,
In an environment (by blocks): Quasistationary hypothesis,
Residual: Non-stationary hypothesis with a known drift, from which residuals are derived and ordinary Krigging can be performed,
Universal: Non-stationary hypothesis and polynomial form with a drift set *a priori*.

2.2. MOS case study

The Mancha Oriental System (MOS) is located in the SE of the Iberian Peninsula and is one of the largest aquifers in Spain (7,260 km^2) (Fig. 4). The area has a semiarid Mediterranean climate. Average rainfall is 350 mm/year and mean annual temperature is 13-15°C; the continental nature of the climate is clear from the extreme temperatures that occur.

The area is characterized as a high plain (700 masl mid-altitude) surrounded by gentle relief, interrupted only by a valley which was carved by the action of the Júcar River. From a hydrogeologic perspective, the MOS is formed by the superposition of three limestone aquifer hydrogeologic units (UHs): UH2: Tertiary, UH3: Upper Cretaceous and UH7: Middle Jurassic. These HUs are separated by aquitards/aquifuges that comprise UH1 (upper and lower), UH4, UH5 and UH6 (Sanz et al., 2009). The impermeable base and southwest boundary of the area of study is composed of marl, clay and gypsum from the Lower Jurassic, belonging to HU8 (Fig. 4).

Over the last 30 years the progressive transformation of approximately 100,000 ha from dry to irrigated farmland has translated into an acceleration of socioeconomic development due to widespread use of groundwater resources. Groundwater abstractions in the MOS exceed 400 Mm^3/yr, of which 98% is used for irrigated agriculture and the rest to supply a population of 275,000 Inhabitants (Estrela et al., 2004). Groundwater pumping is not sustainable with the amount of available resources, estimated at 320 Mm^3/yr by the Júcar Water Authority. Therefore, two major impacts are occurring: (a) the quantity of available groundwater is descending, noted as a continuous decline in the regional water level and a decrease in aquifer discharge to the Júcar River; and (b) the quality is also affected, as researchers have found a significant increase in nitrate concentrations in groundwater (Moratalla et al., 2009).

In this context, the MOS is an ideal case study for testing and validating the usefulness of GIS Techniques for understanding the aquifer system and planning for sustainable management. Following is a description of the interpolation methods applied to these variables: a) The elevations of the top and bottom of the aquifer units, b) Hydraulic parameters, c) Groundwater level data d) Groundwater chemistry. The approach is to explore the variable data with histograms and spatial trend analysis in order to understand the behaviour of the variable in space as well as to establish whether the data are consistent or

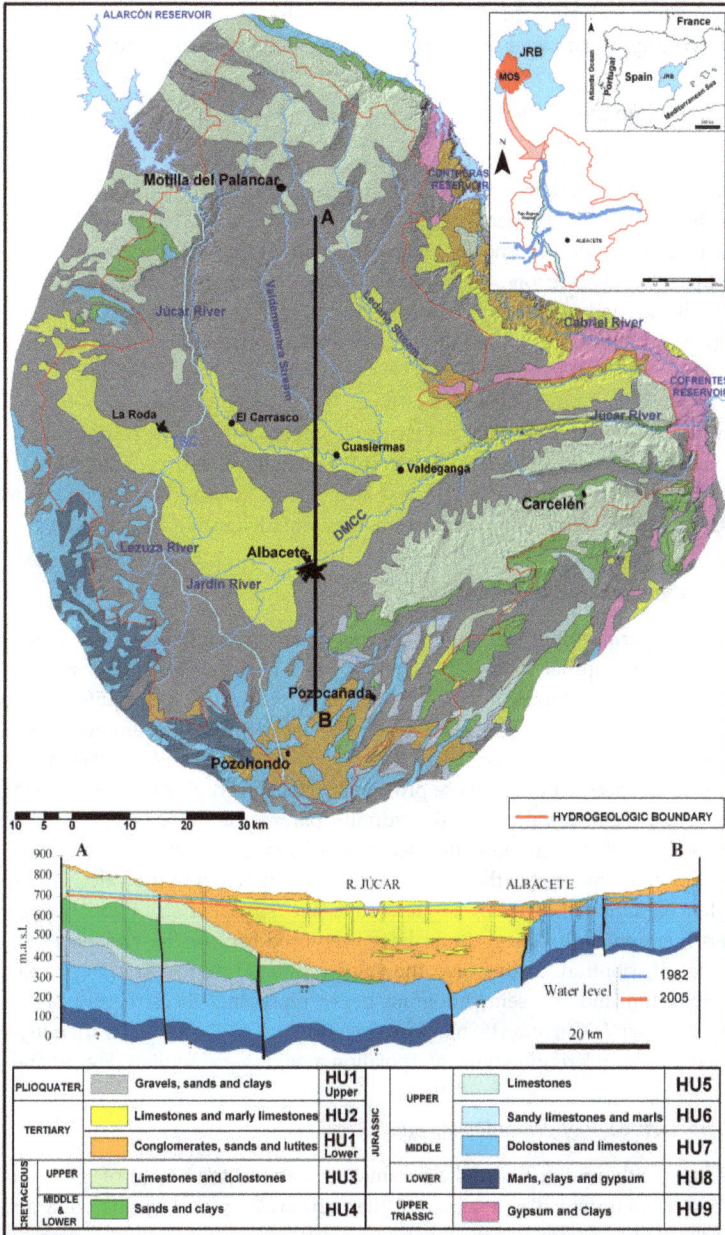

Figure 4. Simplified hydrogeological map of the MOS. Taken from Sanz et al. (2009). JRB Jucar River Basin. Graphical output from GIS software.

anomalous. After analyzing the data, a variable is selected to be interpolated. The type of interpolation to apply for raster maps or continuous spatial entities is chosen.

2.2.1. Hydrostratigraphic framework

Any attempt at making a coherent hydrogeological model should be approached by first understanding, with a certain amount of precision, the geometric configuration. In addition to information on the surface geology, lithologic columns should be analyzed and gathered from water sampling points, and tested materials should be classified within the defined hydrogeologic units (Murray & Hudson, 2002). Using this information, layers of geographically located points (X, Y coordinates) as well as the topographic height of the superior (top) and inferior (bottom) limits were made into attributes of each hydrogeologic unit that behaves as an aquifer. Using geostatistical interpolation models developed on theoretic and applied foundations, GIS software (i.e. ArcMap ® 9.3) was used to determine the continuous geographic entities, ie: raster maps of the surfaces corresponding to the top and bottom of each hydrogeologic aquifer unit. The result is the three-dimensional structure of the hydrogeologic system (Fig. 5A). These 3D geologic models (Fig. 5), constructed using GIS tools, became the foundation for the numeric simulation models in later steps.

2.2.2. Hydrodynamic characterization

Transmissivity, permeability and storage coefficient are hydraulic parameters that must be quantified for an aquifer because they are needed to estimate the progression of groundwater levels, groundwater flow through a section of the aquifer, contaminant transport time, the degree of aquifer homogeneity and the numeric parameterization of the groundwater flow models (Mace, 2000). Generally, estimating these parameters requires pumping tests in specific points. These provide specific geographic entities defined by their coordinates, and attribute values for the hydraulic parameters in the well. It is also useful to have previous knowledge on the spatial behavior of the variable and establish a relationship for the interpolation model (ie: the type of distribution function of the variable). In the case of the Mancha Oriental System, to determine the spatial distribution of any of the parameters mentioned, these logarithms have been used because the variable tends to have a log-normal distribution. In this case, the value estimated by the Krigging method is the absolute optimum and the semivariogram better represents the structure of the spatial variability (Samper & Carrera, 1996). Once the structure of the spatial variability of log-T was studied, ordinary Krigging type interpolation models were applied (Fig. 5B and 3A).

2.2.3. Characterization of groundwater flow

As is the case with aquifer hydraulic parameters, data on the height of the groundwater levels are also point data. The attribute of the groundwater level in this point geographic entity is compiled in the inventory of water points by subtracting the topographic height of the point from the depth of the water level in the well. These measurements should be performed for a specific date and in the least amount of time possible.

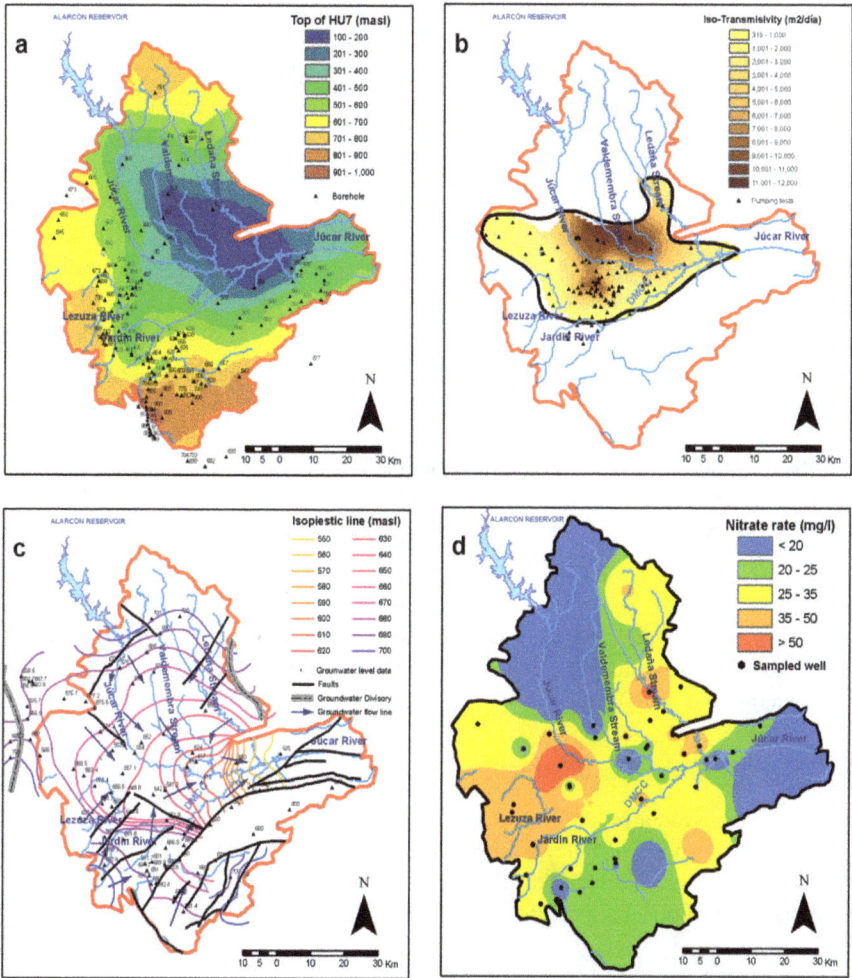

Figure 5. Examples of hydrogeologic variable interpolation. a) The elevations on the top of HU7 Mid-Jurassic aquifer, b) Hydraulic parameters (Transmissivity of HU2 Miocene Aquifer,) c) Groundwater level data for 1975 d) Groundwater chemistry (Nitrate rate-2008).

The raster maps obtained using the data on groundwater level height are called groundwater contour maps (isopiestic lines). These maps serve to determine how groundwater flow functions, where are the recharge and/or pumping areas in addition to indicating gradient calculations, flow and permeability (Fig. 5C). By crossing groundwater contour maps for different dates, contour descent maps can be obtained for that period. In addition, Variation in Saturated Thickness (VST) can be calculated between those dates.

2.2.4. Hydrochemical characterization

The chemical composition of groundwater is conditioned by a multitude of factors. Among those, the most important are: a) chemical composition and disposition of materials with which the water is in contact, b) time of contact with these materials, c) temperature, d) pressure, e) presence of gases, and f) level of water saturation in relation to distinct incorporated salts (Custodio & Llamas, 1983). Although the composition of groundwater is continually changing, the anthropogenic factors can significantly influence the composition. In fact, changes in land use are considered the most influential factor in groundwater pollution. Ions such as NO_3, SO_4, Na and Cl can come from agricultural fertilizers, livestock waste and waste from industry and urban centers. Nitrate is accepted as the most common contaminant in groundwater (Gulis et al., 2002; Jalali, 2009).

In Europe, the objective is for waterways to achieve "good" chemical and ecological status according to Directive 2006/118/EC of the European Parliament and Commission (DOCE, 2006). This directive describes the protection of groundwater from pollution and deterioration and the establishment of a pollution prevention and reduction plan by 2015. In addition, water bodies should be in good quantitative and qualitative status, especially in reference to nitrate content, which should not exceed 50 mg/l.

Thus, establishing the spatial distribution of NO_3 concentrations in groundwater within the aquifer is of vital importance. To accomplish this, point analyses of groundwater in wells and springs must be performed. Using advanced interpolation capacities provided by GIS tools, a complete geostatistical study can be performed to establish the most contaminated areas in terms of nitrate (Fig. 5D).

3. GIS & groundwater abstractions

Intensive use of groundwater for irrigation in arid and semiarid regions has often been the main driver of socioeconomic development over the past four decades (Shah, 2005). However, poor management of pumped volumes of water has led to negative consequences in terms of quality and quantity of available groundwater resources and associated ecosystems.

Controlling the groundwater withdrawals from a wide area of intensive irrigation is not easy. The largest volume of water used for agriculture has been extracted through tens of thousands of pumping-wells which generally have no measurement system and, in many cases, do not meet legal requirements or are unknown even in their location. Various methods of calculating groundwater abstractions have been known for years, but all of them are very expensive or inaccurate in their application to large areas (Brown et al., 2009).

In this scenario, the data provided by satellites (remote sensing) and the computerized processing of these geo-referenced data (GIS) represent a new approach to monitoring and quantifying groundwater abstractions, with the following characteristics: instantaneous observations are available over large areas, there are several images throughout the year, there is information not visible to the naked eye, data distributed in both space and time is available, the information is not conditioned by the legal or administrative characteristics of the pumping wells, and satellite image acquisition and processing is very low-cost compared to traditional methods (Castaño et al., 2009).

3.1. Theoretical foundations

The methodology for determining groundwater pumping for irrigation follows these steps: First, the irrigated crops are identified and classified by the multitemporal analysis of images obtained by multispectral sensors on satellite platforms, comparing the phenological evolution of the crops with the evolution of the Normalized Difference Vegetation Index (NDVI; González-Piqueras, 2006). Then, the area covered by crops is quantified by introducing the data into a Geographic Information System (GIS) and overlay them with the areas or required limits. Based on the surface area of each crop and the knowledge of their water requirements, the theoretical amount of water needed for those crops to reach the stage of development seen in the images is calculated. When agricultural practices are known, a correction factor is applied to translate the theoretical amount of water applied to each crop. Finally, all the information generated is integrated (spatially and temporally distributed) in a Geographic Information System (see Figure 1) and is used to establish relationships among all elements of the water balance (Brown et al., 2009).

3.1.1. Multitemporal analysis of satellite images and cross with vector cartography

The term "Remote Sensing" has different definitions, but the most commonly used is "a group of techniques that analyze data obtained by multispectral sensors located on airplanes, spatial platforms or satellites." The sensors (on satellites) that observe the surface of the Earth are instruments that register the radiation from Earth and the atmosphere and transform it into a signal that can be managed in analog or digital format (Calera et al., 2006). The sensors do this by detecting the electromagnetic signal from the Earth and the atmosphere of a certain wavelength and converting them into an established physical magnitude. The energy values detected, quantified and coded from the sensors are usually in a two-dimensional number matrix or digital image (raster). Each element of the matrix, called a picture element or a pixel, has a digital value assigned to it (digital levels) which is usually registered in a byte or binary code (2^8 values, from 0 to 255). These represent the energy associated with the wavelength to which the detector is sensitive.

According to Chuvievo (2002), each satellite scene can be used to extract four types of information, each with its respective resolution (Table 1):

1. Spatial, derived from the organization and presence of elements on the surface of Earth in three dimensions,
2. Spectral, dependent upon the observed and measured energy,
3. Temporal, associated with changes over time in a specific spatial location and
4. Radiometry, related with the conversion of voltage collected by the apparatus that receives the signal sent from quantifying entities and later on digital levels.

Landsat Satellite TM sensor				NOAA Satellite AVHRR sensor			
Spatial resolution, pixel size	Spectral resolution	Temporal resolution	Radiometric resolution	Spatial resolution, pixel size	Spectral resolution	Temporal resolution	Radiometric resolution
30 x 30 meters	7 bands	15 days	8 bits	1x1 Km	5 bands	12 hours	10 bits

Table 1. Type of resolution of two kinds of sensors, such as Landsat and NOAA satellite. (see http//:www.landsat.org and www.noaa.gov/).

The information referred from the sensor is treated digitally to obtain a geo-referenced representation of the land. Once the interactions of the atmosphere are removed from it, the radiation values received correspond exactly with those measured on the surface (see more detailed information in Chuvieco (2006) or Calera et al., (2006).

The source of radiant energy is solar radiation on the land surface after moving through the atmosphere. The radiation the sensor obtains is that which emerged from the land surface to the proper region of the spectrum when the emissions due to temperature are considered null. Therefore, the electromagnetic spectrum is the continuous succession of these frequency values (wavelengths). Conceptually it can be divided into bands in which electromagnetic radiation has a similar behaviour (Fig. 6).

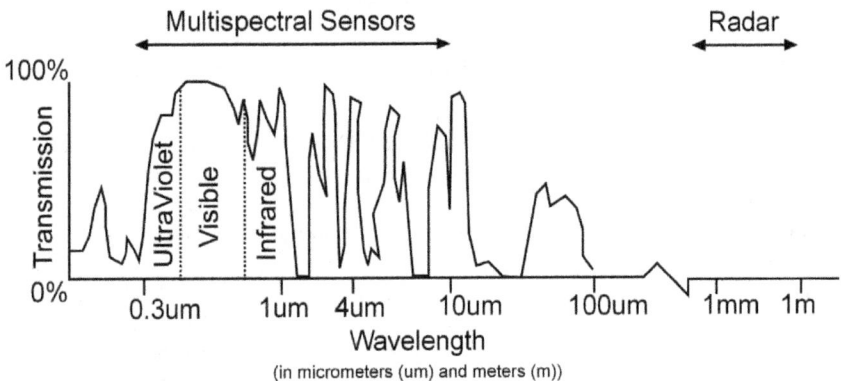

Figure 6. Main features of the electromagnetic spectrum.

Three basic elements can be distinguished as the components comprising all forms of the landscape on the Earth's surface: soil, water and vegetation. The behaviour of these elements in different regions of the electromagnetic spectrum can be observed in Figure 7. The energy emitted (reflectivity) from the ground in the solar spectrum has a uniform response, showing a flat curve and ascending to greater wavelengths. It is important to know that bare soil can present different curves according to the chemical composition, humidity content, organic material content, etc. In the optical spectrum, water can be observed as a strong contrast between the reflectivity of the visible (5%) and the infrared, where water absorbs almost all the radiation in these wavelengths (Fig. 7). This effect is used to separate the water-soil limit.

Similarly to soil, the characteristic reflectivity curve for water can vary as a function of factors such as depth, suspended materials, roughness, etc. (Calera et al., 2006).

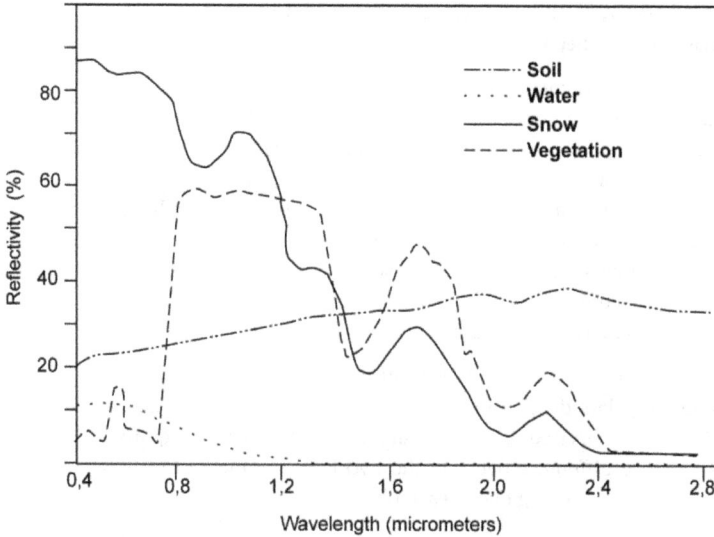

Figure 7. Reflectivity curve characteristic of soil, water and vegetation.

The morphology of the reflectivity curve against the wavelength that vegetation has is well defined (Fig. 7). It has low reflectivity (10%) with a maximum relative to the region of green, high reflectivity in the near infrared which is gradually reduced to the middle of the infrared spectrum. The strong contrast between the reflectivity of red and near infrared indicates that the higher the contrast is, the more vigorous the vegetation is, either due to greater land cover or greater photosynthetic activity (Calera et al., 2006). This spectral behaviour is the foundation for the development of certain indices with an objective to highlighting active vegetation from other components (soil, water, dry vegetation). From the reflectivity of each band (quantitative information distributed and geo-referenced in space) a relationship with the biophysical characteristics can be established (biomass, fraction of plant cover, etc.). This allows for quantitative, spatial-temporal monitoring of the processes on the Earth's surface (Bastiaanssen et al., 2000; Calera et al., 2001; González-Piqueras, 2006). Nonetheless, reflecting the spatial and temporal variability of plant cover is complicated if different spectral bands with the reflectivity values are used. To unify this process, the Vegetation Indices have been developed, one of the most important being the NDVI (Rouse et al., 1973).

The NDVI is:

$$NDVI = \frac{(NIR - R)}{(NIR + R)} \tag{5}$$

where:

NDVI: Normalized Difference Vegetation Index,
NIR: Near Infrared reflectivity (spectrum range in micrometers),
R: reflectivity in red reflectivity.

GIS tools have the capacity to establish dynamic processes if they contain spatially referenced information which is repeated over time in addition to the ability to study spatial changes over the land surface. Mathematical operations can be used between the different sensor bands (digital images) to obtain quantitative information of each satellite scene obtained for a specific date. In this way, a temporal series is available for establishing the progression of a variable, for example NDVI. Multitemporal analysis stems from the availability of a time sequence of images, so these scenes must meet a set of requirements such as geometric coregistration (ability to superimpose images with the highest precision possible and radiometric normalization; Calera et al., 2005).

With this information and digital classification tools each pixel of the image from each date can be assigned a class defined through an automated process. There are two methods for classification: a) supervised and b) unsupervised (does not require intervention of an "interpreter"). The difference between the two is in the method of obtaining the spectral reference classes for assigning one to each pixel.

Supervised classification stems from *a priori* knowledge on specific land uses located in space, which are called training plots. These serve to establish spectral reference classes. There are several methods and a procedure for assigning a class to each pixel, but the most commonly used is an algorithm of maximum probability. Without getting into the details of this method, the algorithm is based on multivariate statistical analysis of components that identify each pixel in terms of their closest resemblance.

Other classification methods that could be used as alternatives or complimentary methods are decision tree (expert systems). These procedures are based on separating the pixel values of a layer into homogeneous groups and subgroups. Another method, called contextual filters, can also be applied. This not only considers the spectral characteristics of an individual pixel, but also considers neighboring pixels (Calera et al., 2006).

Once the classified map has been obtained, the spectral classes can be used to select the classes that are of interest from a hydrogeologic point of view. In our case study, this is crops irrigated with groundwater. Therefore, it is important to know the area of irrigated crops and their spatial distribution. One of the most commonly used GIS techniques for this is overlay vectorial and raster cartography, which is the only way to obtain this information for rasterized areas. There are two types of overlays that depend on the pixel value (ESRI, 1997). When the pixel has a real value (for example a precipitation map, groundwater level, NDVI, etc.) a statistic is calculated for areas by obtaining statistical values from the raster within the selected polygons (mean, minimum, maximum, etc.). The other case is when the attribute of each pixel is a discrete value defining a series of classes (i.e. raster map of the classified land uses). In this occasion (tabulate areas) the result is the surface of each class within the vectorial polygons selected (ESRI, 1997). Once the areas of each irrigated crop have been determined and the amount of irrigation water supply is known, it is possible to calculate the volume of water used to irrigate crops in the area on an annual basis (see summary in Castaño et al., (2009).

3.2. MOS case study

The use of groundwater resources of the MOS above its recharge capacity has led to several quantitative impacts: a steady decline in groundwater level, reduced aquifer discharge to the Júcar River and aquifer pollution. In fact, the quantitative analysis performed on the Júcar River Basin (Estrela et al., 2004) for the European Water Framework Directive (EU, 2000) clearly indicates that the environmental objectives set are not being reached at the present time and there is a certain risk of not meeting them by 2015.

In this situation, quite common in a semi-arid river basin, it is particularly important to precisely quantify the groundwater balance in order to determine aquifer sustainability. The information provided by the multispectral images becomes critical because these data sets are the only consistent and objective information on crops and can replace the data on agricultural statistics. In this regard, the MOS is an ideal case study for testing and validating the adequacy of remote sensing and GIS techniques for calculating groundwater abstractions in agricultural basins in semi-arid climates (Castaño et al., 2009).

Following is a description of several studies in the MOS to classify irrigated crops and quantify the ground water consumption required for ideal phenological development.

3.2.1. Calculation of groundwater withdrawals

The development of a method to calculate groundwater abstractions has been briefly described in the section on the theoretical foundations. In addition to knowing the method, one must have previous knowledge of the study area in order to choose the type of satellite image, for example crops and natural vegetation (phenologic development), soils, climate, relief, etc. In this study, due to the characteristics of the study area, the ideal images for thematic cartography are those from Landsat5 TM and Landsat7 ETM+ (Tables 1 and 2).

	Band 1	Band 2	Band 3	Band 4	Band 5	Band 7
Wavelength (λ μm)	0.45	0.52	0.45	0.52	0.45	0.52
Region	Blue	Green	Red	Near Infrared	Middle Infrared	Far Infrared

Table 2. Bands and wavelengths from the Landsat 5 TM and Landsat 7 ETM+. Information on band 6 from this sensor is omitted because it is not relevant in this study.

Using this information, the number of scenes required can be established as well as the bands to use for differentiating the crops of interest. For example, at least two images are necessary to establish a time series and to identify the non-irrigated crops, one in May or June (maturation process) and another in July (harvest). If more Landsat scenes on specific dates are included spring irrigated crops can be identified. Generally, a minimum of 16 images are used for performing the classification. If the temporal progression of the spectral response is considered a discriminating element for crops (phenologic development), the NDVI spectral band is the most useful. This is obtained by performing mathematical operations with the images for the same dates using bands 3 and 4 of the Landsat sensor (Table 2 and Figure 8).

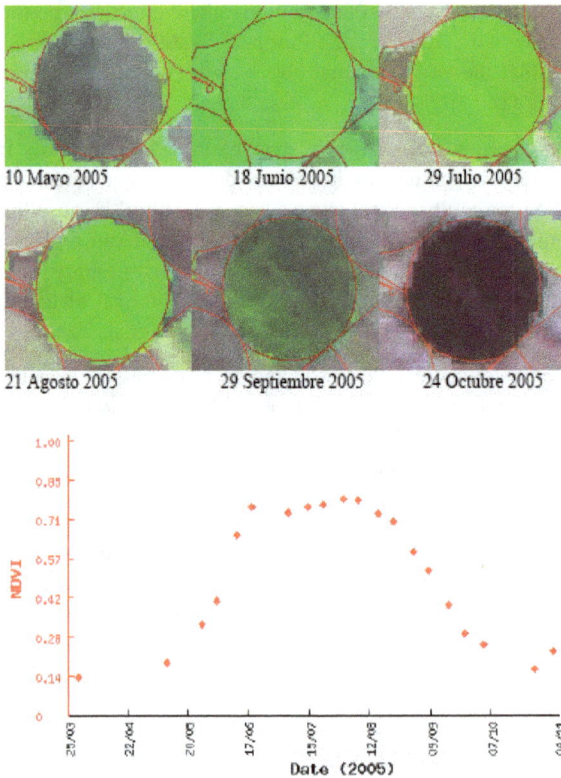

Figure 8. Example of the RGB composition for a maize plot and the temporal progression of NDVI during the crop growth cycle in the year 2005. (From Calera, et al., 2005).

The next step is to choose the classes to use in the classification as a function of those which can be differentiated using the spectral band in the images used such that they meet the objectives of the study (Spring Irrigation, Summer Irrigation, Spring-Summer Irrigation, Alfalfa, Bare soil, Dry farming crops, Shrubs, Forest). For classification, a study is required on the training plots called "true land."

This information is used to perform the classification using the maximum probability algorithm, tree decision and contextual filters. These classifications will be capable of discriminating between the sources of error that can be generated, e.g. in dry farmed crops because several images can have similar spectral responses. For example, in a rainy spring, cereals grown in irrigated or dry farming can be difficult to differentiate. Therefore, using other scenes with tree decision can help in the classification process. Contextual filters can be used to eliminate error on the plot boundaries or isolated pixels that belong to a different class than the rest of the plot. The result is a raster map with the classification of all irrigated crops (Fig. 9).

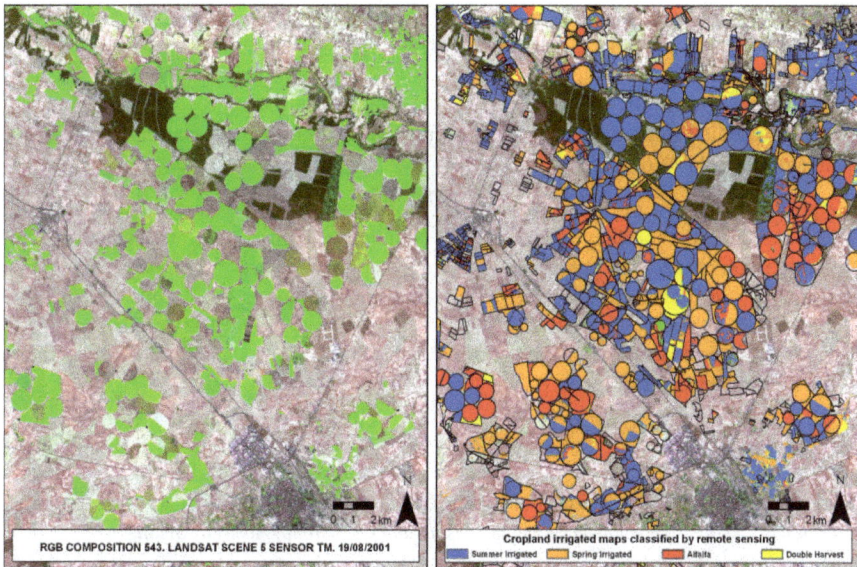

Figure 9. Cropland irrigated with groundwater map, classified by remote sensing techniques for a 2001 in the MOS from LANDSAT-TM scenes.

Once the maps have been classified, the spectral classes can be used to select the classes of interest from a hydrogeologic point of view. The area of irrigated crops must be determined (divided into spring, summer and spring-summer irrigated crops) as well as their spatial distribution (overlay tools). Information on the irrigation needs of the crops present in the MOS are provided weekly through the Irrigation Assessment Service (SAR) by the Agronomic Institute of Technology of the Province of Albacete (ITAP) using the method proposed by Allen et al., (1998). For each agricultural year, the institution groups the irrigation needs for each crop and publishes them in the annual monitoring reports (http://www.itap.es). The theoretical irrigation needs represent the minimum water consumption for sustaining the crops of interest. To determine the true water needs, correction coefficients must be applied to the theoretical irrigation volumes (Castaño et al., 2009). Field work to quantify the agricultural practices applied in the region should be done to perform this calculation. Thus, the irrigation efficiency can be used to calculate the correction coefficient that transforms the theoretic quantity of water necessary into the true values applied to each crop in the area. Generally the true amount of groundwater abstraction is higher than the theoretical irrigation needs.

Therefore, the calculation of water consumption for the different types of irrigated crops by applying the following equation:

$$Vr = A_i \times D \qquad (6)$$

Where:

Vr = the annual volume of irrigation water for each type of crop (m^3).
A = the area of each type of irrigated crop (ha).
D = the irrigation needs for each type of crop, applying the correction coefficient (m^3/ha).
i = Hydrogeologic Domain, Municipality...In this way groundwater consumption could be calculated for the MOS or any part of it.

Estimating the amount of water required for irrigation is critically important in times of water shortage and especially in the current situation of increasing water demands with increasing populations. The use of GIS tools in this endeavor greatly increases the accuracy and efficiency of these types of study. This chapter is meant to be a summary of some methods used in the case study of the Mancha Oriental System, but they can be applied to other systems worldwide at risk of groundwater overexploitation or as a preventative measure to protect natural resources.

Author details

David Sanz, Santiago Castaño and Juan José Gómez-Alday
University of Castilla - La Mancha / Remote Sensing and GIS Group, Albacete, Spain

Acknowledgement

This study was funded by the Spanish Government under research grant CGL2008-06394-C02-02/BTE. Special thanks go to the Jucar Water Authority (CHJ) and Stakeholders (JCRMO) in the Mancha Oriental System for providing information. The findings presented therein belong to the authors. Special thanks go to Dr. A. Moratalla (UCLM) for providing some of the chemistry data and to Dr. A. Calera and Dr. M. Belmonte (UCLM), for providing the multitemporal classification of the crops in the study area. We would also like to thank S.A. Kroll for improving the English text.

4. References

Allen, RG.; Pereira, LS.; Raes, D. & Smith. (1998). Crop evapotranspiration. Guidelines for computing crop water requeriments. *FAO Irrigation and Drainage*. Paper 56. FAO, Rome 300pp.

Bastiaanssen , W.G.M.; Molden, D. & Makin, IW. (2000). Remote sensing for irrigates agriculture: examples from research and possible applications. *Agric Water Manage* 46: 137-155.

Calera, A.; Jochum, AM.; Cuesta, A.; Montoro, A. & López, P. (2005). Irrigation management from space: Towards user-friendly products. *Irrig Drain*. 19. p. 337-353.

Calera, A.; Martínez, C. & Meliá, J. (2001). A procedure for obtaining green plant cover. Its relation with NDVI in a case study for barley. *Int J Remote Sens*. 22. p. 3357-3362.

Calera, A; Castaño, S. y Quintanilla, A. (2006): Historia y visión general de la Teledetección. In La evaluación del impacto ambiental de Proyectos y Actividades agroforestales. La Información en un SIG. In *La evaluación del impacto ambiental de Proyectos y Actividades agroforestales*. Ed. Ediciones de la Universidad de Castilla-La Mancha. Cuenca. ISBN: 84-8427-416-0.

Cassiraga, E.; (1999). Incorporación de información blanda para la cuantificación de la incertidumbre: aplicación a la hidrogeología. Tesis doctoral, Universidad Politécnica de Valencia.

Castaño, S.; Gómez-Alday, J.J. & Sanz, D. (2008). *Aguas continentales: Gestión de recursos hídricos y calidad del agua. Teledetección y SIG en la gestión de aguas subterráneas*. Centro Superior de Investigaciones Científicas. (CSIC). Madrid. ISBN:978-84-00-08664-0.

Castaño, S.; Sanz, D. & Gómez-Alday, J.J. (2009). Methodology for quantifying groundwater abstractions for agriculture via remote sensing and GIS. *Water Resou Manage* 24:795-814.

Chuvieco, E.; (2006). *Teledetección ambiental*. 2 Ed. Ariel Ciencia, Barcelona. 592p. ISBN: 978-84-34-48072-8

European Committee (2000). *The Water Framework Directive 2000/60/EC of the European Parliament and the Council of establishing a Framework for Community Action in the field of Water Policy*, European Commission, Brussels.

Custodio, E.; & Llamas M.R. (1983). *Hidrología subterránea*. Omega, Barcelona. 2350 pp. ISBN 84-282-0446-2

Environment Systems Research Institute (ESRI) (1997): *Understanding GIS – the ARC/INFO Method*, Self-study Workbook. 4th Ed.

Estrela, T.; (ed). (2004). *Jucar Pilot River Basin. Provisional Article 5 Report pursuant to the Water Framework Directive*. Ministerio de Medio Ambiente, Valencia, Spain.

Goodchild, M.F.; (ed). (1996). *GIS and environmental modeling: progress and research issues*. GIS World Books, Fort Collins, CO.

Gulis, G.; Czompolyova, M. & Cerhan, J.R. (2002). An ecologic study of nitrate in municipal drinking water and cancer incidence in Trnava District, Slovakia. *Environmental Research*; 88(3):182-187.

Jalali, M.; (2009). Geochemistry characterization of groundwater in an agricultural area of Razan, Hamadan, Iran. *Environmental Geology*; 56:1479-1488.

Johnston, K.; Ver Hoef, J.M.; Krivoruchko, K. & Lucas, N. (2001). *Using Arc-Gis Geostatylical Analyst*. Ed. ESRI. USA, 300 p.

Mace, R.E.; (2000). Estimating transmissivity using specific-capacity data, The University of Texas at Austin, *Bureau of Economic Geology, Geological Circular* 01-2

Moratalla, A.; Gómez-Alday, JJ.; De las Heras, J.; Sanz, D. & Castaño, S. (2008) Nitrate in the water-supply wells in the Mancha Oriental Hydrogeological System (SE Spain). *Water Resour Manag* 29:1621-1640.

Murray, K.E.; & Hudson, M.R. (2002). Three-Dimensional Geologic Framework Modeling for a Karst Region in the Buffalo National River, Arkansas. U.S. Geological Survey Karst Interest Group. *Proceedings, Shepherdstown, West Virginia. Eve L. Kuniansky, editor. Water-Resources Investigations Report 02-4174*

González-Piqueras, J.; (2006). *Evapotranspiración de la cubierta vegetal mediante la determ,inación del coeficiente de cultivo por teledetección. Extensión a escala regional: Acuífero Mancha Oriental.* Dissertation, University of Valencia.

Peña Llopis, J.; (2006). *Sistemas de Información Geográfica aplicados a la gestión del territorio.* Ed. Club Universitario. Alicante, 310 p. ISBN 84-8454-493-1.

Samper, F.J.; & Carrera, J. (1996). *Geoestadística: aplicaciones a la hidrogeología subterránea.* 2ª Ed. Mundi-Prensa. Barcelona, 484 p. ISBN 84-404-6045-7.

Sanz, D.; Gómez-Alday, J.J.; Castaño, S.; Moratalla, A.; De las Heras, J. & Martínez Alfaro P.M. (2009). Hydrostratigraphic framework and hydrogeological behaviour of the Mancha Oriental System (SE Spain). *Hydrogeol J* 17:1375-1391

Shah, T.; (2005). Groundwater and Human Development: Challenges and Opportunities in Livelihood and Environment, *Water Science and Technology* 8:27-37

Monitoring Land Suitability for Mixed Livestock Grazing Using Geographic Information System (GIS)

Fazel Amiri, Abdul Rashid B. Mohamed Shariff and Taybeh Tabatabaie

Additional information is available at the end of the chapter

1. Introduction

Combining land and land use in a land evaluation procedure defines land suitability, which is the fitness of a land unit for a land use type assessed by comparing land use requirements of each land utilization type with the land (FAO, 1976; 2007). Land suitability analysis is an important tool in making locational and sitting decisions in planning studies. Broadly defined, land-use suitability analysis aims at identifying the most appropriate spatial pattern of future land use according to specified requirements, preferences, and predictors of specific activities (Collins, Steiner, Rushman, 2001; Hopkins, 1977).

In Ghara-Aghch region, center Iran, the need for rangeland suitability evaluation is due to increasing livestock population, which causes an increased demand for forage. In This area, livestock and pasture is a very important business for the community to sustain living. Livestock (Sheep and goats) are fed from pasture. Land management, therefore, is a real issue that requires proper attention from the authorities to ensure sustainability of the rangeland sector in the state. The regeneration rate of rangeland resources is very slow, so it is not able to cope with the ever increasing livestock population growth; hence this imbalance situation leads to regional economic development problems. Proper evaluation based on land planning is essential to solve this problems (Sonneveld, Hack-ten Broeke, van Diepen, Boogaard, 2010).

Definition of the term "mixed livestock grazing" was first used in rangeland forage grazing of livestock by Cook (1954) and subsequently by Smith (1965). They defined the term "mixed livestock grazing" as the use of a pasture's forage for more than one variety of livestock (cattle, sheep and goats) with the aim of achieving maximum productivity. Holechck et al. (1995) explained the rationale for stability improvement of rangelands against the mixed livestock grazing as follows: better distribution of livestock in the pasture, harvesting more than one plant species, and more uniform use of pasture lands. In terms of the economy of the rangeland, "mixed livestock grazing" can be studied in three aspects: firstly, with 'mixed livestock grazing' there is an increase in livestock products and the income will increase; secondly, the risk hazard will decrease; and thirdly, the invading species will be controlled. On the other hand, with "mixed livestock grazing" the preservation costs will increase, and rangeland management becomes more difficult (Coffey, 2001). Heady (1975) reported that with mixed livestock grazing, the efficiency of forage use will increase due to combined use of the grasses, forbs, and shrubs. However, Smith (1965) observed that topography, water resources, and priority of management goals are among the factors determining the success or failure of management of "mixed livestock grazing" rangelands. Coffey (2001) noted that selective grazing species by the livestock in 'mixed livestock grazing' is very important. The cattle prefer grasses to the forbs and shrubs, while the sheep prefer the forbs to the grasses and the goats prefer the shrubs and small branches compared to the grasses and forbs. Therefore, the common grazing of cattle, sheep and goats on rangelands results in all vegetation being grazed and as a result the woody plants and shrubs which form a large part of the rangeland will be grazed in large quantities with common grazing. Luginbuhl et al. (2000) observed that by adding goats to a pasture being grazed by cattle showed a decrease in shrubs and provision of sufficient time for regeneration of the grasses. In fact, by adding the goats to the pasture grazed by the cattle controlled woody plants without influencing the cattle's grazing preference, and thus grazing capacity was increased with a rise in income. Adding the sheep in a pasture which is being grazed by the cattle showed similar results, although sheep in comparison with goats consume fewer woody species; however, the sheep can be used to control the woody species with suitable grazing pressure and thus cause an improvement in the rangeland. Several studies have reported model suitability of the rangelands for livestock grazing (Alizadeh, Arzani, Azarnivan, Mohajeri, Kaboli, 2011; Amiri, 2009a; b; Arzani, Jangjo, Shams, Mohtashamnia, Fashami, Ahmadi, Jafari, Darvishsefat, Shahriary, 2006; Bizuwerk, Peden, Taddese, Getahun, 2005; Gavili, Ghasriani, Arzani, Vahabi, Amiri, 2011; Javadi, Arzani, Farahpour, Zahedi, 2008; Thornton, Herrero, 2001). The allocation of limited rangeland resources to various land uses, lack of sufficient environmental policies for sustainable use of rangelands as well as degradation of these areas have caused increasing concern among managers and revealed the importance of land suitability analysis. However, no research has been reported on the mixed livestock grazing of sheep and goats. Therefore, the objectives of this study, while recognizing important factors affecting model suitability for 'mixed livestock grazing' of the rangelands, was also designed to determine the kind and rate of the limitations and factors reducing the suitability for an adequate plan for grazing.

As complexity of decisions increases, manual processes become time consuming and are liable to errors, resource managers may increasingly lack the necessary expertise, and, therefore, capacity to make resource management decisions that integrates the range of issues involved. One of the reasons is that the decision may be based on very little information. Other reasons may be the lack of module with flexible user interface (Barbari, Conti, Koostra, Masi, Guerri, Workman, 2006).

A number of technological developments have facilitated the implementation of land evaluation principles and models. In order to incorporate the different land attributes that differ spatially and to identify the best suitable land use, GIS has proved to be the best tool (Bizuwork, Taddese, Peden, Jobre, Getahun, 2006). The powerful query, analysis and integration mechanism of GIS makes it an ideal scientific tool to analyze data for land use planning. Management of natural resources based on their potential and limitation is essential for development of rangeland on a sustainable basis. GIS technology is being increasingly employed by different users to create resource database and to arrive at appropriate solutions/strategies for sustainable development of rangelands (Venkataratnam, 2002). Today, GIS is a tool that can assist a community to plan and to support the information management during the rangeland production process, while at the same time ensures the proper balance between competing resource values. It can enhance the accessibility and flexibility of information and can improve the linkages and understanding of relationship between different types of information (Baniya, 2008).

2. Methodology

2.1. Study area

The study area is located in the Ghara-Aghch catchment in Isfahan province (10 kilometers northeast of Semirom), in the central part of Iran. The area under study (51º, 34′, 54″ to 51º, 45′, 53″ E and 31º, 26′, 19″ to 31º, 03′, 28″ N) comprises of 8962.25-hectares of which 79.9% is rangeland (Figure.1). The climate is semi-arid with an average annual rainfall of 358 mm yr⁻¹, falling mainly in the autumn and winter. The average minimum and maximum temperatures are 3.1 and 16.7 ºC, respectively. The Mean annual temperature is about 10 degrees Celsius) and the climate based on the classification using the Dumbarton method is semi-arid. Sheep and goats were the two main sources of animal production. In Ghara-Aghch, the rangeland area is negatively affected by inappropriate land management practices, e.g. over utilization. Uncontrolled utilization of the vegetation of the rangelands affects forage quality because of the transition from a plant community with a higher nutritive value to one with lower nutritional value.

2.2. Vegetation type

Site evaluation and data collection was carried out during the spring until autumn of 2010. Vegetation segments, pasture boundaries, agricultural lands, fruit gardens, urban settlements, bare soils, out cropped rocks and stony areas were mapped in field studies using 1:50,000 scale map and aerial photographs. Preliminary vegetation types were distinguished with the physiognomic-floristic-ecologic method. Meanwhile with the determination of the pasture types, the boundaries were checked on the map according to the features of vegetation entities and dominant species. In this study, a visual scoring method of the available dominant species was used to report the vegetation cover map, botanical composition, and forage production in 17 vegetation types (Figure2; Table1). Overstocking and extended grazing periods are characteristics of inappropriate

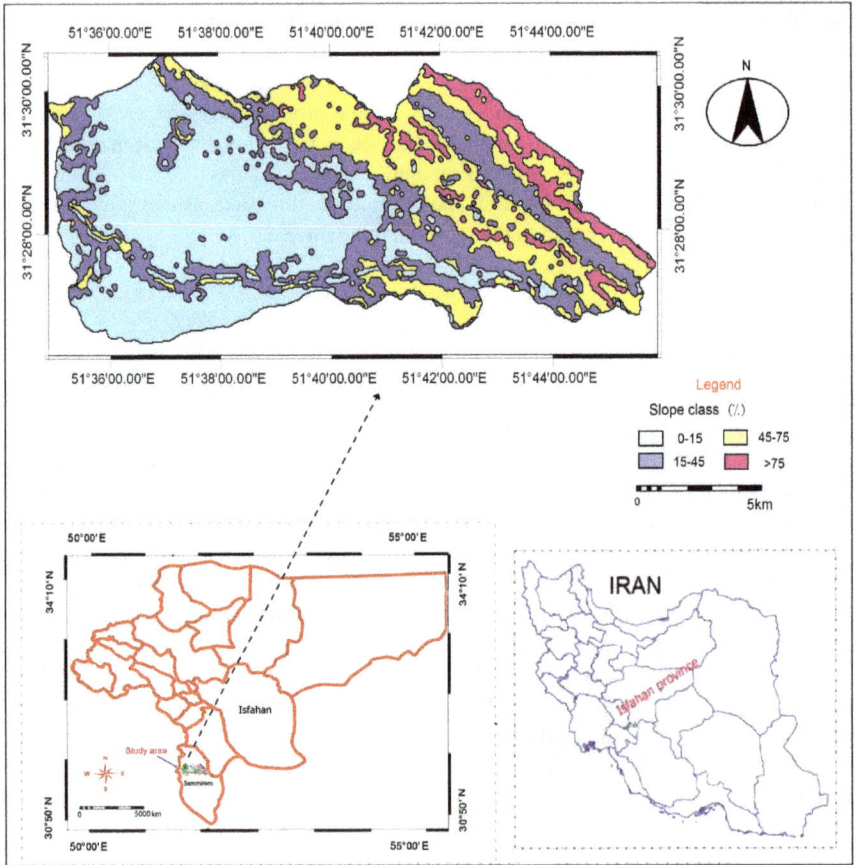

Figure 1. Location of study area within the Ghara-Aghch District

management practices in the study area. In this study, 182 plant species in ten major vegetation types were identified in the rangelands in Ghara-Aghch showing negative and poor trends and conditions.

No.	Abbreviations	Vegetation type	Area (ha)
1	Ag.tr	*Agropyron trichophoum*	122.77
2	Ag.tr-As.pa	*Agropyron trichophoum-Astragalus parroaianus*	305.59
3	Ag.tr-As.ca-Da.mu	*Agropyron trichophoum- Astragalus canesens- Daphne macronata*	898.36
4	As.ad-Ag.tr-Da.mu	*Astragalus adsendence-Agropyron trichophoum-Daphne macronata*	385.59
5	As.pa-Ag.tr	*Astragalus parroaianus-Agropyron trichophoum*	162.77
6	As.ly-Ag.tr-Da.mu	*Astragalus lycioides-Agropyron trichophoum-Daphne macronata*	237.51
7	As.ca-Br.to-Co.cyl	*Astragalus canesens-Bromus tomentellus-Cousinia cylianderica*	2029.68
8	As.br-Br.to-Da.mu	*Astragalus brachycalyx-Bromus tomentellus-Daphne macronata*	116.2
9	As.go-Co.cyl	*Astragalus gossipianus-Cousinia cylanderica*	362.66
10	As.pa-Co.cyl-Da.mu	*Astragalus parroaianus-Cousinia cylanderica-Daphne macronata*	-
11	As.cy-Fe.ov	*Astragalus cyclophylus-Ferula ovina*	105.7
12	Br.to-As.pa	*Bromus tomentellus-Astragalus parroaianus*	373.11
13	Co.ba-As.go	*Cousinia bachtiarica-Astragalus gossipianus*	188.52
14	Co.ba-Sc.or	*Cousinia bachtiarica-Scariola orientalis*	499.07
15	Fe.ov-Br.to-As.za	*Ferula ovina-Bromus tomentellus-Astragalus zagrosicus*	212.33
16	Ho.vi-Po.bu	*Hordeum bulbosum-Poa bulbosa*	36.76
17	Br.to-Sc.or	*Bromus tomentellus-Scariola orientalis*	153.58
Total rangeland area			7158.81

Table 1. Vegetation communities in Ghara-Aghch rangelands

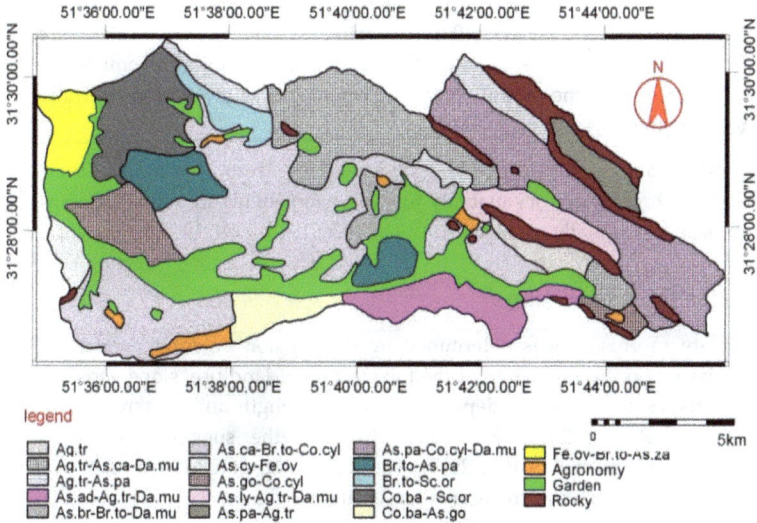

Figure 2. Vegetation type (VT) of Ghara-Aghch rangelands

2.3. Factors of livestock model

The livestock grazing model suitability comprises of three measures: the capacity and production of forage, the soil sensitivity to erosion, and physical factors (water resources and slope). The components of the suitability model for livestock grazing are illustrated in Figurer 3.

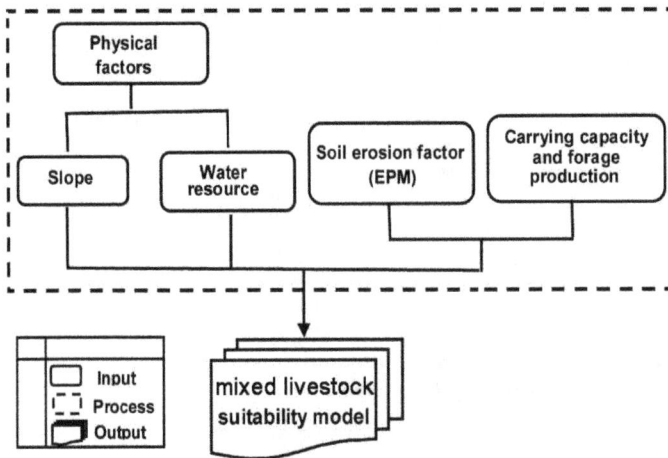

Figure 3. Components of mixed livestock grazing suitability model

The method introduced by FAO (1991) for range suitability classification used ILWIS version 3.6 as the GIS Software. Land evaluation normally requires a comparison between the inputs required and the outputs obtained when each relevant land utilization type is applied to each land unit.

Two orders of range suitability for livestock grazing were considered: suitable (S) or unsuitable (N). Three classes of suitability were determined: highly suitable (S1), moderately suitable (S2), and marginally suitable (S3) (FAO, 1976; 1983; 1984; 1985; 1991; 2002; 2007).

2.4. Soil sensitivity to erosion

Soil sensitivity to erosion was determined by the Erosion Potential Model (EPM). This model was based on the evaluation of the four factors of land use, slope, erosion potential, soil characteristics, and geology, depending on the strength and weakness of each factor (Ahmadi, 2004; Rafahi, 2004). Figurer 4 illustrates the suggested factors and their relationships in this model (Amiri, 2010). The slope map and EPM model were used to calculate erosion potential and create erosion sensitivity classes.

According to this model:

$$Z = Y.Xa \, (\Psi + I \, 0.5) \tag{1}$$

where, Z is the erosion severity index, Y is the sensitivity of soil and bedrock to erosion, Xa Is the land use index, Ψ is the erosion index of the watershed, and I is the average gradient of the slope (Amiri, 2010).

Soil depth, type, texture, gravels, structure, rocky outcrops, and groundwater were the characteristics used to categorize each group (Figure 4). Sensitivity to erosion in the sub-model for each vegetation type was created and classified by integrating range condition, land use, slope, erosion potential, soil characteristics, and geology (Table 2). The lower erosion class was placed in suitability category S1, low and medium erosion class in S2 suitability category and high and very high erosion classes were placed in the suitability categories of S3 and N respectively (Table 2).

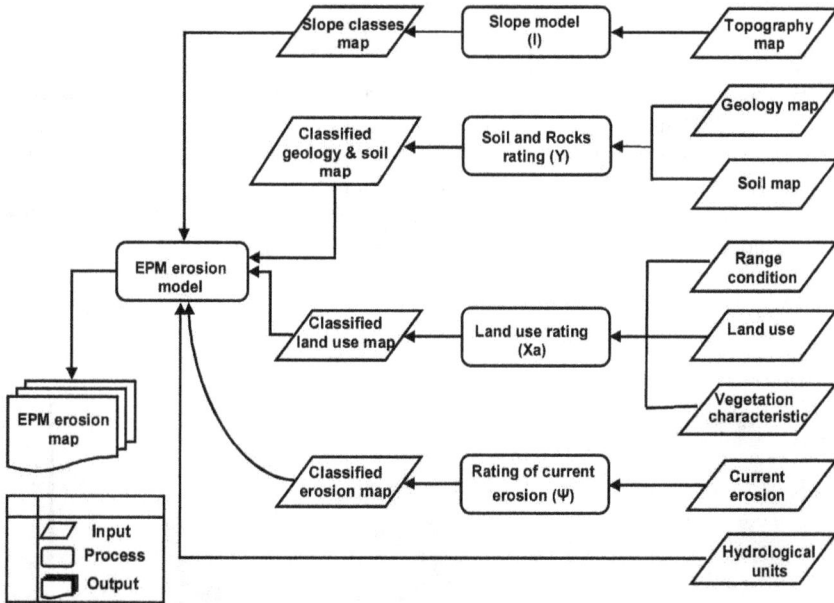

Figure 4. EPM model for soil erosion

Symbol	Range of Z	classes	Suitability classes
1	< 0.2	Low	S1
2	0.2-0.7	Medium	S2
3	0.7-1	High	S3
4	>1	Very High	N

Table 2. Classes of sensitivity to erosion (Amiri, 2009a)

2.5. Grazing capacity and forage production factors

In terms of data relevant to the conditions of livestock breeding, the percentile herd combination in each Samman unit [In Iranian rangelands, the livestock can only use water in Samman unit] was used to determine the livestock grazing capacity (Amiri, 2009b). First the Samman unit plan was adopted with the vegetation types of the region so that the percentile of the herd combination in the vegetation types located in the boundaries of each Samman unit could be determined. The vegetation parameters recorded in April-May and May-June 2009 and April-May and May-June 2010 were used in the study.

The grazing capacity and the suitability of the forage production in vegetation types was first determined. The existing plant species in the vegetation types were listed and the percentage of canopy cover of each variety was determined separately based on the percentage derived from total plots sampled. The production of entire plant varieties edible to sheep and goats were separately determined by cutting and weighing of samples in each plot at the end of the active growth period (Milner, Hughes, Gimingham, Miller, Slatyer, 1968). Samples were taken at random in the 10 vegetation types (determined via floristic-physiognomic method) within one-square-meter plots with three 200-meter transects. Based on field visits and interviews with experts from the Natural Resources Institute (NRI) the palatability classes of the species separately for sheep and goats were classified into one of the three palatability classes (I, II, and III) and the proper use factors (PUF) in vegetation types were determined based on the soil sensitivity to erosion suitability class adapted from the EPM model with respect to the range conditions and range trends in vegetation types (Table 3).

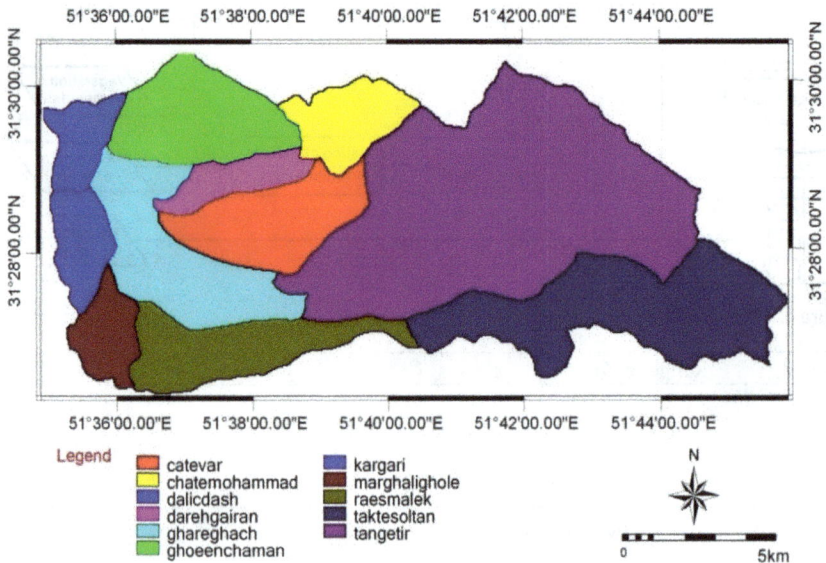

Figure 5. Utilization units in Ghara-Aghch rangeland

Then the available forage of the existing varieties in the vegetation types for sheep and goats in livestock use was calculated from the product of palatability or Proper Use Factor (PUF) (each one, which is lesser) of each variety and herd combination percentage (sheep and goats) and by adding up the available forage products of all varieties of a type (Smith, 1965).

Figurer 6 illustrates the components of capacity and suitability of forage production for livestock use. The diagram derived from the livestock grazing capacity model will then be applied in the next stage as input for the water resource model.

Soil Erosion sensitivity (SE)	Range condition (RC)	Range trend (RT)	Proper Use Factor (PUF)
Low and Medium (S1 or S2)	Good or Excellent	Up or Static	50
Low and Medium (S1 or S2)	Good or Excellent	Down	40
Low (S1)	Fair	Up or Static	40
Medium (S2)	Fair	Up or Static	35
Medium (S2)	Fair	Down	30
High (S3)	Fair	Up or Static	30
High (S3)	Fair	Down	25
Medium (S2)	Poor	Up or Static	30
Medium (S2)	Poor	Down	25
High (S3)	Poor	Up or Static	25
High (S3)	Poor	Down	20

Table 3. Palatability coefficients and proper use factor rates used in the calculation of available forage [When the erosion suitability class is S3 and the pasture is in a poor condition and the tendency is negative, the allowed exploitation limit for goats is considered zero and the production suitability class is considered N (unsuitable)] (Amiri, 2009a)

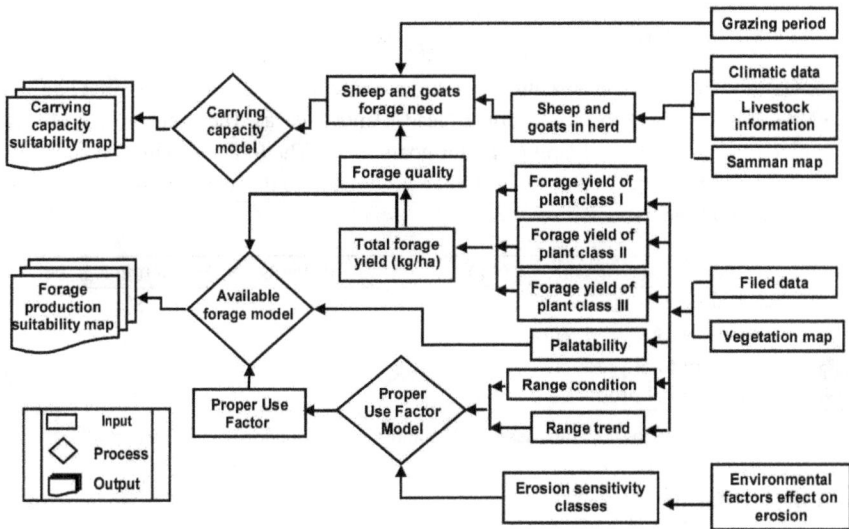

Figure 6. Components of carrying capacity and suitability of forage production in livestock use model

As illustrated in Figurer 6 the components of the livestock grazing capacity model comprises of four sub-models which include the amount of available forage for the sheep and goats, gazing period, forage needed for livestock use, and the area of the vegetation types (ha). In order to create the available forage, the relevant information was integrated for each vegetation type using Equation 2:

$$DLNN = GP + T + FQ \tag{2}$$

where, DLNN = Daily Livestock Nutrition Need, GP = Grazing Period, T = Topography, and FQ = Forage Quality (Amiri, 2010a).

The average daily requirement of a 50 kg sheep and 37 kg goat consuming quality forage was determined as 1.35 kg dry matter. Available forage (AF, kg/day) for livestock was calculated as:

$$AF = \Sigma(Y + (P/PUF)) \tag{3}$$

where; Y= yield (kg/ha), P = palatability, and PUF = proper use factor (Guo, Liang, Liu, Niu, 2006); while PUF was determined by combining information on trends in range condition and erosion sensitivity (Amiri, Shariff, 2011).

The livestock grazing capacity model as described earlier comprises of four sub-models which include the amount of available forage for the sheep and goats, gazing period, forage needed for the sheep and goats, and the area of the vegetation types. The grazing capacity was calculated using Equation 4 (Guo, Liang, Liu, Niu, 2006).

$$GC = \frac{AF}{DLNN} \tag{4}$$

where GC id the for grazing capacity, AF is the available forage (Kg/ha) in the area (ha), and DLNN is the Daily Livestock Nutritional Need (Amiri, 2009a). The number of goats was determined using the Animal Unit (A.U) for goats as 0.8. The forage production suitability class, based on the ratio of the available forage production to the total products of that type was determined from Table 4.

State	Available forage production (AF) *	Production classes
1	%40 (of total production)	S1
2	%30-40 (of total production)	S2
3	%20-30 (of total production)	S3
4	< %20 (of total production)	N

Table 4. Forage production suitability classes [* Minimum production lower then 100 (kg/h)]

2.6. Physical factors

The suitability class of this model was determined via the combination of the two measures of slope and water resources.

2.7. Slope

The slope suitability categories in livestock use were determined from the slope suitability classes (Table 5).

Slope (%)	0-10	10-30	30-60	60<
Suitability classes	S1	S2	S3	N

Table 5. Slope suitability classes (Neameh, 2003)

2.8. Water resources

The suitability categories of this model were determined via the combination of three sub-models of quality, quantity and distance from water sources (Figure 7). The distance from water sources suitability classes in livestock use are illustrated in Table 6 (Figure 8).

Slope class (%) / Suitability class	0-10	10-30	30-60	>60
S1	0-3400	0-3000	0-1000	N
S2	3400-5000	3000-4800	1000-3600	N
S3	5000-6400	4800-6000	3600-4100	N
N	>6400	>6000	>4100	N

Table 6. Water resource distance and its suitability classes

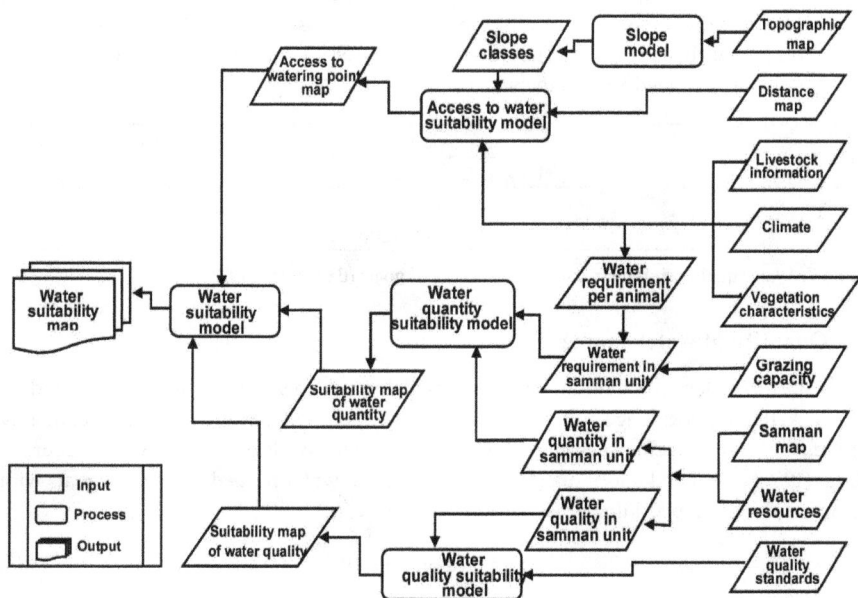

Figure 7. Model for classification of water resource suitability (Amiri, 2009a; b)

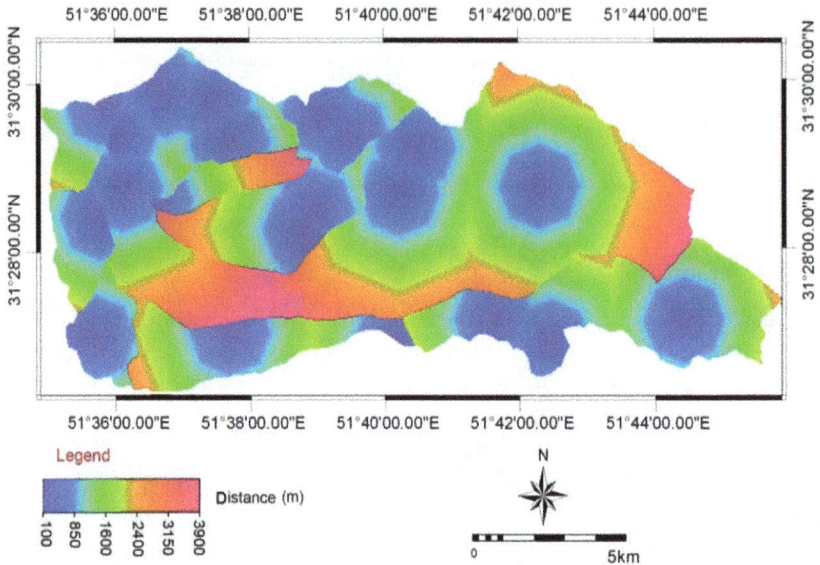

Figure 8. Water resource distance in Ghara-Aghch rangeland

Khan and Ghosh (1982) studied tolerance of goats and sheep to saltiness under difficult environmental conditions in the Rajasthan desert, and observed that the tolerance of the goats was higher than the sheep. The water suitability precincts of the region were classified based on Total Dissolved Salts in the water (TDS) (Table 7).

Suitability class	Total Dissolved Salts (TDS; ppm)			
	S1	S2	S3	N
Sheep	<3000	3000-6000	6000-10000	>10000
goats	<3000	5000-7000	7000-10000	>10000

Table 7. Water quality suitability classes for sheep and goats (Bagley et al 1997)

2.9. Quantity of water sources

Many factors affect the amount of water used by livestock which include the kind of livestock, the livestock's age and breed, the regions topography, available forage and quality of the forage, the grazing season, the quantity, quality and distance from water resources. King (1983) developed a formula (5) for the amount of water needed by African goats with an average weight of 37 kilograms:

$$a \ l/kg^{0.82}/day =? \ lit /day \tag{5}$$

In this formula 'a' is the coefficient which is to be calculated based on local investigations. The '?' is the amount of water needed by the livestock, and 'kg' is the live weight of the livestock on the kilogram basis. Ferreira et al. (2002) calculated the amount of water needed by Merino breed sheep with an average weight of 50 kilograms via the following formula (6):

$$37 \text{ ml/kg}^{0.82} \tag{6}$$

Therefore, based on all the factors involved in the calculation of the water needed in each Samman unit and information from local pastoral farmers (Formulas 5 and 6), the water needed for a mature sheep (Ghashghaei Turkish breed) was calculated as five liters per day and for a mature goat (Ghashghaei Turkish breed) it was estimated as four liters per day (Amiri, 2009b).

The grazing capacity map of each vegetation type was overlaid with the map of the Samman unit and via weight averaging based on the area of each Samman unit, the quantity of the water resources was determined and the number of permitted livestock (sheep and goats) was calculated for each Samman unit. The suitability categories were then determined by comparison of the available water in each Samman unit with the water needed by the livestock in each Samman unit (Table 8).

Available water in pasture ration to livestock need (%)	>76	51-75	26-50	< 25
Suitability classes	S1	S2	S3	N

Table 8. Water resource suitability classes

3. Results

3.1. Erosion sensitivity model

The erosion sensitivity model on vegetation types showed that about 3.5% of the regions rangeland surface (254.25 hectares) was classified in as erosion Class II (low sedimentation intensity), 64% (4585.98 hectares) as Class III (medium sedimentation intensity), and 32.4% (2318.95 hectares) was classified as erosion Class IV (high sedimentation intensity). Furthermore, the results of suitability categories of soil sensitivity to erosion revealed that 4585.98 hectares (64%) of the rangeland surface was classified in the S2 suitability category and 2572.84 hectares (36%) was placed in the S3 suitability category. The map of suitability categories of the EPM model are shown in Figurer 9.

3.2. Forage production and livestock grazing capacity

The results of the suitability model on forage production of vegetation types in the study area under investigation are illustrated in Table 9. According to the forage production model, none of the vegetation types fall into the S1 suitability category. About 1352.46 hectares (18.89%) of the rangeland fell into the S2 suitability category, around 4837.74 hectares (67.57%) of the rangeland fell in the S3 suitability category on forage production, and finally 968.61 hectares (10.8%) of the rangeland fell in the N suitability category (Figurer 10). The results on the livestock grazing capacity in the study area are shown in Table 10.

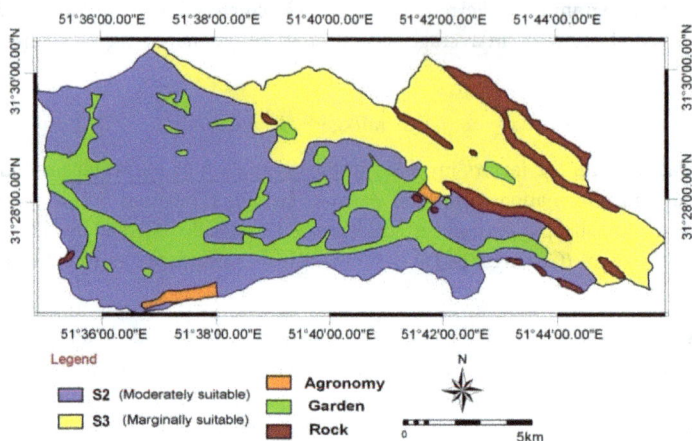

Figure 9. Erosion class properties in Ghara-Aghch

Number	Vegetation type	Available forage based on herd composition (kg/h)		Ratio of available forage to total production	Forage suitability classes
		sheep	goats		
1	Ag.tr	88.9	31.8	31.7	S2
2	Ag.tr-As.pa	60.7	34.3	27.6	S3
3	Ag.tr-As.ca-Da.mu	66.6	23	27.7	S3
4	As.ad-Ag.tr-Da.mu	66.7	24.2	30.8	S2
5	As.pa-Ag.tr	58.2	22.4	25.8	S3
6	As.ly-Ag.tr-Da.mu	60	20.8	28.2	S3
7	As.ca-Br.to-Co.cyl	42.1	16.8	25.1	S3
8	As.br-Br.to-Da.mu	67.8	23.2	32.3	S2
9	As.go-Co.cyl	35.2	24.2	23.1	S3
10	As.pa-Co.cyl-Da.mu1	-	-	-	N
11	As.cy-Fe.ov	56.1	31.2	30.01	S2
12	Br.to-As.pa	58.5	19.8	30.2	S2
13	Co.ba-As.go	44.2	23.8	27.5	S3
14	Co.ba-Sc.or	33.9	19.5	23.3	S3
15	Fe.ov-Br.to-As.za	81.3	30.2	33.4	S2
16	Ho.vi-Po.bu	152	53.8	31.8	S2
17	Br.to-Sc.or	53.6	29.8	28.4	S3

Table 9. Suitability classes based on forage production and available forage in Ghareh Aghach [Since this type is classified in the S3 erosion suitability class and is of a poor condition with a downward trend it is unsuitable for livestock grazing]

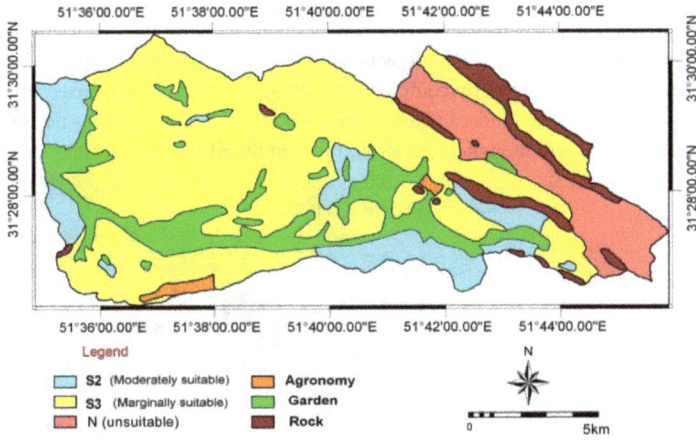

Figure 10. Suitability map of forage production at Ghara-Aghch

Number	Vegetation type	Available forage (kg/h)		Area (h)	Daily forage need (kg)		gazing period (day)	Livestock grazing capacity (on AUM)	Livestock grazing (number of head)	
		sheep	goats		sheep	[2]goats			sheep	goats
1	Ag.tr	88.9	31.8	122.77	1.36	1.165	120	95	67	35
2	Ag.tr-As.pa	60.7	34.3	305.59	1.38	1.179	120	186	112	93
3	Ag.tr-As.ca-Da.mu	66.6	23	898.36	1.4	1.19	120	500	356	180
4	As.ad-Ag.tr-Da.mu	66.7	24.2	385.59	1.41	1.2	120	217	152	81
5	As.pa-Ag.tr	58.2	22.4	162.77	1.44	1.234	120	79	55	30
6	As.ly-Ag.tr-Da.mu	60	20.8	237.51	1.4	1.194	120	119	85	42
7	As.ca-Br.to-Co.cyl	42.1	16.8	2029.68	1.49	1.272	120	700	477	279
8	As.br-Br.to-Da.mu	67.8	23.2	116.2	1.42	1.217	120	64	46	23
9	As.go-Co.cyl	35.2	24.2	362.66	1.54	1.315	120	125	69	70
10	[1]As.pa-Co.cyl-Da.mu	-	-	-	-	-	120	-	-	-
11	As.cy-Fe.ov	56.1	31.2	105.7	1.43	1.222	120	57	34	29
12	Br.to-As.pa	58.5	19.8	373.11	1.55	1.322	120	158	113	56
13	Co.ba-As.go	44.2	23.8	188.52	1.52	1.3	120	75	46	36
14	Co.ba-Sc.or	33.9	19.5	499.07	1.63	1.397	120	145	87	73
15	Fe.ov-Br.to-As.za	81.3	30.2	212.33	1.37	1.168	120	151	105	57
16	Ho.vi-Po.bu	152	53.8	36.76	1.52	1.3	120	44	31	16
17	Br.to-Sc.or	53.6	29.8	153.58	1.39	1.19	120	82	50	40

Table 10. Livestock grazing capacity of vegetation types [[1]This type is unsuitable for livestock graze. [2] The weight of the sheep (a livestock unit) was 50 kilograms and the average weight of the goats was 37 kilograms, 50/37 = 0.8, as a result the ratio of each sheep to a goats in the region is 0.8]

3.3. Suitability model of water resource quality

The water resource quality sub-model was determined by examination of the effective factors on the water quality and by comparison with specific standards. Based on the water resources quality sub-model and considering the water quality, there were no limitation in the region in question, and the whole region fell within the S1 suitability category (Table 11).

No.	Water code	Samman name	Q mean	Q max	Q min	pH	Ec/25 c mimhos/Cm	T.D.S mgL⁻¹	Na%	Cl mel⁻¹	Co₃ mel⁻¹	Hco₃ mel⁻¹	So₄⁻² mel⁻¹	Ca²⁺ mel⁻¹	Mg²⁺ mel⁻¹	K⁺ mel⁻¹	S.A.R	Suitability class of water quality
1	A	Takhte soltan	3.79	4.1	3.48	7.8	315	203	3.00	3.51	0.00	193.5	5.25	62.25	2.59	0.38	0.07	S1
2	B	Tange tir	9.59	13.38	5.8	7.2	366	238	2.00	3.54	0.00	244.04	0.48	70.14	6.08	0.39	0.07	S1
3	C	Tange tir	1.08	1.8	0.35	7.2	366	238	2.00	3.54	0.00	244.04	0.48	70.14	6.08	0.39	0.07	S1
4	D	Tange tir	0.79	1.41	0.18	7.2	366	238	2.00	3.54	0.00	244.04	0.48	70.14	6.08	0.39	0.07	S1
5	E	Reismelek	0.27	0.35	0.2	7.8	230	150	12.00	3.54	0.00	134.22	10.08	34.06	2.43	0.39	0.28	S1
6	F	Takhte soltan	1.82	3.54	0.1	7.8	314	204	3.00	3.54	0.00	195.23	5.28	62.12	2.43	0.39	0.07	S1
7	G	Chat mohammad	0.71	1.1	0.32	7.6	387	252	6.9	7.09	0.00	189.12	43.6	54.2	14.6	0.39	0.21	S1
8	H	Chat mohammad	0.95	1.2	0.7	7.6	387	252	6.9	7.09	0.00	189.12	43.6	54.2	14.6	0.39	0.21	S1
9	I	Ghoen chaman	1.12	1.31	0.94	7.3	388	252	7.00	7.09	0.00	189.12	43.7	54.1	14.59	0.39	0.21	S1
10	J	ketivar	0.72	1.3	0.15	7.9	374	343	4.9	7.08	0.00	225.6	10.0	60.12	10.94	4.6	0.39	S1
11	K	ketivar	0.71	1.5	0.5	374	230	150	7.08	0.00	0.00	134.21	10.0	34.0	10.94	4.6	0.39	S1
12	L	Reis malek	0.86	1.2	0.52	7.8	373	343	12.00	3.54	0.00	225.7	10.0	60.12	6.02	0.38	0.28	S1
13	M	ketivar	0.88	1.12	0.65	7.9	313	204	4.9	7.08	0.00	194.8	5.35	61.8	10.94	4.6	0.39	S1
14	N	Takhte soltan	0.38	0.65	0.12	7.8	387	353	2.9	3.51	0.00	188.9	43.5	54.3	2.52	0.37	0.08	S1
15	O	Delig dash	1.03	1.12	0.94	7.3	386	251	7.00	7.0	0.00	189.12	43.2	53.8	14.7	0.36	0.22	S1
16	P	Ghoyein chaman	0.27	0.35	0.2	7.26	387	253	7.01	7.03	0.00	188.9	43.5	54.3	14.49	0.33	0.2	S1
17	Q	Ghoyein chaman	1.15	1.3	1.0	7.31	315	204	7.00	7.00	0.00	195.2	5.28	54.3	14.7	0.36	0.22	S1
18	R	Mergh aligholi	0.75	1.5	0.3	7.8	386	251	3.00	3.54	0.00	189.12	43.2	62.12	3.43	0.39	0.21	S1
19	S	Mergh aligholi	1.53	2.2	0.86	7.8	387	204	3.00	3.54	0.00	195.2	5.28	53.8	3.43	0.39	0.22	S1
20	T	Ghoyein chaman	1.5	2.4	0.6	7.2	315	251	7.01	7.03	0.00	189.12	43.2	53.8	14.49	0.33	0.22	S1
21	U	Ghoyein chaman	0.79	1.4	0.18	7.2	386	251	7.01	7.03	0.00	225.7	43.2	62.12	14.49	0.33	0.39	S1
22	V	Ghare aghach	0.4	0.68	0.12	8.00	374	343	5.00	7.09	0.00	188.9	10.08	60.12	10.94	4.6	0.22	S1
23	W	Khar gari	0.52	0.91	0.14	7.3	387	353	7.00	7.00	0.00	225.7	43.5	54.3	14.7	0.36	0.39	S1
24	X	Dareh jeiran	0.48	0.85	0.11	7.9	373	343	4.9	7.08	0.00	225.7	10.08	60.12	10.94	4.6	0.39	S1

Table 11. Ghareh Aghach water resources quality and quantity

3.4. Suitability model of water resources quantity

The results achieved from the sub-model on water resource's quantity are presented in Table 12. The results of the sub-model, revealed that there were no limitations on the amount of water in the Samman units in question and that all fell into the S1 suitability category.

Samman unit	water content (lit/day)	Carrying capacity in each Samman unit (A head of livestock in 120 day)		Water need (lit/day)		Suitability classes of water quantity	
		sheep	goats	sheep	goats	sheep	goats
Catevar	224640	389	122	1945	488	S1	S1
Chatemohammad	143424	328	166	1640	664	S1	S1
Dalicdash	88992	99	55	495	220	S1	S1
Darehgairan	41472	249	143	1245	572	S1	S1
Ghare-aghach	34560	125	84	625	336	S1	S1
Ghoeenchaman	417312	107	81	535	324	S1	S1
Kargari	44927	27	24	135	96	S1	S1
Marghalighole	196992	326	192	1630	768	S1	S1
Raesmalek	97632	258	152	1290	608	S1	S1
Taktesoltan	517536	91	50	455	200	S1	S1
Tangetir	990144	166	94	830	376	S1	S1
Total	2797632	2165	1163	10825	4652	S1	S1

Table 12. Quantity suitability of water resource in each Samman unit

3.5. Distance from water resources suitability

The results of the sub-modal on the distance from water resources suitability revealed that 6385.17 hectares of the rangeland area (89.2%) fell in the S1 suitability category, 530.04 hectares (7.4%) of the rangeland of the region in question fell into the S2 suitability category, and only 243.6 hectares (3.4%) of the rangeland fell into the unsuitable (N) category; in addition, no rangeland area fell into the S3 suitability category. The final outcome of the model on water resources is illustrated in Table 13. The region in question had no problems regarding the quantity and quality of the water resources; it was only the distance from the resources that mainly determined the suitability of the rangeland with respect to water resources.

	Livestock grazing
Suitability classes	Area (ha)
S1	6,385.17 (89.2%)
S2	530.04 (7.4%)
S3	-
N	243.6 (3.4%)
Total land area in study	7,159 ha

Table 13. Categorization of land area into suitability classes based on water resources model

3.6. Final application for livestock grazing

The final outcome of the suitability model for livestock grazing was derived from the combination of three suitability sub-models involving soil sensitivity to erosion, forage production suitability, and the suitability of the water resources (Table 14).

Sub-model	S1	S2	S3	N
Erosion	0	4,078 (57.0%)	386 (5.4%)	2,696 (37.6%)
Water Resources	0	4,519 (77.1%)	859 (12.0%)	478 (6.7%)
Forage production	0	979 (13.7%)	5,211 (72.8%)	969 (13.5%)
Integrated model	0	1,126 (15.7%)	4,918 (68.7%)	1,116 (15.6%)
Total land area in study = 7,159 ha				

Table 14. Model-based categorization of land area into suitability classes

4. Discussion

Iran is the second largest country in the Middle East, but has limited natural resources such as fertile soil and water, resulting in limited opportunities to expand and/or intensify arable farming. Extensive animal husbandry, on the other hand, including nomadic, transhumant and sedentary forms, is widespread over the rangelands of the country. Rangelands and animal husbandry have been important in Iran for a very long time, as witnessed by the teachings of Zoroaster. More recently, many people have died in the defence of their rangelands after land nationalization, when only the right of use was at stake. The degree of importance attached to a specific rangeland area reflects its productivity, land scarcity and the availability of alternative sources of income. In Iran, as in most parts of the world, animal husbandry is the most productive use of semi-arid zones bordering the desert. However, overgrazing is a major problem in most of these areas. Therefore, the objectives of this paper, was to apply the concept of range inventory in the recognition and evaluation of potential and actual production for optimal utilization of this valuable natural resource for domestic livestock production.

The degree of importance attached to a specific rangeland area reflects its productivity, land scarcity and the availability of alternative sources of income. In Iran, as in other parts of the world, animal husbandry is the most productive use of semi-arid zones bordering deserts (Breman, De Wit, 1983; Reed, Bert, 1995). Farahpour et al. (2004) had estimated that 80 to 90% of the livestock production in Iran, equivalent to 168,000 - 180,000 ton y^{-1} of meat, was associated with the rangelands. Annual dry matter production of rangelands was estimated at more than ten million tons per hectare. In addition to forage production, mining, fuel wood collection, industrial use of rangeland, e.g. as source of medicinal plants and recreation are other rural enterprises in the rangelands of Iran.

Several researchers have reported that with livestock use and increasing grazing evenness (Forbes, Hodgson, 1985), will in the long term result in increased grazing capacity and livestock production (Abaye, Allen, Fontenot, 1994; Meyer, Harvey, 1985; Pringle,

Landsberg, 2004), as well as increase plant diversity and income of the livestock-farmers. Livestock use can also enhance the consumption of poisonous and invasive plants by the livestock which are not sensitive to these species, and thus increase livestock production. For example, the leaves of plants such as Spurge and Larkspur are poisonous to cattle, but safe for sheep. Thus, sheep grazing will indirectly protect the cattle on the rangeland (Taylor, Ralphs, 1992). In the livestock use model adapted in this study no area fell in the S_1 suitability category (limitless), while the main part of the rangeland area (75.9%) fell within the S_3 suitability category (with high limitation). Among all the factors considered in the lands surveyed, factors related to the vegetation and forage production were the most significant in decreasing the region's rangeland livestock use suitability.

In adapting the grazing suitability model for the rangeland due consideration was given to the climatic conditions, vegetation, soil, the status of the current utilization, and topography, and the factors were found to be effective to different degrees. Therefore, recognizing the factors effective in the model and determining the amount of limitations they impose was important for analyzing and assessing the rangeland. Arzani et al. (2006), Amiri (2009a) and Alizadeh et al. (2011) determined rangeland suitability for sheep and goats grazing using a livestock grazing model with the three components of forage production, water resources, and the soil sensitivity to erosion. In the present study via application of the FAO method (1991) the same three measures were employed to determine the final livestock grazing suitability model for the rangelands.

This research describes the use of a geographical information system (GIS) to construct land suitability models for livestock grazing in the Ghara-Aghch region, Iran. Based on FAO method and the source data, sub models were created focusing on three different themes: sensitivity of the soil to erosion, water resources and available forage. Models recognized the important factors affecting model suitability for livestock use of the rangeland, and also determining the kind and rate of the limitations and factors reducing the suitability with the aim of gaining an adequate plan for grazing. In assessing site considerations these general models identified wider resource management options and solved conflicts of rangeland allocation and livestock grazing between pastoral and rancher.

The results of the final suitability outcome of the model revealed (a) none at the S1 suitability category (unlimited), (b) 694.36 hectares (9.7%) in the S2 suitability category (with minor limitation), (c) 5439.35 hectares (75.9%) in the S3 suitability category (major limitation), and (d) 1025.81 hectares (14.3%) in the N suitability category (unsuitable). The most important reducing factors in model suitability model were: (a) land use and the vegetation cover (in relation to sensitivity of the soil to erosion), (b) the amount of the available forage in comparison with the total production and (c) the existence of less palatability plants among the pasture plants (forage production suitability). In general, no serious difficulty was observed for the livestock's water source, but in some areas the considerable distance from the water source and the precipitous slope resulted in a decrease or limitation in the graze suitability. Among all parameters studied, the specifications on vegetation and forage production were determined as the most significant factors in reducing the suitability of the rangeland for livestock grazing of sheep and goats.

4.1. Soil sensitivity of the erosion model

The most important erosion reducing factors in the study area were determined by land and vegetation use. In the present study the factors affecting erosion were in compliance with reports by many similar studies. Factors of land use, surface cover, run off, and the current erosion in the region are among the most important factors influencing erosion in the Ghara-Aghach region. Amiri (2009a) stated that the important factors in increasing erosion are soil sensitive to the erosion, unsuitable vegetation cover and the lack of proper management in land use. Neameh (2003) had also mentioned unsuitable land use (plowing the rangeland and changing them into farmlands) as the main factor in reducing the suitability of Roozeh Chay rangelands in Uromieh. The negative effects of over grazing and early grazing on the reduction of infiltration and increased run off (and consequently, increased erosion) were clearly specified.

4.2. Water resources model

The results of the study showed that the quantity (number of permanent water resources), quality and the distance from the water resources did not impose much limitations on the rangelands suitability for grazing livestock. However, the steep slopes along the livestock path to the water resources resulted in the formation of an 'unsuitability' category for livestock. Valentine (2001) reported on the importance of the slope factor in reaching the water resources, and declared that by increasing the slope the ability to graze decreases and increases the livestock demand to expend lots of energy. Steep slopes are not recommended for grazing, but instead they can be applied for other purposes (such as wild life and tourism). The quality and quantity of the water resources in the region did not impose any limitations. This study demonstrated that the slope factor in the rangelands of Semirom region was the major factor decreasing and limiting rangeland suitability with respect to the distance from water resources. The outcome of the research indicates the slope as the reducing and sometimes limiting factor in the range suitability. Hence, the slope factor is of considerable importance in determining the suitability of the pasture for grazing. As slope increases the water retention time on the ground decreases, the rate of penetration decreases, and the amount of water run-off increases. The possibility of retaining mature soils on steep slopes is reduced.

Grazing on steep slopes will cause movement of the soil and consequently, will make it difficult for plants to remain stable. Furthermore, the livestock will spend lots of energy in walking on the steep slopes (for grazing and reaching water sources) and as a result their function will decrease. Cook (1954) explained that on slopes of more than 60 degrees little forage is grazed. Amiri (2009b) and Gavili et al. (2011) defined the slopes with more than 60 percents as useless for all kinds of livestock, while Holechek et al. (1995) reported slopes of more than 60 percent, and Arzani et al. (2006) defined slopes of more than 60 percent as useless for livestock grazing. On such steep slopes wild animals would graze better than livestock.

4.3. Forage production model

The major factors reducing suitability of the rangelands in the study area were improper use or exploitation limit, the existence of Class II and III plants in the forage combination, and the decrease in available forage for livestock. It must be noted that factors which cause reduction of proper use factor exploitation limit in the region, are themselves deemed as the reducing factors of the suitability of the rangeland. The effects of previous usage (changing the rangeland into farmlands and leaving them, or over grazing), the low vegetation cover, and the existence of low palatability class plants among the vegetation (perennial forbs and annual grasses) are among the factors reducing the suitability of forage production in the study area. Plowing rangelands with the aim of developing un-irrigated cultivation in the regions is one of the factors responsible for the destruction of the rangeland, although the annual rainfall allows for rain-watered cultivation. It must be noted that these rangelands with deep and good soils are among the best rangelands in the country. As the climatic conditions in the study area facilitates un-irrigated cultivation the region's rangeland has in the past been plowed and cultivated, wherever the soil depth and the slope were not limiting. During the early years of neglect of the un-irrigated-farms, the invader plants (most of the annual grasses and forbs) had become established in the region. The annual forbs and grasses make up a temporary vegetative ground cover (during the growing season), while for much of the year the ground has no vegetative cover and hence is defense-less against erosion. The present study revealed that changing the rangeland to rain-watered farms and neglecting them, over grazing, early grazing, low vegetation cover, and presence of fewer palatable species as the most important factors reducing suitability of the study area in terms of forage production. Amiri (2009a) had also observed low vegetative cover as among the most important factors in reducing production suitability of a region.

The results of the final range suitability model revealed that the most important factor in reducing the rangeland suitability of the study area was the low amount of the available forage in comparison to the total herbage production. It must be noted that other factors responsible for reducing the suitability of the region's rangeland include low vegetations cover, lack of proper vegetative ground cover to protect the surface soil, surface run-off, slope, the sensitivity of the soil to erosion, climatic conditions, plant combination, the condition and trend in vegetation types, over grazing, and finally invasion of the rangeland areas determine the suitability of the region's rangeland. Furthermore, an important factor in limiting grazing is the steep slope of the region (more than 60 degree).

Farahpour et al. (2004) reported that early and over grazing as the main causes of the reduction of the suitability of the rangelands of Shadegan in Isfahan, but in the Ghara Aghach district, due to the limitations imposed on early grazing by the Institute of the Natural Resources (I.N.R.) of Isfahan Province and Semirom City, early grazing was not the suitability limiting factor in the region's rangelands.

Guenther et al. (2000) in determining the suitability of a region in Australia noted the two factors of slope and water resources as the suitability limiting factors of rangeland for grazing cattle. Due to the existence of numerous permanent water resources in the Ghara-Aghach rangelands, the water resources factor does not impose much limitation on the suitability of the rangeland of the region. However, the slope factor in reaching the water resources in limited areas of the region's rangeland was a suitability limiting factor. Fitumukiza (2004) on determining the suitability of the rangeland of the Gaza Province in Mozambique for cattle grazing, considered such parameters as rainy and growing seasons, soil characteristics, vegetative cover, the needed and available forage, reaching the water resources and slope, and expressed the major suitability limiting factors in the region's rangeland as: firstly, the lack of accessibility to the water resources; then, low palatability of the plant species, low production of the forage and the slope. It must be noted that the results reported by Guenther et al. (2000) and Fitumukiza (2004) were similar to that observed in the present study. Arzani et al. (2006) studied sheep grazing in four regions of Siahrood and Lar in the Alborz mountain range, Ardsetan in central area, and Dasht-e Bakan in Zagros region, and observed that in the Siahrood region, the variety of the poisonous plants, the steep slope, temporary water resource, and the components sensitive to erosion were the main factors limiting the suitability of the region. The factors limiting the suitability of the rangeland in Lar region in order of their importance were: the steep slope, the sensitivity of the soil to erosion, and the manner of exploiting the lands. The factors limiting the suitability of rangeland in the Ardedstan region were: low productivity, the existence of invasive plants, greater distance from the water resources, the manner of exploiting the lands, and the current erosion. In the Dasht-e Bakan region, the slope, distribution of the water resources, and lack of permanent water resources were the factors limiting the suitability of the rangeland for grazing sheep.

The outcome of the present study also showed that due to low productivity of palatable forage as a result of constant utilization of the rangeland, the shortage or lack of palatable plants on one hand and the existence of numerous un-palatable and thorny plants in the vegetation composition on the other, effective grazing of livestock in the rangeland will be limited.

5. Conclusion

Assessment of rangelands is an activity that frequently challenges those involved in the livestock industry, environmental protection, and in land and rangeland management. The main objectives of an integrated land, forage and livestock resources suitability assessment are to quantify the resource endowment, understand interrelationships between resource components, predict environmental impact, estimate livestock support capacity, and evaluate development options.

Geographic Information Systems (GIS) have experienced rapid growth in recent years. GIS is a technology using a computer programme which aids in managerial, policy and development decisions, primarily by modeling suitability of land and forage resources for planning livestock grazing, taking system complexity into account.

In this chapter, recent developments of using GIS as a smart tool in supporting the ranchers and pasture owners for monitoring land suitability for livestock feeding purposes is challenged. This research aims at developing a module based on GIS for predicting the physical suitability of land for livestock feeding. It can help decision makers to monitoring the level of land suitability for livestock grazing. It gives clear indicator for the suitability of land and limitation factors to be applied to practical land management with greater success. This study was carried out on a regional scale to examine limitations and opportunities for extensive grazing. While we may present a comprehensive attitude towards extensive grazing, one should know that grazing is one of the uses readily available for rangelands. As FAO argues, different land units have different qualities for certain utilizations. As might be understood, rangelands' utilizations comprise certain qualities and criteria that the model must consider in assessing suitability. However, mixed livestock grazing could be substituted with single utilization in order to gain sustainability of these resources and gain ultimate but sustainable benefits.

Author details

Fazel Amiri
Corresponding Author
Spatial and Numerical Modeling Laboratory, Institute of Advance Technology (ITMA),
Faculty of Engineering, Universiti Putra Malaysia, Malaysia

Abdul Rashid B. Mohamed Shariff
Spatial and Numerical Modeling Laboratory, Faculty of Engineering,
Universiti Putra Malaysia, Malaysia

Taybeh Tabatabaie
Environnemental Science, Faculty of Engineering, Islamic Azad University Bushehr Branch, Iran

Acknowledgement

I dedicate this chapter to my mother, and to all people that contributed towards its successful completion.

6. References

Abaye, A., Allen, V., Fontenot, J., (1994). Influence of grazing cattle and sheep together and separately on animal performance and forage quality. Journal of Animal Science 72, 1013-1022.

Ahmadi, H., (2004). Applied geomorphology (Vol.1, water erosion). Tehran University Publication, Tehran, Iran.

Alizadeh, E., Arzani, H., Azarnivan, H., Mohajeri, A., Kaboli, S., (2011). Range Suitability Classification for Goats using GIS: (Case study: Ghareaghach Watershed- Semirom). Iranian Journal of Range and Desert Research 18, 353-371.

Amiri, F., (2009a). A model for classification of range suitability for sheep grazing in semi-arid regions of Iran. Journal of Livestock Research for Rural Development 21, 68-84.

Amiri, F., (2009b). A GIS model for determination of water resources suitability for goats grazing. African Journal of Agricultural Research 4, 014-020.

Amiri, F., (2010). Estimate of Erosion and Sedimentation in Semi-arid Basin using Empirical Models of Erosion Potential within a Geographic Information System. Air, Soil and Water Research 3, 37-44.

Amiri, F., Shariff, A.R.B.M., (2011). An Approach for Analysis of Integrated Components on Available Forage in Semi-Arid Rangelands of Iran. World Applied Sciences Journal 12, 951-961.

Arzani, H., Jangjo, M., Shams, H., Mohtashamnia, S., Fashami, M., Ahmadi, H., Jafari, M., Darvishsefat, A., Shahriary, E., (2006). A model for classification of range suitability for sheep grazing in Central Alborz, Ardestan and Zagros regions. Journal of Science and Technology of Agriculture and Natural Resources 10, 273-290.

Baniya, N., (2008). Land suitability evaluation using GIS for vegetable crops in Kathmandu Vally/Nepal, Agriculture and Horticulture. University of Berlin.

Barbari, M., Conti, L., Koostra, B., Masi, G., Guerri, F.S., Workman, S., (2006). The use of global positioning and geographical information systems in the management of extensive cattle grazing. Biosystems engineering 95, 271-280.

Bizuwerk, A., Peden, D., Taddese, G., Getahun, Y., (2005). GIS Application for analysis of Land Suitability and Determination of Grazing Pressure in Upland of the Awash River Basin, Ethiopia, International Livestock Research Institute (ILRI), Addis Ababa, Ethiopia.

Bizuwork, A., Taddese, G., Peden, D., Jobre, Y., Getahun, Y., (2006). Application of geographic information systems (GIS) for the classification of production systems and determination of grazing pressure in uplands of the Awash River Basin, Ethiopia. Ethiop. Vet. J. 10.

Breman, H., De Wit, C., (1983). Rangeland productivity and exploitation in the Sahel. Science 221, 1341-1347.

Coffey, L., (2001). Multispecies grazing. Appropriate Technology Transfer for Rural Areas (ATTRA).-www. attra. ncat. org.

Collins, M.G., Steiner, F.R., Rushman, M.J., (2001). Land-use suitability analysis in the United States: historical development and promising technological achievements. Environmental Management 28, 611-621.

Cook, C.W., (1954). Common use of summer range by sheep and cattle. Journal of Range Management 7, 10-13.

FAO, (1976). A framework for land evaluation. Food and Agriculture Organization of the United Nations, Soils Bulletin 32.FAO, Rome. .

FAO, (1983). Guidelines: land evaluation for rainfed agriculture. Food and Agriculture Organization of the United Nations, Soils Bulletin 52. Rome, Italy.

FAO, (1984). Land evaluation for forestry. Forestry Paper 48. Food and Agriculture Organization of the United Nations, Rome, Italy.

FAO, (1985). Guidelines: land evaluation for irrigated agriculture. Food and Agriculture Organisation of the United Nation, Rome.

FAO, (1991). Guidelines: land evaluation for extensive grazing, Issue 58 of FAO soils bulletin. Food & Agriculture Org., 158 pp.

FAO, (2002). Agricultural drainage water management in arid and semi-arid areas. Food and Agriculture Organization of United Nations., Rome.

FAO, (2007). Land evaluation Towards a revised framework Food and Agriculture Organization of the United Nations., Rome, Italy.

Farahpour, M., Van Keulen, H., Sharifi, M., Bassiri, M., (2004). A planning support system for rangeland allocation in Iran with case study of Chadegan Sub-region. The Rangeland Journal 26, 225-236.

Ferreira, A., Hoffman, L., Schoeman, S., Sheridan, R., (2002). Water intake of Boer goats and Mutton merinos receiving either a low or high energy feedlot diet. Small Ruminant Research 43, 245-248.

Fitumukiza, D.M., (2004). Evaluating rangeland potentials for cattle grazing in a mixed farming system, International institute for geo-information science and earth observation. ITC, Netherlands.

Forbes, T., Hodgson, J., (1985). The reaction of grazing sheep and cattle to the presence of dung from the same or the other species. Grass and Forage Science 40, 177-182.

Gavili, E., Ghasriani, F., Arzani, H., Vahabi, M., Amiri, F., (2011). Determine water resources accessibility for sheep grazing by GIS technology (Case study: Feraidun Shahr rangeland in Isfahan Province). Journal of Applied RS and GIS Technology in Natural Resource Science 1, 89-99.

Guenther, K.S., Guenther, G.E., Redick, P.S., (2000). Expected-use GIS maps. Rangelands 22, 18-20.

Guo, Z.G., Liang, T.G., Liu, X.Y., Niu, F.J., (2006). A new approach to grassland management for the arid Aletai region in Northern China. The Rangeland Journal 28, 97-104.

Heady, H.F., (1975). Rangeland management. McGraw-Hill New York.

Holechek, J.L., Pieper, R.D., Herbel, C.H., (1995). Range management: principles and practices. Prentice-Hall, USA, 526 pp.

Hopkins, L.D., (1977). Methods for generating land suitability maps: a comparative evaluation. Journal of the American Institute of Planners 43, 386-400.

Javadi, S., Arzani, H., Farahpour, M., Zahedi, A., (2008). Evaluating Rangeland Suitability for Camel Grazing Using GIS (Case study: Tabashalvan range). Rangeland 2, 46-62.

Khan, M., Ghosh, P., (1982). Comparative physiology of water economy in desert sheep and goats, Proceedings of the 3rd International Conference on Goat Production and Disease, Dairy Goat Publishing Co., Scottsdale, USA.

King, J.M., (1983). Livestock water needs in pastoral Africa in relation to climate and forage. International Livestock Centre for Africa, Addis Ababa, Ethiopia.

Luginbuhl, J., Green Jr, J., Poore, M., Conrad, A., (2000). Use of goats to manage vegetation in cattle pastures in the Appalachian region of North Carolina. Sheep & Goat Research Journal 16, 124-135.

Meyer, H.H., Harvey, T., (1985). Multispecies livestock systems in New Zealand. In: Baker, F.H.J., R.K. (Ed.), Proceedings of a conference on multispecies grazing; June 25-28 Winrock International Institute for Agricultural Development, Morrilton, Ark. (USA), pp. p. 84-92.

Milner, C., Hughes, R.E., Gimingham, C., Miller, G., Slatyer, R., (1968). Methods for the Measurement of the Primary Production of Grassland. International Biological Programme (London and Oxford and Edinburgh). p.70, 70 pp.

Neameh, B., (2003). Land evaluation for Land Use Planning with especial attention to sustainable fodder production in the Rouzeh Chai catchment of Orumiyeh area, Iran. MS unpublished Thesis, International Institute for Geo-information Science and Earth Observation., Nederland. ITC, Netherlands, pp. 95.

Pringle, J.H., Landsberg, J., (2004). Predicting the distribution of livestock grazing pressure in rangelands. Austral Ecology 29, 31-39.

Rafahi, H.G., (2004). Wind erosion and conservation. Tehran University Publication, Tehran, Iran, 671 pp.

Reed, J., Bert, J., (1995). The role of livestock in sustainable agriculture and natural resource management. Livestock and sustainable nutrient cycling in mixed farming systems of sub-Saharan Africa 2, 461-470.

Smith, A.D., (1965). Determining common use grazing capacities by application of the key species concept. Journal of Range Management 18, 196-201.

Sonneveld, M.P.W., Hack-ten Broeke, M.J.D., van Diepen, C.A., Boogaard, H.L., (2010). Thirty years of systematic land evaluation in the Netherlands. Geoderma 156, 84-92.

Taylor, J., CA, Ralphs, M., (1992). Reducing livestock losses from poisonous plants through grazing management. Journal of Range Management 45, 9-12.

Thornton, P., Herrero, M., (2001). Integrated crop–livestock simulation models for scenario analysis and impact assessment. Agricultural Systems 70, 581-602.

Vallentine, J.F., (2001). Grazing management. Elsevier, 659 pp.

Venkataratnam, L., (2002). Remote sensing and GIS in agricultural resources management., First National Conference on Agro-Informatics, INSAIT.

Chapter 14

Effects of Population Density and Land Management on the Intensity of Urban Heat Islands: A Case Study on the City of Kuala Lumpur, Malaysia

Ilham S. M. Elsayed

Additional information is available at the end of the chapter

1. Introduction

The increased size of urban areas in terms of their population and their land consumption has intensified adverse urban environmental impacts. The increased capacity of the human race provokes adverse environmental change on a truly global scale. In the last two decades all over the globe rapid changes in technology and in the re-location of population from rural to urban areas have altered local natural environments beyond recognition, now the global environment is at risk. Most people would argue that changes in the location and concentration of commercial activities, especially in large cities, have produced the greatest visual impact on the built environment (Tamagno et al., 1990). In many developing countries, towns are expanding and an increasing proportion of the land is being taken up for urban land uses, replacing fields, farms, forests and open spaces. As a result, distinctive and often unpleasant climatic conditions are experienced by the majority of urban inhabitants in the world today (Shaharuddin, 1997). Urban settlements provide one of the best examples of change in human activities and perceptions. Residential areas are constantly undergoing modification and expansion into areas that were formally occupied by agriculture and the natural environment. Residential lands were reclaimed or will be reclaimed from the sea or swampland if the demand for land is sufficiently high. By 1950, approximately 30% of the world's population lived in urban areas. That number is now nearing 50%, with a current urban population estimated at 2.9 billion people. By the year 2030, the global population is predicted to rise by two billion (Streutker, 2003), a growth expected to occur almost entirely in urban areas. The increased capacity of the human race provokes adverse environmental change on a truly global scale, something to which urban populations make a major contribution. Atmospheric modifications through urbanization have been noted. Climatically (Sham, 1987), one obvious consequence of urbanization is the creation of the heat island. (Streutker, 2003) focused on one of the primary effects of urbanization on weather and climate, the urban heat island; he found that the urban temperature depends on population density.

Several factors result in temperature difference between the urban and rural areas, stemming from changes in the thermal properties of surface materials to alterations of the topography and man activities in cities. Large urbanized regions have been shown to physically alter their climates in the form of elevated temperatures relative to rural areas at their periphery (Brain, 2001). The effect of metropolitan regions is not only confined to horizontal temperatures but also to those in the vertical direction with far-reaching consequences, studies have shown that the thermal influence of a large city commonly extends up to 200-300 m and even to 500 m and more (Sham, 1993).

The study aims to study the level of urbanization in terms of population density and land management and its effect on the intensity of the urban heat island of the city of Kuala Lumpur.

The measurements for level of urbanization vary from country to another. Usually, national procedures followed for such measurements based on specific criteria that may include any/ some/ all of the following:

a.   The concentration or size of populations.
b.   The process in which the in-migration of people to cities blends into an urban lifestyle.
c.   The process in which urban culture spreads to agricultural villages.
d.   The predominant type of economic activity.
e.   The development of urban areas and their urban characteristics such as specific services and facilities.
f.   The process in which the proportion of people living in an urban area increases.

Of these definitions, the last one is the most quantitative. Therefore, for the purposes and limitations of this study, the last definition is used to defining and measuring the level of urbanization. Thus, the level of urbanization depends solely on density of population per acres and land use for the city.

2. Methodology

The data related to the population density and land management of the city of Kuala Lumpur was gathered from Malaysian Governmental sources, specifically, from the City Hall of Kuala Lumpur. On the other hand, two major sources of data are used to study the UHI of the city.

2.1. Population density

The 1970 and 1980 censuses in Malaysia classified urban areas into three categories: "metropolitan," with a population in excess of 75,000; "large town," with a population size of 10,000 and over; and "small town," with a population size of 1,000 to 9,999 persons. "Small towns," however, are excluded from the consideration of urbanization levels. Based on this definition, Malaysia has 14 metropolitan areas and 53 towns with a population of 10,000 to 75,000. Kuala Lumpur city is the capital city of Malaysia with a population of 1504300 persons. It is recognized as the greatest metropolitan area within the country (Elsayed, 2006). Table 1. and Fig. 1. below illustrate the changes in the population densities for Kuala Lumpur City and its City Centre in 1980, 2000 and 2004 respectively.

Year	Population		Area in sq km		Population density	
	KL	City centre	KL	City centre	KL	City centre
1980	156980	93800	234	18.13	670	5174
2000	1423900	128720	234	18.13	6085	7100
2004	1504300	121655	234	18.13	6429	6710

Table 1. Changes in the population densities for KL and City centre in1980, 2000 & 2004

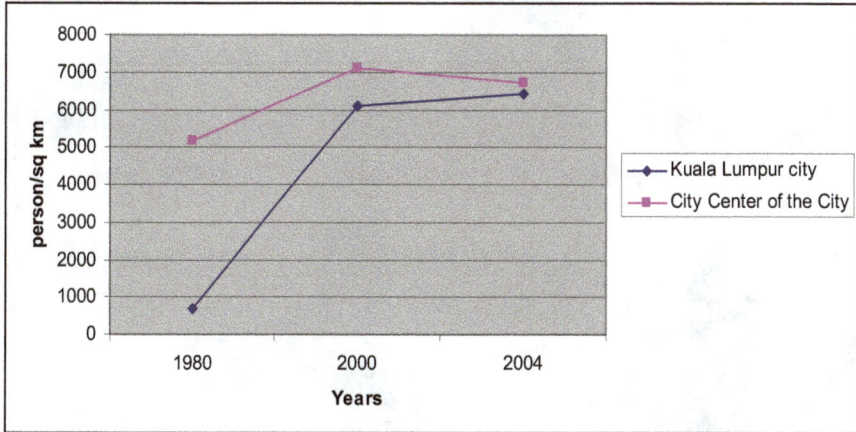

Figure 1. Changes in population densities for KL and City centre in 1980, 2000 & 2004

2.2. Land management

There is a tremendous change in the land use of the city of Kuala Lumpur since 1980 to 2004. Map1. and Map 2 below depict the land use of KL city in 1980 and 2004 respectively.

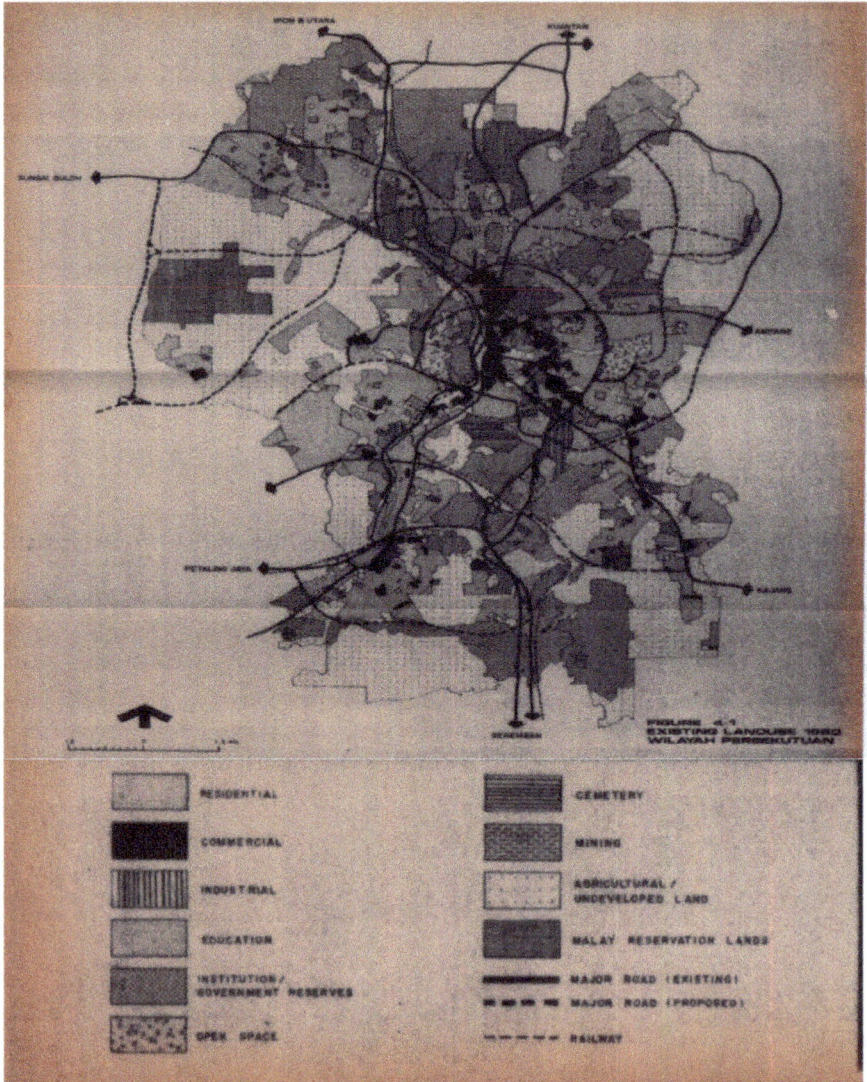

Map 1. Land use for the city of Kuala Lumpur 1980

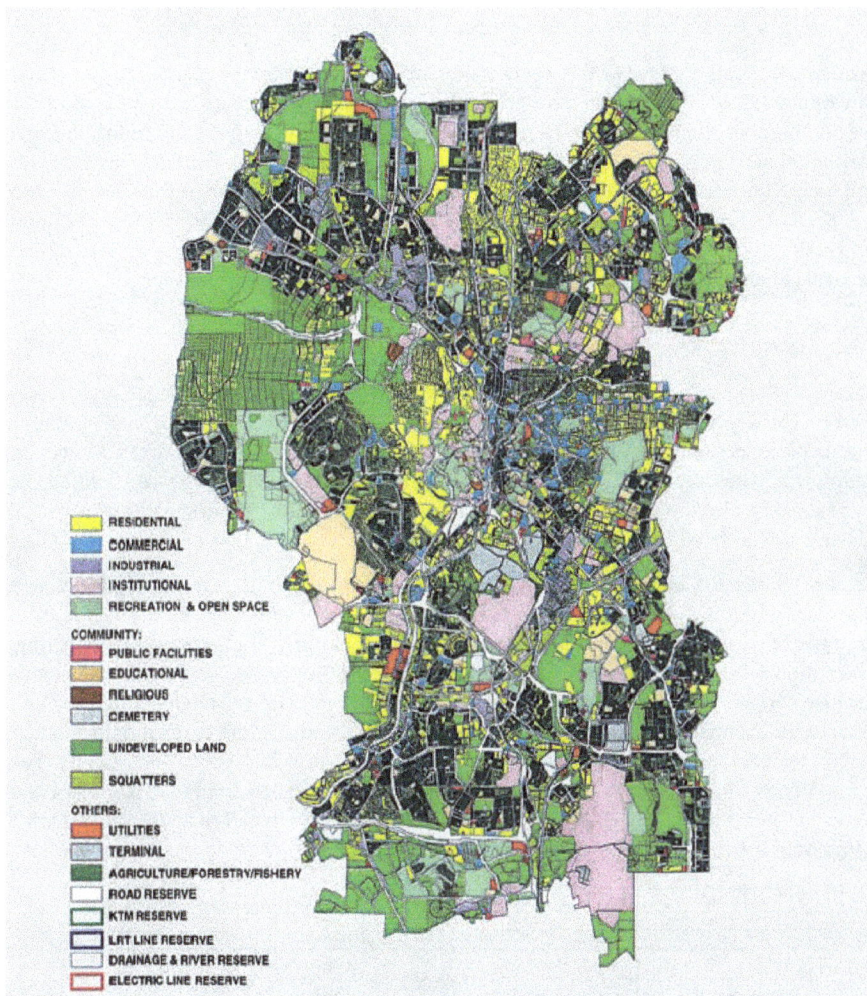

Map 2. Land use for the city of Kuala Lumpur 2004

2.3. Urban heat island

Secondary and Primary sources of data are used to study the UHI of the city. The Secondary data is collected from the relatively longer records of meteorological data provided by specific weather station networks, while the Primary data is collected through an intensive fieldwork done with the collaboration of number of assistants and field observers. These two methods were combined and used to study and measure the urban heat island of the city:

2.3.1. Measuring the urban heat island through weather station networks

Two weather station networks cover the City of Kuala Lumpur and its periphery; Governmental weather station network and private one. According to the case study, a specific number of stations are selected to be involved in the study. Concerning the first weather station network, which is under Malaysian Ministry of Science and Environment and called the Malaysian Meteorological Services (MMS), the stations selected to be used are: Kuala Lumpur International Airport (KLIA), Petaling Jaya, Subang, Sungai Besi, and University Malaya. While for the private weather station network, the stations selected are: Combak, Shah Alam, Cheras, Contry Height, Klang, Nilai, and Petaling Jaya.

2.3.2. Measuring the urban heat island through traverses surveys

This method is used in a specific confined area within the study area for this research. It was used for the city center of Kuala Lumpur city and four major Gardens within Kuala Lumpur and its periphery, and that because of the lake of weather station in those areas. Moreover, within the city center of the city no weather station is located. The area was confined not only because of lack of data in that areas, it is moreover because of equipments and financial constraints that faced the researcher during that period.

Because of the difficulty of making simultaneous measurements, a number of eighteen observers took measurements and readings. They are senior undergraduate students from College of Architecture and Environmental Design and College of Engineering, International Islamic University Malaysia. With the help of these observers, an intensive traverse surveys were carried out for measuring the air temperature, relative humidity and air velocity during one week period in December 2004, starting in 20th of the month and end by 26th for one-hour duration per day from 21:00-22:00 Local Malaysian Time (LMT). The study area is divided into several sectors. Each sector is assigned to one or two observers according to the area and complexity of the sector. The total number of sectors is 12. (Table 2. and Map 3. below).

No.	Name of the station	No of observers
1-	KLCC	Two
2-	Bukit Bentang	One
3-	Time Square	One
4-	Chow kit	One
5-	Sogo	One
6-	Central Market	One
7-	Puduraya	One
8-	Hang Tuah	Two
9-	KLCC Park	Two
10-	Main Lake Garden	Two
11-	Titiwangsa Lake Garden	Two
12-	National Zoo	Two

Table 2. Stations used for Traverses Surveys Method

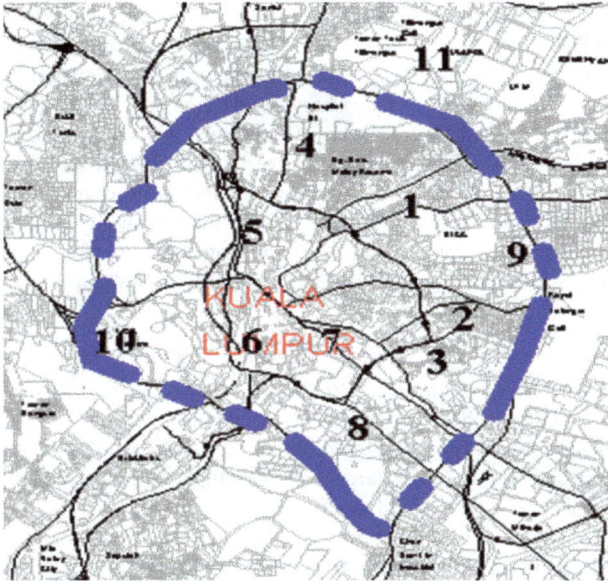

Map 3. Location of the Stations with the City Center of Kuala Lumpur City

3. Results and analysis

The results and analysis of the level of urbanization in terms of population density and land use, and the urban heat island are detailed below.

3.1. Population density

The population densities in 2000 for the city centre of Kuala Lumpur city, Kuala Lumpur City (KL) and Kuala Lumpur Metropolitan Region (KLMR) are 6085, 7100 and 1052 (persons/sq km) respectively. While by 2004 these population densities become 6710 for the city center and 6429 for the city of Kuala Lumpur. Furthermore, the expected population densities for 2020 are 1750 for KLMR, 9402 and 13547 for Kuala Lumpur city and the city centre of the city respectively. Thus, the highest population density is located in the city center of the city, then Kuala Lumpur city, while the less population density is in KLMR. The population density of the city of KL has been increasing from 670 in 1980 to 6085 in 2000 to 6429 in 2004 due to the increasing levels of urbanization of the city compare to its periphery. It rose because of the increasing number of migrants searching for better working opportunities, services, and facilities.

3.2. Land management

Using Charts 1, 2 & 3 below, a tremendous change in the residential, commercial, open space and recreational, road and rail reserves, and undeveloped land of the city from 1980 to

2004 is recognized. The residential and undeveloped land use of the whole city both decreased from 25.7% to 22.66% and from 27.7% to 23.7% respectively. Under the undeveloped land use the agricultural/ fishery/ forest land use is categorized. There is a recognized decrease in the agricultural/ fishery/ forest land use. By 2004 it occupied only 0.07% (16.13 acres) of the total area of the city. Conversely, the commercial, open space and recreational, and road and rail reserves land increased from 2.1% to 4.51%, 1.3% to 6.52%, and from 14.0% to 23.42% correspondingly. Almost there is no change in the industrial, institutional, cemetery, and educational land use of the whole city. The industrial and institutional lands decreased from 2.3% to 2.28% and from 7.2% to 6.69% respectively. While the cemetery, and educational lands increased from 3.3% to 3.98% and from 1.1% to 1.13 %respectively.

The changes in the land use of the city center are almost following the same manner of the city of Kuala Lumpur. The commercial, road and rail reserves land increased from 254.88 to 318.99 hectares and from 498.69 to 566.68 hectares respectively. While the residential, industrial, and institutional land use reduced from 390.58 to 287.6 hectares, from 4.12 to 0.93 hectares, and from 266.04 to 163.06 hectares correspondingly. In converse to the city increase in the open space and recreational land use, the city centre open space and recreational land use decreased from 179.28 to 170.25 hectares. While the undeveloped land use of the city center increased from 0.0 to 137.89 hectares.

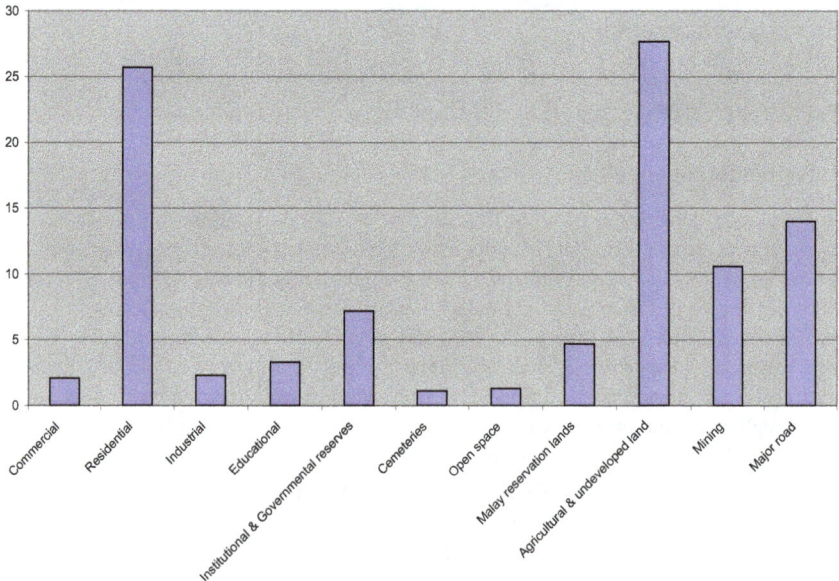

Chart 1. Land use in percentage for city of Kuala Lumpur in 1980

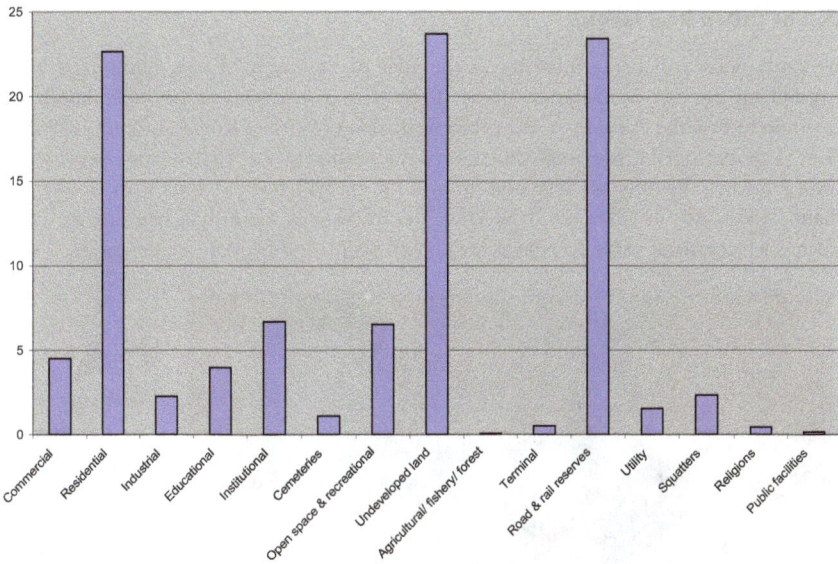

Chart 2. Land use in percentage for city of Kuala Lumpur in 2004

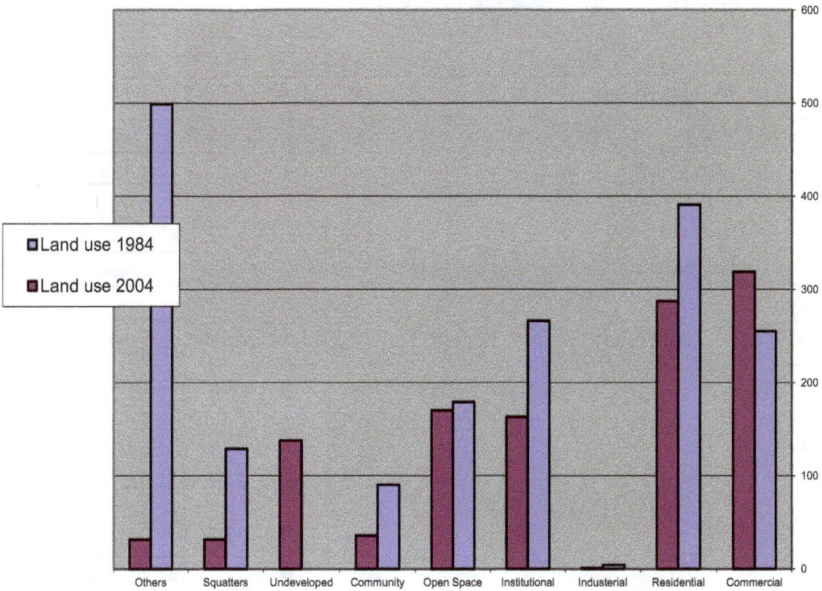

Chart 3. Land use in Hectares for the city center of Kuala Lumpur in 1984 & 2004

3.3. The urban heat island

The study shows that, the intensity of the UHI of the city of Kuala Lumpur is 5.5 ° C recorded on Sunday 26 December 2004 (Map. 4 & 5 below). On the other hand, from previous studies, the intensity of the urban heat island of city of Kuala Lumpur in 1985 was 4.0 °C. Comparing the previous values of the intensity of the UHI to this recent valued (Table 3 below), the intensity increased from 4.0 °C in the latest previous work done in 1985 (Sham, 1986, 1987) to 5.5 °C in 2004. Thus, the increase is more than one degree Celsius, which is a recognized value whenever the human health and comfort are the issues.

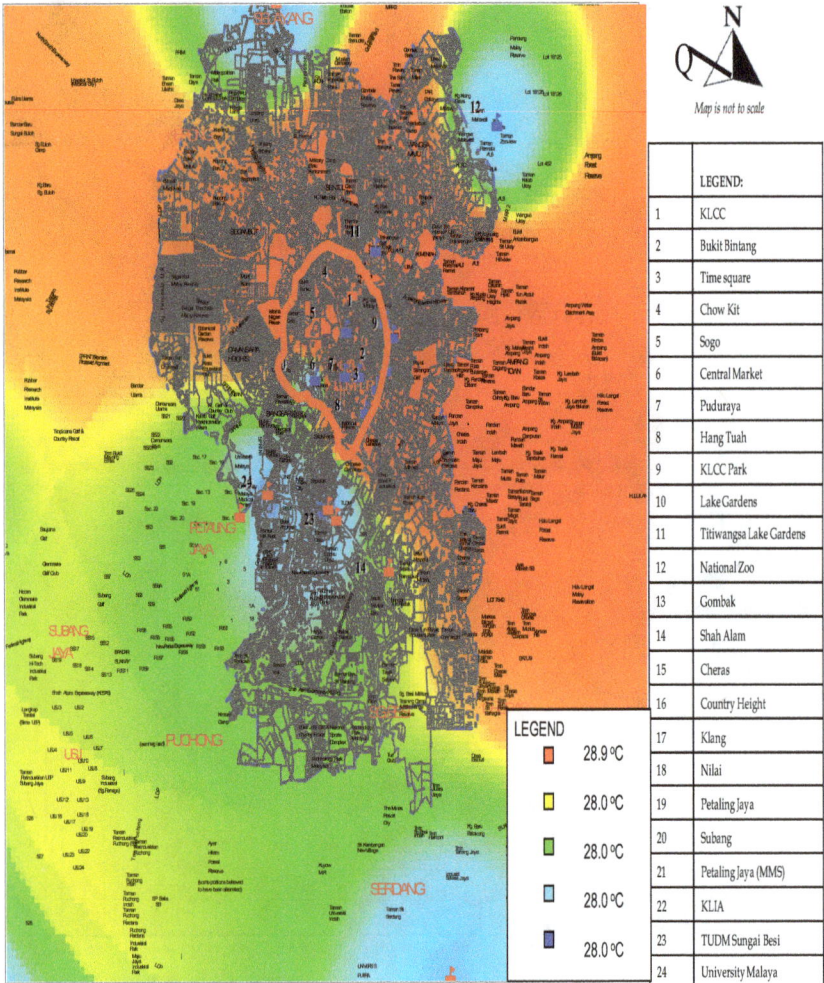

	LEGEND:
1	KLCC
2	Bukit Bintang
3	Time square
4	Chow Kit
5	Sogo
6	Central Market
7	Puduraya
8	Hang Tuah
9	KLCC Park
10	Lake Gardens
11	Titiwangsa Lake Gardens
12	National Zoo
13	Gombak
14	Shah Alam
15	Cheras
16	Country Height
17	Klang
18	Nilai
19	Petaling Jaya
20	Subang
21	Petaling Jaya (MMS)
22	KLIA
23	TUDM Sungai Besi
24	University Malaya

LEGEND
- 28.9 °C
- 28.0 °C
- 28.0 °C
- 28.0 °C
- 28.0 °C

Map 4. The UHI of the sity of Kuala Lumpur on Sunday 26 December 2004

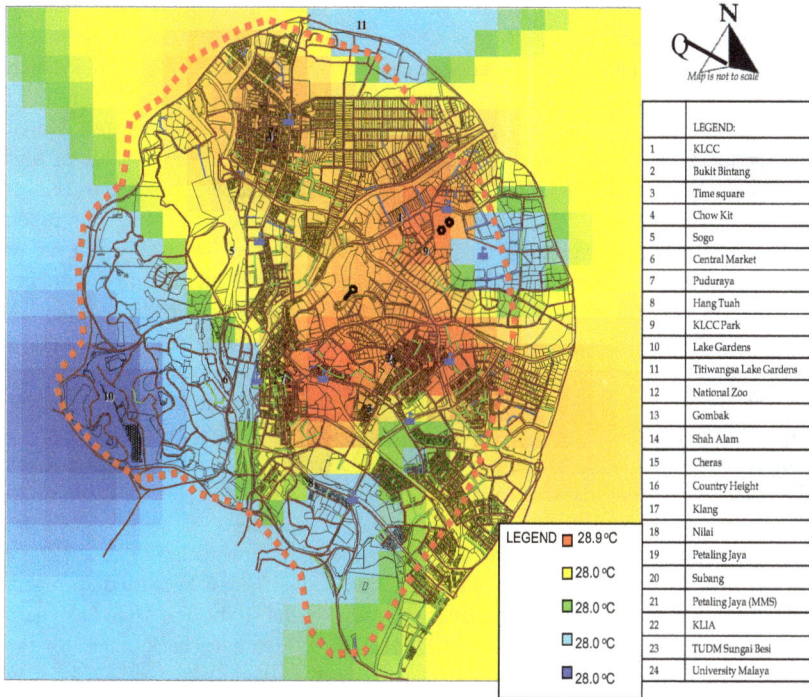

	LEGEND:
1	KLCC
2	Bukit Bintang
3	Time square
4	Chow Kit
5	Sogo
6	Central Market
7	Puduraya
8	Hang Tuah
9	KLCC Park
10	Lake Gardens
11	Titiwangsa Lake Gardens
12	National Zoo
13	Gombak
14	Shah Alam
15	Cheras
16	Country Height
17	Klang
18	Nilai
19	Petaling Jaya
20	Subang
21	Petaling Jaya (MMS)
22	KLIA
23	TUDM Sungai Besi
24	University Malaya

LEGEND 28.9 °C
28.0 °C
28.0 °C
28.0 °C
28.0 °C

Map 5. The UHI of the sity center of the sity on Sunday 26 December 2004

Year	UHI Intensity (°C)	Highest Temp. (°C)	Location of highest temperature	Lowest Temp. (°C)	Location of lowest temperature
1985	4.0	28.0	City centre	24.0	Outside the city centre
2004	5.5	29.2	City center	23.7	Outside the city centre

Table 3. Intensity and location of the UHI of the city of Kuala Lumpur in 1985 & 2004

The study finds that as the intensity of the UHI of the city increased the residential, industrial, institutional, undeveloped and agricultural/ fishery/ forest lands decreased. Conversely the commercial, open space and recreational, road and rail reserves, cemetery, and educational lands increased. In addition to that, as the city centre get warmer and its temperature increased its commercial, undeveloped, road and rail reserves land increased, while its open space and recreational, residential, and institutional land decreased.

There is no contradiction between recent and previous findings of the first published similar work concerning UHI of the city that reported by (Sham, 1973a). The city centre still is the hottest area of the city of Kuala Lumpur. Such finding is due to continuous human activity and development within the city centre of KL. In the last two decades the city centre of KL experienced rapid changes in concentration of commercial activities and in the re-location of population. The results of the study show that, the records of temperature for most of the stations located within the city center are recorded as the highest temperatures, while the records for the stations located within KL but outside the city centre are that of higher temperatures. On the other hand, the less heat and the high temperatures are register only for the stations located outside KL. Therefore, the higher the level of urbanization in terms of population density, the higher the temperature value recorded. The City center has now been occupied by multi stories and tall buildings. These multi-storied buildings found in the city centers dominate the skyline, and have a dramatic effect on the microclimates of the city centre. Man, through his constant constructions, has affected the exchange of energy and moisture within the system by altering the physical qualities and materials of the earth's surface with in the city centre. He has continually replaced vegetation and greenery with buildings. Furthermore, he has become a primary source of heat production from his transportation systems, industrial plants, and HVAC systems. Therefore, the city centre is still the hottest area of the city of Kuala Lumpur. On the other hand, the study shows that, all gardens and parks have relative low temperatures regardless of their locations, in or outside KL. Furthermore, the lowest temperature is recorded for a station located within the city centre of the city, which is the Main Lake Garden station. That is because of the age and area of the garden compared to other gardens included in the study. The Main Lake garden is the largest lake park in the city (Hamidah, 1984). This garden dates back to the 1890s with an area of 73 hectares. While Titiwangsa Lake garden is the second lake park in the city with an area of 44.5 hectares. The garden is even different from other gardens in terms of its type and age of plants.

Recent studies (Elsayed, 2006, 2009) show that, although the dependence of the intensity of the urban heat island of the city of KL on population density is significant, the population density at the city centre area is decreasing. It might be of interest to urban planners that, although the temperature is likely to rise with the increase of population density, the situation at the city centre is different. This is due to the intensive human activity and development within the city centre of KL. That indicates that, the management of those lands is highly affecting the intensity of the urban heat island of such land. The city centre experiences rapid changes in concentration of commercial activities and constructions. Man through his constructions has affected the exchange of energy and moisture within the system by altering the physical qualities and materials of the earth's surface with in the city centre. The city centre has been occupied by multi stories and very tall buildings e.g. Petronas Twin Towers. These multi-storied buildings found in the city centers dominate the skyline, and have a dramatic effect on the microclimates of the city centre. Man replaces vegetation and greenery by buildings and becomes a primary source of heat produce. Therefore, the city centre is still the hottest area of the city of Kuala Lumpur regardless of the reduction happened in its population density. This fact should help in convincing urban planner and design makers in placing more emphasis on the strategies that relates the land management to the mitigation of urban heat island.

4. Conclusion

The effects of population density on the intensity of the urban heat island of the city of Kuala Lumpur could be concluded from Table 1, Figure 1 and Table 3 above which illustrate the changes in the population densities for KL and City centre in1980, 2000 & 2004, and the intensity and location of the UHI of the city of Kuala Lumpur in 1985 & 2004 respectively.

The study shows that, the population density of the city is proportional to the records of temperature taken during the survey. The population density of the city of KL has been increasing from 670 in 1980 to 6085 in 2000 to 6429 in 2004. Consequently the intensity of the UHI of the city increased from 4.0°C in 1985 to 5.5°C in 2004. Thus, there is a proportional relationship between the population density and the UHI of the city of KL. Therefore, the study concludes that, the UHI of the city of Kuala Lumpur is proportional to the population density of the city. Accordingly, the study concludes that, the population density affects the urban heat island of the city and contributes to the increase in the intensity of the urban heat island of the city of Kuala Lumpur, Malaysia.

The study shows that, although the overall population density of the city increases, that of the city centre decreases, while the nucleus of this UHI is the city centre. Therefore it is difficult to conclude that the intensity of the UHI is inversely proportional to the population density of the city centre. Nevertheless, it is possible to conclude that, the increase in the intensity of the UHI is not only related to the population density of the city centre, it is actually affected by other different factors and human activities. The study finds that, the commercial, road and rail reserves lands of the city is proportional to the intensity of the UHI, while the open space and recreational, residential, institutional, and agricultural/ fishery/ forest lands is inversely proportional to the intensity of the UHI of the city. Therefore, utilizing these findings and literature reviewed the study concludes that, the intensity of the urban heat island could be reduced if the land of the city of Kuala Lumpur managed in such ways that:

- Trees should be planted to shade the hot tarmac of city roads or at least low-level bushes and greenery. Within the city of KL, Many open areas are covered with blocks of marble, granite or tiles. Although these are better than black tarmac, these areas still absorb a lot of heat in direct sunlight and release the heat at late afternoons, evenings and early nights. Again, the author recommends that, such open areas should be turned into green areas or even very small parks. Furthermore, trees should be planted to shade the hot tarmac of inner city roads like Jalan Tuanku Abdul Rahman, Chow Kit…etc; or low level bushes planted along the covered drains in such areas. In addition to that, some roads and highways, which take up an increasing proportion of the urban area, should also be creatively designed to include green shade. The large masses of concrete in new flyovers that are continuously being built all over the city, capture and store large quantities of solar heat, should also take into consideration some plant cover, like overhanging creepers which can shield or block absorption of the heat and reduce the air temperature significantly.

- Roads and highways, which take up an ever-increasing proportion of the urban area, should also be creatively designed to include green shade, at the very least along the medians. The large masses of concrete in new flyovers continuously being built all over the city, which can capture and store large quantities of solar heat, should also take into consideration plant cover, like overhanging creepers which can shield or block absorption of the heat.

- Urban car parks should comply with a minimum of 50% shade requirement. Previous studies ([Eliasson, 1993; Sham, 1987, 1990/1991; Shashua, 2000) show that shade trees contribute significantly to temperature reduction, hence the reduction on the intensity of the UHI. Therefore, the author suggests that, urban car parks should comply with a minimum of 50% shade requirement by plantation of trees or/and low level bushes.

- Tree planting programs should be reintroduced for all housing estates. Incentives and subsidies should be part of the long term planning.

- Many commercial buildings, almost all (Ahmad, 2004) are having flat roofs in Malaysia either to accommodate air-conditioning equipment or water tanks, or for another purposes. Such buildings should green their roofs and planted them with shrubs and low level bushes. This means cultivating greenery on the flat roof surfaces to absorb the heat. This will not only help the city to counter UHI but building owners will also benefit in terms of savings in air-conditioning power consumption. As proven in previous studies; please check chapter 2 for more details.

- The creation of as many cities parks as possible will improve the situation and help significantly in reducing the intensity of the UHI of the city. Therefore, tree planting programs should be reinforced in the city of KL, and incentives and subsidies should be part of the long term planning for the city. Previous studies (Eliasson, 1993; Sham, 1987, 1990/1991; Shashua, 2000) prove that green areas moderate urban temperatures. The results of this study confirm this theory; it shows that, the green areas are relatively low in temperature than the non-green areas.

- Reduce summer solar radiation by managing the land covered by critical surfaces, for example, pedestrian walks, waiting areas, and busy streets. Reduce the abundance of concrete and asphalt, and increase the amount of vegetation and open water. This will increase higher volumetric heat capacities and greater rates of latent heat influx, thereby lowering air temperatures.

- Increase airflow at ground level to flush heated and polluted air away from the city and that could be achieved by managing the land cover and building design.

Author details

Ilham S. M. Elsayed
University of Dammam, College of Engineering, Saudi Arabia

Acknowledgement

The author acknowledges the financial support provided by Sudan University of Science and Technology, Ministry of Higher Education, Sudan, and the Centre for Built Environment, International Islamic University Malaysia, for field works and surveys.

5. References

Abdul Samad, H. (2000). *Malaysian urbanization and the environment: sustainable urbanization in the new millennium*, Akam print, Malaysia

Ahmad, F. E. & Norlinda, B. M. D. (2004). *Urban heat islands in Kuala Lumpur*, Kuala Lumpur, Department of Irrigation and Drainage Malaysia

Brain, S. J. (2001). *Remote Sensing Analysis of Residential Land Use, Forest Canopy Distribution, and Surface Heat Island Formation in Atlanta Metropolitan Region*, Ph. D. Thesis, Georgia Institute of Technology, Atlanta

Chan, K. E. ; Abdullah, N. & Tan, W. H. (1984). *Population and demographic characteristics in Kuala Lumpur*, Proceedings of Seminar on Urbanization and ecodevelopment: with special reference to Kuala Lumpur, University of Malaya, Institute of Advance Studies, Malaysia

Eliasson, I. K. (1993). *Urban Climate Related to Street Geometry*, Ph. D. Thesis, Goteborgs Universitet, Sweden

Elsayed, I. S. (2006). *The Effects of urbanization on the Intensity of the Urban Heat Island: a Case Study on the City of Kuala Lumpur*, Ph. D. Thesis, International Islamic University Malaysia

Elsayed, I. S. (2009). *Land Management and its effects on the Intensity of the Urban Heat Island: a Case Study on the City of Kuala Lumpur*, Proceedings of The IASTED International Conference on Environmental Management and Engineering, Alberta, Canada

Ghani, S. (2000). *Urbanization & regional development in Malaysia*, Utusan Publications & Distributors, Malaysia

Hafner, J. (1996). *The Development of Urban Heat Islands in the Southeast Region of the United States in the Winter Season (Global Warming)*, Ph. D. Thesis, Huntsville, University of Alabama

Hamidah, K. (1984). (ed.). *Kuala Lumpur: the city of our age*, City Hall of Kuala Lumpur, Malaysia

Hoong, Y. Y & Sim, L. K. (1984). (ed.). *Urbanization and ecodevelopment: with special reference to Kuala Lumpur*, Proceedings of Seminar: PRO, 2, Institute of Advance Studies, University of Malaya Press, Kuala Lumpur

Khoo, S. G. (1996). *Urbanization and urban growth in Malaysia*, Jabatan Perangkaan Malaysia

Kok, K. L. (1988). *Patterns of Urbanization in Malaysia. National Population & Family Development Board*, Kuala Lumpur, Proceedings of the conference on Urbanization in Malaysia: Patterns, Determinants and Consequences, pp. 20-55

Orville, R. E. (2001). Enhancement of cloud-to-ground lightning over Houston, Texas, *Geographical Research*, 28, pp. 2597-2600,

Shaharuddin, A. (1997). Urbanization and human comfort in Kuala Lumpur-Petaling Jaya, Malaysia, *Ilmu Alam*, 23, pp. 171-189

Shahruddin, A. & Norazizah, A. (1997). *The essential usage of air conditioning system in Petaling Jaya, Selangor, Malaysia*, Proceedings of Symposium on Population, Health and the Environment, International Geographical Union Commission on Population and the Environment, Chiang Mai, Thailand

Sham, S. (1973a). Observations on the city's form and functions on temperature patterns: a case study of Kuala Lumpur, *Tropical Geography*, 36, No. 2, pp. 60-65

Sham, S. (1973b). The urban heat island: its concept and application to Kuala Lumpur, *Sains Malaysiana*, 2, No.1, pp. 53-64

Sham, S. (1980a). Effects of urbanization on climate with special reference to Kuala Lumpur-Petaling Jaya area, Malaysia, In: *Urbanization and the atmospheric environment in the low tropics: Experiences from the Klang Valley Region, Malaysia,* Penerbit Universit Kebangsaan Malaysia, pp. 264-268

Sham, S. (1980b). *The climate of Kuala Lumpur, Petaling Jaya area, Malaysia: A study of the impact of urbanization on local within the humid tropic,* Universiti Kebangsaan Malaysia

Sham, S. (1984a). Inadvertent atmospheric modifications through urbanization in the Kuala Lumpur area, Malaysia, In: *Urbanization and Ecodevelopment with Special Reference to Kuala Lumpur,* Universiti Malaya, Institute of Advanced Studies, Malaysia

Sham, S. (1984b). Urban development and changing patterns of night-time temperatures in the Kuala Lumpur – Petaling Jaya area, Malaysia, *Journal Teknologi*, 5, pp. 27-36

Sham, S. (1985). Post-Merdeka development and the environment of Malaysia, In: *Urbanization and the atmospheric environment in the low tropics: Experiences from the Klang Valley Region, Malaysia*, Penerbit Universit Kebangsaan Malaysia, pp. 41-75

Sham, S. (1986). *Temperatures in Kuala Lumpur and the merging Kelang Valley conurbation*, The Institute of Advanced Studies, Universiti Malaya, Malaysia

Sham, S. (1987). *Urbanization and the atmospheric environment in the low tropics: Experiences from the Klang Valley Region,* Penerbit Universit Kebangsaan Malaysia

Sham, S. (1990/1991). Urban climatology in Malaysia: An overview, *Energy and Buildings Journal*, 15-16, pp. 105-117

Sham, S. (1993). Environment and Development in Malaysia: Changing Concerns and Approaches", *ISIS Malaysia*, Malaysia

Shashua-Bar, L. & Hoffman, M. E. (2000). Vegetation as a climatic component in the design of an urban street: An empirical model for predicting the cooling effect of urban green areas with trees, *Energy and Building*, 31, pp. 221-235

Smoyer, K. E. (1997). *Environmental risk factors in heat wave mortality in St. Louis*, Ph. D. Thesis, University of Minnesota, Minnesota

Streutker, D. R. (2003). *A Study of the Urban Heat Island of Houston, Texas*, Ph. D. Thesis, , Rice University, Taxes

Tamagno, B. et al. (1990). *Changing environments*, Cambridge University Press, London, UK, pp. 1-104

Thong, L. B. (1990). *Urbanization Strategies in Malaysia and the Development of Intermediate Urbanization Centers*, Proceedings of the IGU Regional Conference of Asian Pacific Countries on Urban Growth and Urbanization, Beijing, China

Valazquez-Lozada, A. (2002). *Urban heat island effect analysis for San Juan, Puerto Rico*, M. Sc. Thesis, University of Puerto Rica

Wan, M. N. & Abdul Malek, M. (2004). *Utilizing satellite remote sensing and GIS technologies for analyzing Kuala Lumpur's urban heat island,* Faculty of Architecture and Environmental Design, International Islamic University Malaysia

Demand Allocation in Water Distribution Network Modelling – A GIS-Based Approach Using Voronoi Diagrams with Constraints

Nicolai Guth and Philipp Klingel

Additional information is available at the end of the chapter

1. Introduction

In water distribution network modelling the topology of the network is commonly mapped as a (directed) graph consisting of nodes and links (Walski et al., 2001). Pipe sections with constant parameters are modelled as links. Intersections, water inlets, points of water withdrawal and locations of pipe parameter changes are modelled as nodes. Coordinates related to the nodes define the spatial location. To calculate the hydraulic system state (pressures and flows) the water demand of the consumers is assigned to the relevant nodes as the driving parameter (demand driven simulation). Figure 1 schematically shows the abstraction of a water distribution network.

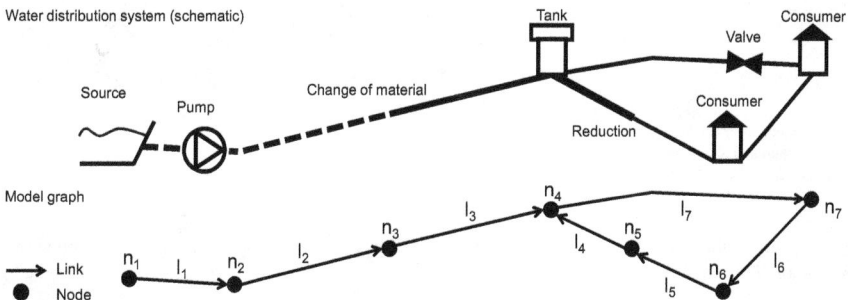

Figure 1. Illustration of a water distribution system and the corresponding model graph (Klingel, 2010)

To reduce the model size typically not every known demand (e.g. demand at a house connection) is modelled as a node. Besides, location and demand of consumers are often not known, especially in developing countries.

To allocate demands to the nodes Voronoi diagrams might be used. The Voronoi diagram is also known as Thiessen polygon and dual to the Delaunay triangulation (c.f. fig. 2, left sketch). The Voronoi diagram is based on a set of nodes and defines one area for each node with every point within the area being closer to the originating node of the area than to any other node. In the following these areas are referred to as Voronoi regions. They represent the catchment areas of the nodes. The borders of the Voronoi regions constitute of lines called bi-sectors. Thus, known demands (value and location) can be allocated to the closest nodes according to the Euclidean distance. This approach is also commonly applied to determine, for example, the rainfall catchment areas of inlets of sewer networks.

If demands and their location are not known, they can be determined based on the size of the Voronoi regions, the population distribution and density and a consumption per capita value. Alternatively, building areas can be allocated to the nearest nodes based on the Voronoi diagrams if the correspondent data is available. In this case the actual demand is calculated by using a coefficient relating the population to a building area unit and a consumption per capita value. The sketch on the right in figure 2 schematically shows the allocation of demands (in this case building areas) to nodes using a Voronoi diagram.

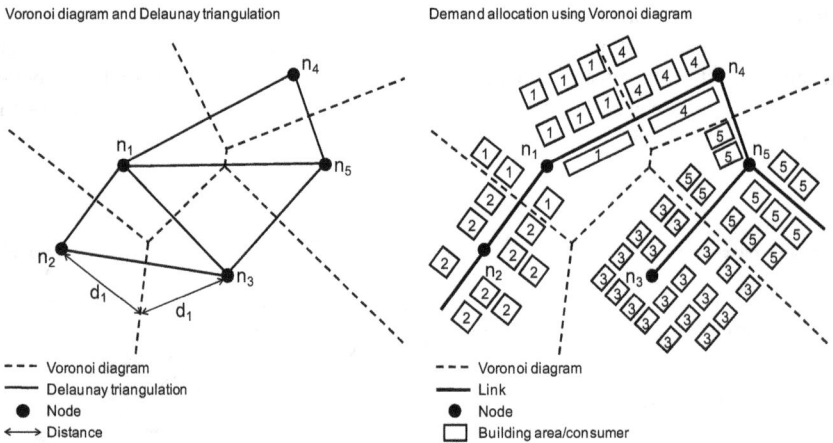

Figure 2. Illustration of a Voronoi diagram and a Delaunay triangulation and demand allocation using a Voroinoi diagram

Pipe network data, population distribution and building data are commonly stored and managed in geographic information systems (GIS). Thus, a GIS-application to allocate water demand values to the network model nodes is self-evident and commonly applied.

The calculation of common Voronoi diagrams is based on a set of nodes and the Euclidean distance between the nodes and the points of the areas. Various well known and often used algorithms to generate such Voronoi diagrams exist, for example the sweep-algorithm (Fortune, 1986; Preparata and Shamos, 1985) and the divide-and-conquer-approach (Shamos, 1975).

The approach to allocate demands using such common Voronoi diagrams follows the assumption that a consumer is connected to the closest existing pipe and to the closest existing node respectively. Boundaries and obstacles are not considered as constraints. In this context, areas with different population densities, for example, constitute boundaries and railway trails and lakes within the supply area, for example, constitute obstacles which cannot be passed by distribution pipes. Thus, with common Voronoi diagrams consumers, areas or building areas are possibly allocated to nodes which are closest according to the Euclidean distance but are not closest if boundaries or obstacles are considered. This constitutes a significant limitation of the application of Voronoi diagrams based on the Euclidean distance for the allocation of water demand.

If boundaries are convex, no obstacles are located in between two originating nodes of a bi-sector. The Voronoi diagram can still be calculated based on the Euclidean distance. The convex boundaries can be considered easily by intersecting the Voronoi diagram with the convex boundary. Conversely, sections of non-convex boundaries or obstacles might be located in between two originating nodes of a bi-sector. In these cases the determination of the bi-sectors based on the Euclidean distance is not sufficient any more. The shortest way to bypass the non-convex section or obstacle respectively has to be considered instead.

Okabe et al. (2000) describe this problem as "Voronoi diagrams with obstacles" and introduce the "shortest path Voronoi diagram". Thereby, obstacles are supposed to be considered by calculating Voronoi diagrams based on the shortest path instead of the Euclidean distance. A satisfactory solution of this approach would result in an increased model accuracy and therefore higher reliability of the simulation calculations.

Figure 3 illustrates the difference between a Voronoi diagram, which is determined based on the Euclidean distance (left sketch) and based on the shortest path (right sketch). Not considering (convex sections of) the boundaries, for example, results in an allocation of areas to node n_1 which can only be part of the Voronoi regions of the nodes n_4 and n_5 if the boundaries given in the example are respected. Not considering the shortest path to bypass non-convex sections of the boundaries leads, for example, to an allocation of an area to node n_2 instead of node n_1. Accordingly, an area is allocated to node n_3 instead of node n_2 if the shortest path to bypass the obstacle given in the example is ignored and the Voronoi diagram is determined based on the Euclidean distance.

Several authors have tackled this problem. The approach of Aronov (1987) considers non-convex boundaries but not obstacles within the Voronoi regions. The approach of Papadopoulou and Lee (1998) does consider obstacles as constraints but only if specific conditions are given. None of the known approaches solves the problem satisfactorily for an application in the given context. The consideration of both, non-convex boundaries and obstacles would be desirable.

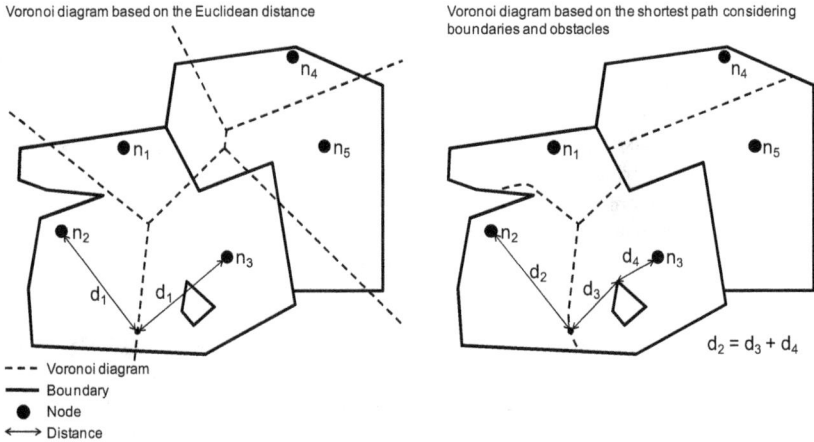

Figure 3. Comparison of a Voronoi diagram based on the Euclidean distance and the shortest path (Klingel, 2010)

In this context, the additively weighted Voronoi diagram is of particular interest. Johnson and Mehl (1939) invented the additive weights of originating nodes of Voronoi diagrams to model the development of crystals, which grow with different speed into different directions. Every node is weighted with a positive or negative constant (additive constant) which is subtracted from the distance function when calculating the Voronoi diagram. Thus, contrary to a common Voronoi diagram based on two nodes, the bi-sector is not a straight line in the geometric centre between the nodes if the two nodes are weighted with different constants. The bi-sector constitutes a hyperbola being closer to the node with the smaller weight (Aurenhammer, 1991).

This chapter presents a novel approach to calculate Voronoi diagrams based on the shortest path instead of the Euclidean distance considering non-convex boundaries and obstacles, which Guth (2009) has developed in his Bachelor thesis. At critical points of boundaries and obstacles, the so called subordinated nodes are temporarily defined and weighted with an additive constant to consider the shortest path as distance function for the calculation of the Voronoi diagram. The approach is presented in section 2 of this chapter. The resultant algorithm is expounded in section 3. The testing of the algorithm and the implementation of the algorithm as a GIS-tool based on the software package ESRI ArcGIS (Ormsby et al., 2004) are demonstrated in sections 4 and 5.

Demand Allocation in Water Distribution Network Modelling – A GIS-Based
Approach Using Voronoi Diagrams with Constraints

287

2. Approach

The basis of the novel approach described herein is a geometric and iterative method to calculate Voronoi diagrams. Thereby, the pairs of originating nodes of a bi-sector (the immediate neighbours) have to be determined first. Each node of a pair is defined to be the origin (centre) of concentric circles. Circles with the same radius are intersected. The intersection points constitute points of the bi-sector. Thus, the Voronoi diagram based on the Euclidean distance can be calculated.

As discussed in section 1, to consider non-convex boundaries and obstacles, the determination of the Voronoi diagram has to be based on the shortest path. Thus, the shortest paths from every point of a bi-sector to each of the originating nodes of the bi-sector have the same length. This leads to a linear form of the bi-sector if the points of the bi-sector are 'visible' from both originating nodes (the length of the shortest path is equal to the length of the Euclidean distance) and to a hyperbolic form if a non-convex boundary section or an obstacle leads to a shortest path which is longer than the Euclidean distance (c.f. fig. 3).

When considering boundaries and obstacles they are assumed to be represented by polygons with vertices, which is the case in the context of the development of a GIS-solution. The critical points of bounding polygons are the extreme vertices of non-convex sections. Accordingly, the critical points of obstacles are the extreme vertices of convex sections. These critical points are identified and added to the set of given nodes (primary nodes) as subordinated nodes. A subordinated node is always related to the closest primary node. An additive constant which represents the shortest path length to that primary node is assigned to each subordinated node. In figure 4 an example of three primary nodes (n_1, n_2 and n_3), a bounding polygon and three obstacles is given. Subordinated nodes are defined at all critical points of the boundary and the obstacles. For example, the three subordinated nodes n_{11}, n_{12} and n_{13} are related to node n_1. The additive constant of the subordinated node n_{12} is also shown exemplarily in figure 4.

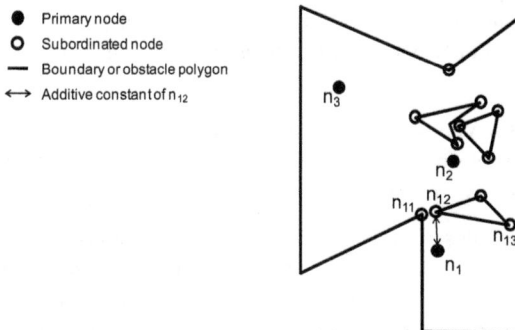

Figure 4. Illustration of subordinated nodes of a given configuration in a bounding polygon with three primary nodes and three obstacles

Even though the principle idea of using additively weighted nodes was invented by Johnson and Mehl (1939), their approach to calculate the Voronoi diagram cannot be applied successfully to solve the given problem. Transferred to the given context, the approach does only solve configurations where the additive constant of a subordinated node is shorter than the distance between the primary node and the related subordinated node (Okabe et al., 2000). In the given case the additive constant is always equal to the distance between primary and subordinated node. Thus, the development of a novel solution is needed.

Like the computation of Voronoi diagrams based on a set of primary nodes by intersecting circles, the determination based on primary and subordinated nodes can be regarded as the projection of the intersection of cones. Each node (located in the x-y-plane) represents the tip of a cone which is opened upwards. All cones have the same cone angle. The cones at the subordinated nodes are moved vertically up (z-direction) by the value of their additive constant. At a certain height (z-value), the intersection of the cones can be represented by the intersection of circles with the radius of the cones at that specific height (z-value). Cones related to primary nodes have the same radii at the same height. Thus, the projection (to the x-y-plane) of the intersection of cones related to primary nodes results in a linear shaped section of the according bi-sector. Accordingly, the intersection of the cones related to a primary and a subordinated node results in a hyperbolic shape due to the different radii at a certain height of the cones.

To simplify the calculation, the projected cones are treated (discretised) as concentric circles. The height above the plane is reflected by the circle's radius. Circles with the same radii are intersected to calculate bi-sector points. For the subordinated nodes' circles, the radii are decreased by the respective additive constant. Resultant radii below zero are not admissible.

For the calculation of the Voronoi diagram the radii of the circles are increased successively. For each radius (iteration), the circles which are related to the primary nodes or their subordinated nodes are intersected in several calculation steps. In one calculation step the circles of two nodes are intersected. If there is an intersection point which is visible from both circles' centres (originating nodes) and is not located within another circle, the point is stored as bi-sector point. Visibility may be limited by the bounding polygon or an obstacle. The difference between the radii of two sequent iterations constitutes the discretisation of the iterative approach.

The result of the approach is a Voronoi diagram consisting of polygons with discretely determined vertices. As GIS do not store mathematical functions, this is not a drawback. The resulting polygons can be directly stored without any further processing.

Figure 5 shows a schematic illustration of the approach. The given configuration comprehends two primary nodes (n_1 and n_2). A boundary polygon constitutes a critical point. The pertinent subordinated node n_{21} is related to the closest primary node, node n_2. In the area visible from both primary nodes (right of the 'visibility boundary' of node n_2) the bi-sector between the nodes is linear. The bi-sector vertices (black squares in the figure) are determined by intersecting concentric circles with the same radii. In the area not visible from

Demand Allocation in Water Distribution Network Modelling – A GIS-Based
Approach Using Voronoi Diagrams with Constraints

289

node n_2 (left of the 'visibility boundary' of node n_2), the bi-sector has a hyperbolic shape. The vertices (black rhombi in the figure) are determined by intersecting the concentric circles of the primary node n_1 and the subordinated node n_{21}. The radii of the circles with the subordinated node n_{21} as origin are always decreased by the additive constant, which is the length of the shortest path between the nodes n_2 and n_{21}. In the figure, circles which are intersected, are plotted in the same colour.

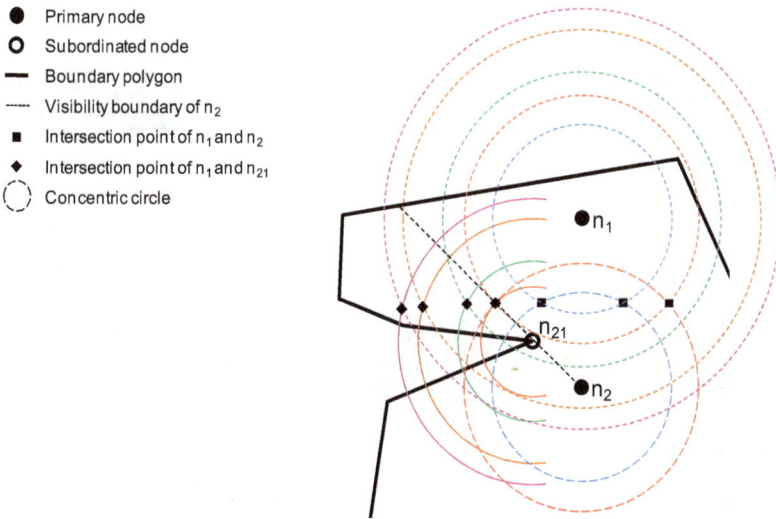

Figure 5. Schematic illustration of the approach

3. Algorithm

3.1. Overview

For the realisation of the approach described in the previous section, an algorithm was developed, tested (c.f. section 4) and finally implemented in a GIS-tool (c.f. section 5). An overview of the individual steps of the algorithm is given in figure 6.

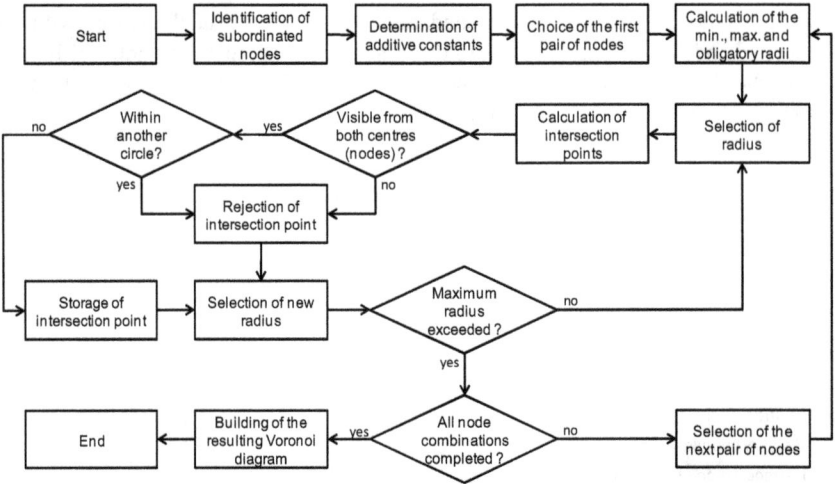

Figure 6. Flowchart of the algorithm

First, the subordinated nodes of the given configuration are identified and weighted with an additive constant. Next, points of the bi-sectors are calculated by intersecting circles with corresponding radii and the nodes as origin. Finally, based on the calculated set of bi-sector points, the Voronoi diagram is determined. Thus, the algorithm can be divided into three principle parts: initialization, computation and creation of the resulting Voronoi diagram. Each of the following subsections describes one of the three parts.

3.2. Initialization

Identifying the subordinated nodes constitutes the first step of the initialization. Therefore, the bounding polygon and the polygons which describe the obstacles are analysed regarding the exterior angle at each vertex. For the bounding polygon this is facilitated by calculating the azimuths of the current vertex to the predecessor (t_1 in equation 1) and the successor (t_2 in equation 1) vertices. The difference between the azimuths is subtracted from the full circle. Thus, the outer angle β is calculated by equation 1:

$$\beta = 2\pi - (t_1 - t_2) \tag{1}$$

The result is normalised to $\beta \in [0;2\pi)$ by successive subtraction of 2π where required. Assuming that the polygon is defined clockwise, the vertex is marked as a subordinated node if it holds that $\beta < \pi$. The angles β of all vertices are added ($\sum \beta$) and compared with the theoretical angular sum W. W is computed by equation 2, where n is the number of vertices:

$$W = (n + 2) \cdot \pi \tag{2}$$

If $\sum \beta = W$, the polygon is defined clockwise and the marks are correct. Otherwise, the order of the vertices is switched and the algorithm is executed again. For the obstacle polygons, the same algorithm is used but the marks are switched in the end. Thus, only the vertices which stick out of the obstacle (convex sections) are considered as subordinated nodes. The identified subordinated nodes are added to the list of originating nodes.

In the succesive step, with the algorithm of Dijkstra (1959), the relation of subordinated nodes and primary nodes is determined and the additive constant for each subordinate node is calculated. The Dijkstra algorithm analyses the graph based on all subordinated and primary nodes with links between pairs of nodes which are visible to each other considering the given boundaries and obstacles. From each subordinated node the possible paths to all primary nodes are calculated. Thus, the shortest path determines the primary node related to the subordinated node. Furthermore, the length of the shortest path constitutes the additive constant.

3.3. Computation of bi-sector points

As mentioned before, the bi-sectors between the Voronoi regions are made up of intersection points of circle intersections. Node pair by node pair, circles of all nodes are intersected with each other to determine the intersection points. The process starts with a randomly chosen pair of nodes.

For the circles to be intersected, the minimum and maximum radius and obligatory radii are determined first. The minimum radius is where both circles touch. To identify the maximum radius, the maximal possible distances from each originating node (centres of the circles) to a visible point within the boundary are calculated. The shorter distance is used as maximum radius. The obligatory radii for each pair of nodes are those where the intersection points lie on a segment of the boundary or obstacle polygon. When the circles of a primary and a subordinated node are intersected, the obligatory radii have to be determined iteratively.

The iteration method is described here below and illustrated in figure 7. B_1 is the starting vertex of a segment, IP the unknown intersection point of circles with the centres (originating nodes) n_1 and n_2. The vector $\underline{v}$ points into the direction of the segment and is normalised to (length) 1. The unknown vectors $\underline{u}$ and $\underline{w}$ point from the intersection point to the respective centres.

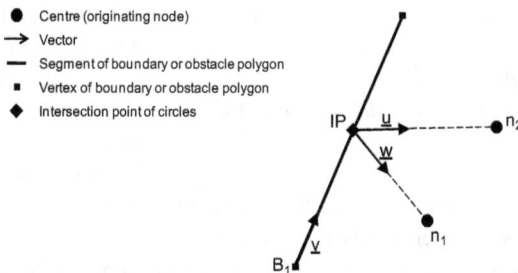

Figure 7. Setup for the iterative calculation of intersection points on polygon segments

To compute the unknown vectors, the following constraints (equations 3 to 7) are used with u_x, u_y, w_x, w_y, v_x, v_y as x- and y-components of the respective vectors, n_{1x}, n_{1y}, n_{2x}, n_{2y}, B_{1x}, B_{1y} as x- and y-coordinates of the respective points and m as length multiplication factor of vector $\underline{v}$, which expresses the distance from B_1 to IP:

$$u_x + n_{1x} - n_{2x} - w_x = 0 \tag{3}$$

$$u_y + n_{1y} - n_{2y} - w_y = 0 \tag{4}$$

$$m \cdot v_x + w_x + B_{1x} - n_{1x} = 0 \tag{5}$$

$$m \cdot v_y + w_y + B_{1y} - n_{1y} = 0 \tag{6}$$

$$\sqrt{w_x^2 + w_y^2} = \sqrt{u_x^2 + u_y^2} \tag{7}$$

Equation 3 and 4 express a closed vector-path from IP via n_1 and n_2 back to IP. Equation 5 and 6 represent the closed path from B_1 via IP and n_1 back to B_1. Equation 7 guarantees, that the vectors $\underline{u}$ and $\underline{w}$ have an equal length. Taking the constraints into account, m can be computed using equation 8:

$$m = \frac{n_{2x}^2 - 2 \cdot n_{2x} \cdot B_{1x} + n_{2y}^2 - 2 \cdot n_{2y} \cdot B_{1y} + 2 \cdot n_{1x} \cdot B_{1x} - n_{1x}^2 + 2 \cdot n_{1y} \cdot B_{1y} - n_{1y}^2}{2 \cdot (n_{2x} \cdot v_x + n_{2y} \cdot v_y - n_{1x} \cdot v_x - n_{1y} \cdot v_y)} \tag{8}$$

Finally, IP can be calculated with equation 9:

$$\underline{IP} = B_1 + m \cdot \underline{v} \tag{9}$$

In the first iteration, the centres are moved in normal direction away from the segment by the value of the additive constant. The normal direction is used as temporary vector for $\underline{u}$ and $\underline{w}$ and IP is computed. In the next iterations, the centres are moved in direction of the last $\underline{u}$ and $\underline{w}$ respectively. After the changes of $\underline{u}$ and $\underline{w}$ fall below a given bound, the final radii are calculated by adding the distance from IP to the centre and the additive constant.

After the minimum, maximum and obligatory radii are identified the intersection points (bi-sector points) related to the pair of nodes are calculated. Beginning with the minimum radii, the circles of the pair of nodes are intersected. For each intersection point it is checked if the point is visible from both centres and if no other centre is closer to the point. If one of the constraints is not fulfilled, the point is dismissed. Otherwise, the point is stored in a list, individually created for each pair of nodes. After the first intersections with the minimum radius, the radius is increased. The new radius is determined with a function, which takes the preceding intersection results into account. For small bends of the resulting bi-sectors the increment is larger than for intersections which form a strong bend. Increments may also be negative if necessary. When the maximum radius is reached, the obligatory radii for the points on the boundary and the obstacles are processed and the next pair of nodes is chosen.

Demand Allocation in Water Distribution Network Modelling – A GIS-Based
Approach Using Voronoi Diagrams with Constraints

293

3.4. Determination of Voronoi diagram

As the intersection points are stored in lists without order, the resulting polygons of the bi-sectors cannot be created directly. For each list a polyline is created first. This is done by picking the first point in the list and searching for the point with the greatest distance to it. This point is defined as the starting point of the polyline. Next, the closest point to the starting node is searched for, defined as the next point and marked. The process is repeated until no unmarked points are left on the list. To form the resulting Voronoi diagram, the individual polylines are merged by comparing the coordinates of the polylines' starting points and endpoints and assembling the nearest ones.

4. Testing

The developed approach and algorithm respectively are tested in order to evaluate the functionality and performance. Several configurations of given sets of nodes and boundaries are analysed by comparing the Voronoi diagram based on the Euclidean distance and the Voronoi diagram calculated with the novel approach. The comparison is mainly based on the size of the Voronoi regions. Two of the tested cases are discussed in the following.

The first case is a synthetic example configuration solely created to test the functionality of the developed algorithm. It consists of only three originating nodes (n_1, n_2, n_3), a non-convex boundary and three obstacles located within the boundary. Figure 8 shows the case and the calculated Voronoi diagrams based on the Euclidean distance and the shortest path. The individual sizes of the areas and the differences are shown in table 1. The differences do not sum up to zero as the approach is approximate and small holes between the individual Voronoi regions may occur. Furthermore, the values in the table are rounded to two decimal places.

Node	Voronoi region based on the Euclidean distance [length²]	Voronoi region based on the shortest path [length²]	Difference [length²]	Difference [%]
n_1	2,067.60	1,889.24	-178.36	-8.6
n_2	2,210.80	2,213.44	2.63	0.1
n_3	2,120.66	2,296.30	175.65	8.3

Table 1. Comparison of the Voronoi region sizes of case 1

The Voronoi region of node n_1 decreases significantly in size (-8.6%) with the new approach compared to the Euclidean Voronoi diagram due to the large area on the left. Considering the bounding polygon, this area is not visible from the primary node n_1. Thus, within the shortest path Voronoi diagram, the specific area is allocated to the Voronoi regions of the nodes n_2 and n_3 instead of n_1. The Voronoi region of node n_2 remains almost the same (0.1%): it gains area from the region of node n_1, but nearly the same amount is allocated to the region of node n_3 in the upper part of the image, where the bisector is bent to the right around the obstacle. Coming closer to the boundary the same bi-sector is bending a little to the left again. This is due to the change of the shortest path to bypass the left obstacle from the way around the obstacle on the left hand side to the way around on the right hand side

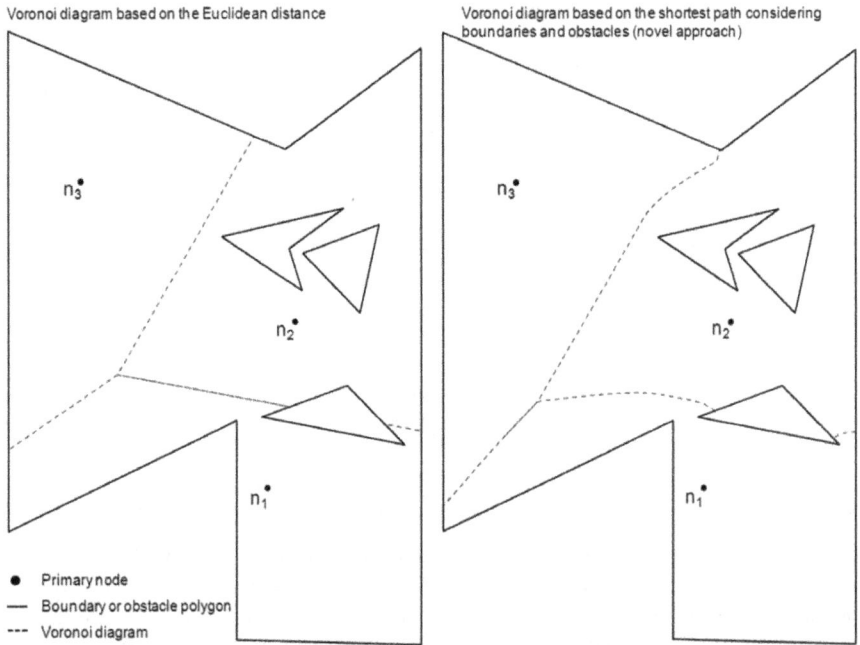

Figure 8. Test case 1 with calculated Voronoi diagrams based on the Euclidean distance and the shortest path (novel approach)

(between the two obstacles). Within the Euclidean Voronoi region of node n_3 no obstacles or non-convex sections of the bounding polygon are located which might have an influence on the bi-sectors if the Voronoi diagram is calculated based on the shortest path. Thus, it gains area of the Voronoi regions of both nodes n_1 and n_2 (8.3%).

The second case is a detail of an existing city in Algeria. It consists of 20 originating nodes (n_1, n_2, …, n_{20}), a non-convex boundary and several obstacles located within the boundary. Figure 9 shows the case and the calculated Voronoi diagrams based on the Euclidean distance and the shortest path. The individual sizes of the Voronoi regions and the differences are shown in table 2. In addition to the explanation given already for table 1, the area differences do not sum up to zero as only a small detail of the data set is shown.

Case 2 clearly shows, that the Voronoi diagram based on the Euclidean distance relates areas to nodes, which are closer to other nodes, if boundaries or obstacles are respected. This is the case, for example, for the nodes n_7, n_{11} and n_{12}. To node n_{11} the area to the right is assigned, which can only be reached by passing through the Voronoi region of node n_7. To the two nodes close to the obstacle in the centre, n_7 and n_{12}, areas on the opposite side of the obstacle are allocated. Contrarily, all Voronoi regions of the shortest path Voronoi diagram have a direct connection to the related originating node. As ascertained in table 2, the differences in the area sizes are up to 13.3% gain and -12,7% loss respectively.

Demand Allocation in Water Distribution Network Modelling – A GIS-Based
Approach Using Voronoi Diagrams with Constraints

295

Node	Voronoi region based on the Euclidean distance [length²]	Voronoi region based on the shortest path [length²]	Difference [length²]	Difference [%]
n_1	33,820.06	38,722.92	-4,902.86	-12.70
n_2	28,519.46	25,179.86	3,339.60	13.30
n_3	20,510.38	20,517.97	-7.59	0.00
n_4	14,185.69	13,991.04	194.65	1.40
n_5	19,486.64	20,731.79	-1,245.15	-6.00
n_6	36,329.44	35,075.23	1,254.21	3.60
n_7	t, the 51	33,492.01	-3,970.49	-11.90
n_8	15,188.57	14,350.52	838.05	5.80
n_9	3,656.39	3,656.57	-0.18	0.00
n_{10}	2,809.11	2,809.19	-0.08	0.00
n_{11}	23,676.38	25,289.44	-1,613.06	-6.40
n_{12}	16,706.28	16,185.26	521.03	3.20
n_{13}	16,916.03	17,686.10	-770.07	-4.40
n_{14}	21,487.65	20,796.47	691.18	3.30
n_{15}	7,085.54	7,085.79	-0.25	0.00
n_{16}	6,603.06	6,603.53	-0.46	0.00
n_{17}	3,779.35	3,831.44	-52.09	-1.40
n_{18}	3,156.40	2,834.48	321.92	11.40
n_{19}	11,720.82	11,494.35	226.47	2.00
n_{20}	23,474.75	22,949.54	525.21	2.30

Table 2. Comparison of the Voronoi region sizes of case 2

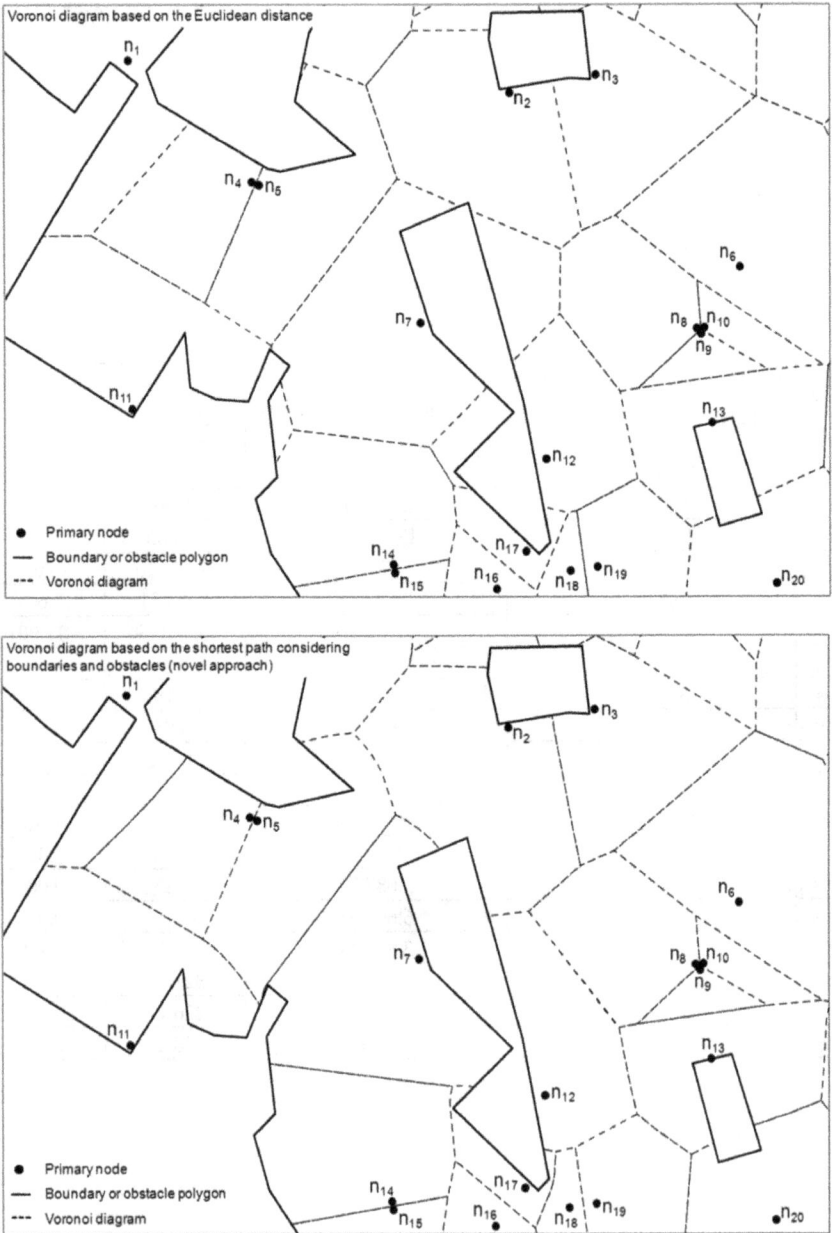

Figure 9. Test case 2 with calculated Voronoi diagrams based on the Euclidean distance and the shortest path (novel approach)

5. GIS-tool

Due to the spatial reference of the water distribution network data and the basic data for demand determination such like population density, population distribution, building areas etc. a GIS-application to determine and allocate water demand values to the network model nodes is advantageous. Following is a description of a GIS-tool implemented as an independent library of functions and integrated into ESRI ArcGIS (version 9).

ArcGIS, a product of the company ESRI, constitutes of several interlinked GIS products. The desktop application *ArcMap* supports visualising, analysing and editing of spatial data and linked attributes. The desktop application *ArcCatalog* enables data management. The data is commonly stored in relational data bases (*geodatabases*) but using single files (e.g. shape files) is possible too. Geometric objects (*features*) are modelled as vector data through defining shape and location of the object. Three principle *feature types* exist: *Point features* (punctual objects), *line features* (linear objects with start point, end point and vertices) and *polygon features* (areas defined by a polygon where the start and end points have the same location). Attributes (without spatial reference) can be linked to *features*. *Features* of the same type and with identical characteristics are stored in object classes (*feature classes*). All ArcGIS products are based on the library *ArcObjects*, developed as Microsoft Component Object Model (COM). *ArcObjects* offers numerous functionalities for the extension of ArcGIS products or the integration of applications into ArcGIS products. For further details on ArcGIS it is referred to the technical literature, c.f. e.g. Ormsby et al. (2004). The application for demand allocation described in this section is based on the ArcGIS data concept and the *ArcObjects* functionalities. The application was implemented as a dynamically linked library (DLL) which is integrated into *ArcMap* as a toolbar with an own user interface.

The demand allocation consists of two principle steps: (1) The calculation of the catchment areas of the nodes (Voronoi regions related to the originating nodes) and (2) the determination of the water demand within the catchment area and its allocation to the corresponding node.

For the calculation of the catchment areas (first step), input parameters are the demand nodes of the network model graph, the boundaries of the populated areas and the obstacles within the populated areas. The nodes have to be provided as *point features* in a *point feature class*, the boundaries and obstacles as *polygon features* in *polygon feature classes*. With the approach described in the previous sections the Voronoi diagram is calculated based on the nodes and stored in another *polygon feature class*. The references between Voronoi regions (catchment areas) and originating nodes are saved as an attribute of the nodes.

For the calculation of the demand per catchment area (second step) three options are developed and implemented considering the commonly available data basis: (A) Values and locations of the demands are given as *point features*, (B) areas with numbers of the population living there are given as *polygon features* or (C) building areas are given as *polygon features* and areas with the population are also given as *polygon features*. Moreover, option B and C need a constant demand per capita value as input parameter

Option A) The demand values [volume/time] are accumulated per catchment area by intersecting the catchment areas (*polygon features*) and the demands (*point features*) according to the spatial references. Finally, the accumulated demands per catchment area are allocated to the originating nodes.

Option B) The population per catchment area [inhabitants] is determined by intersecting the areas with constant values of the population (*polygon features*) and the catchment areas (*polygon features*) according to the spatial reference. To calculate the demand per node [volume/time], the population values [inhabitants] of the catchment area are multiplied with the demand per capita [volume/inhabitant/time].

Option C) The population per building area [inhabitants] is determined by intersecting the areas with constant values of the population (*polygon features*) and the building areas (*polygon features*) according to the spatial reference. The building areas, and thus the population, are accumulated per catchment area by intersecting the building areas (*polygon features*) and the catchment areas (*polygon features*) according to the spatial references. Finally, the demand per node [volume/time] is calculated by multiplying the population values of the catchment areas [inhabitants] and the demand per capita [volume/inhabitant/time].

Figure 10 shows an exemplary detail of the visualised result of a demand allocation with the GIS-tool. In the example, the nodes of the network model (black dots), a building area data set (dark green and orange areas) and a population density data set (transparent blue and transparent green areas) are given (as *point feature class* and *polygon feature classes*). Moreover, a satellite picture of the setting is used as basic layer. Based on the set of nodes and the bounding population density polygons, the Voronoi diagram was calculated applying the novel approach (*polygon feature class*). The Voronoi regions (*polygon features*) constitute the catchment areas of the nodes (red lines). Considering the given data, the demand is determined based on the population density and the building distribution (option C).

For each population density area (*polygon feature*) the number of inhabitants is calculated by multiplying the respective density and area. Next, the determined population is related to the building areas. Therefore, the total building area per population density area is determined by intersecting the two *polygon feature classes* according to their spatial reference. With the number of inhabitants and the total area of buildings per population density area the inhabitants per building area unit are calculated for each population density area. After accumulating the building areas per node according to the node catchment areas the number of population per node can be calculated. Finally, with the specific per capita demand, the demand per node is determined. Non domestic consumers with given demands (orange areas in the figure), such like industries or administrations, are excluded from the process described and allocated to the nodes directly.

Demand Allocation in Water Distribution Network Modelling – A GIS-Based
Approach Using Voronoi Diagrams with Constraints

299

Figure 10. Exemplary solution of demand allocation with the developed GIS-tool (option c)

6. Conclusion

The need for a more precise demand allocation in order to increase the accuracy of water distribution network models led to the development of a novel approach to calculate Voronoi diagrams with constraints. The constraints considered are convex and non-convex boundary and obstacle polygons. Thus, the approach basically solves the problem of generating Voronoi diagrams in arbitrary environments.

By means of a geometric approach, namely successive circle intersection with discrete radii, the bi-sectors which form the Voronoi diagram are determined. Thus, the result is not a mathematically exact solution but an iterative approximation. Hereby, the accuracy can be controlled by the distances between the individual radii. For GIS software, the discretisation does not present a drawback as only polygons with discrete vertices can be stored.

At the moment, the algorithm is not comparable with other algorithms determining Voronoi diagrams based on the Euclidean distance (sweep-algorithm, divide-and-conquer-approach) regarding the computation speed. But the advantages of increased accuracy of the results of an application for demand allocation in water distribution network modelling (and possibly in other applications) outbalance the speed issue. Furthermore, with an enhanced radius controlling mechanism and a reduction of the circle combinations for which intersections are computed, there is still potential for improvements. Moreover, the overall speed could be increased by using highly parallel computations, since each circle combination can be worked at separately. This could be facilitated by means of GPU or multi CPU usage.

Another method to compute the shortest path Voronoi diagram following the described approach is the analytic representation of the bi-sectors. A bi-sector between just two nodes can be defined as a mathematical function. The major challenge is to identify those sections of the functions, which make up the bi-sectors of the total Voronoi diagram based on more than two nodes by intersecting the functions. Due to the mathematical properties the function intersections cannot be computed directly but only iteratively. Thus, in a mathematical sense, the approach is not exact as well. A benefit of the analytically described bi-sectors is the fact that only two points per function have to be calculated. To compute a Voronoi diagram polygon, the relevant vertices can be simply calculated with a free choice of density using the determined functions. Thus, the computation has no impact on the time needed for the actual determination of the mathematical description of the Voronoi diagram. An analytic solution is currently being developed by the authors.

Furthermore, a GIS-tool to allocate the water demand was developed and presented in section 5. The GIS-tool consists of two principle calculation steps. First, the catchment areas of the nodes are computed using the developed approach to calculate Voronoi diagrams. The second step consists of three options to determine the demand. The most precise result is achieved by applying option A due to the superior input data. In this case, values and location of demands are known and directly allocated to the nodes according to the Voronoi diagram. If such input data is not available, the distribution of water demand has to be determined first. Option B offers an approach to determine the water demand based on the population distribution, which is considered to be constant for defined areas. Option C follows the same approach but provides a higher accuracy by relating the population (and thus, the demand) to the building density instead of considering a constant population distribution for specific areas. The developed GIS-tool has been already successfully applied in several network modelling projects of the Institute for Water and River Basin Management of the Karlsruhe Institute of Technology and proofed to be a useful solution for demand allocation.

Finally, it is worth mentioning that an application of the presented approach is not limited to demand allocation in water distribution network modelling. A broad field of applications is conceivable, such as determination of rainfall catchment areas, modelling of growth processes or stochastic analysis.

Demand Allocation in Water Distribution Network Modelling – A GIS-Based
Approach Using Voronoi Diagrams with Constraints

301

Author details

Nicolai Guth and Philipp Klingel

Karlsruhe Institute of Technology (KIT), Institute for Water and River Basin Management, Germany

7. References

Aronov, B. (1987). On the geodesic Voronoi diagram of point sites in a simple polygon. *Proceedings of the third annual symposium on computational geometry, SCG '87*, pp. 39-49, Association for Computing Machinery, ISBN 0-89791-231-4, Waterloo, Ontario, Canada, 1987

Aurenhammer , F. (1991). Voronoi diagrams – A Survey of a Fundamental Geometric Data Structure. *ACM Computing Survey*, Vol.23, No.3, (September 1991), pp. 345-405, ISSN 0360-0300

Dijkstra, E. W. (1959). A note on two problems in connexion with graphs. *Numerische Mathematik*, Vol.1, No.1, (1959), pp. 269-271, ISSN 0029-599X

Fortune, S. (1986). A Sweepline Algorithm for Voronoi Diagrams. *Proceedings of the second annual symposium on computational geometry, SCG '86*, pp. 313-322, Association for Computing Machinery, ISBN 0-89791-194-6, Yorktown Heights, New York, USA, 1986

Guth, N. (2009). *Ermittlung der Einzugsflächen von Knoten mittels Voronoi-Diagrammen unter Berücksichtigung von Nebenbedingungen*, Bachelor thesis (unpublished), Hochschule Karlsruhe Technik und Wirtschaft, Karlsruhe, Germany

Johnson, W. A. & Mehl, R. F. (1939). Reaction kinetics in processes of nucleation growth. *Transactions of American Institute of Mining and Metallurgical and Petroleum Engineers*, Vol.135, (1939), pp. 416-458

Klingel, P. (2010). *Von intermittierender zu kontinuierlicher Wasserverteilung in Entwicklungsländern*, PhD thesis, Karlsruhe Institute of Technology, Department of Civil Engineering Geo and Environmental Sciences, Karlsruhe, Germany, Retrieved from <http://digbib.ubka.uni-karlsruhe.de/volltexte/1000019357>

Okabe, A.; Boots, B.; Sugihara, K. & Nok Chiu, S. (2000). *Spatial Tessellations: Concepts and Applications of Voronoi Diagrams* (2. edition), John Wiley & Sons, Ltd, ISBN 0-471-98635-6, Chichester, England

Ormsby, T.; Napoleon, E.; Burke, R.; Groess, C. & Feaster, L. (2004). *Getting to know ArcGIS Desktop: Basics of ArcView, ArcEditor and ArcInfo* (2. Edition), ESRI Press, ISBN 978-1589482104, Redlands, USA

Papadopoulou, E. & Lee, D. T. (1998). A new approach for the geodesic Voronoi diagram of points and simple polygon and other restricted polygonal domains. *Algothimica*, Vol.20, Nr.4, (1998), pp. 319-352, ISSN 0178-4617

Preparata, F. P. & Shamos, M. I. (1985). *Computational geometry: an introduction* (1. edition), Springer, ISBN 0-387-96131-3, New York, USA

Shamos, M. I. (1975). Geometric complexity. *Proceedings of seventh annual ACM symposium on Theory of computing, STOC '75*, pp. 224-233, Association of Computing Machinery, Albuquerque, New Mexico, USA, 1975

Walski, T. M.; Chase, D. V.; Savic, D. A.; Grayman, W.; Beckwith, S. & Koelle, E. (2003). *Advanced Water Distribution Modeling and Management*. 1, Haestad Press, ISBN 0-9714141-2-2, Waterbury, USA

Chapter 16

A Primer on Recent Advancement
on Freight Transportation

Shih-Miao Chin, Francisco M. Oliveira-Neto, Ho-Ling Hwang,
Diane Davidson, Lee D. Han and Bruce Peterson

Additional information is available at the end of the chapter

1. Introduction

The efficient and reliable flow of urban goods and services is essential for the economic well-being of the United States (U.S.), particularly the majority of Americans who live in urbanized areas. According to recently published estimates (FAF3, 2007), the U.S. freight system moves about 19 billion tons of goods valued at $17 trillion in 2007. These shipments result in a total of 5.7 trillion ton-miles of movements during the same year. The performance of the freight system has economic impacts on national productivity, the nation, the costs of goods and services, and the global competiveness of American industries. While demand for freight transportation has been rising steadily and shows little sign of abating, the capacity of the freight system has only modestly increased. Without an efficient freight system, other critical systems, such as energy supply, can be seriously impacted.

In order to prepare for future transportation challenges, the transportation agencies at federal, state and local levels are heavily engaged in comprehensive transportation planning efforts. Local, regional, and national planners and policymakers are beginning to more deeply understand freight transportation issues as they apply within urban areas in the United States and around the world. Furthermore, freight mobility is a leading factor in the decision of a private business to locate in or near a particular city. But the goods movement supply chain uses regional, state, and national infrastructure while adopting increasingly complicated intermodal and multimodal transport solutions so it is very complex. Therefore, both the public and private sectors conduct planning and analysis upon which to make critical strategic decisions. This requires that freight data, modelling techniques and other computational and visualization tools be available to aid transportation stakeholders in their decision-making and analysis.

For over a decade, the Center for Transportation Analysis (CTA) of the Oak Ridge National Laboratory (ORNL) has provided assistance to federal transportation agencies in the development of comprehensive national and regional freight databases and network flow models. The objective of this chapter is to provide readers with an understanding of recent ORNL advancements in freight analysis and visualization.

For the freight transportation system to move commodities effectively and efficiently, it needs to overcome geographic barriers within specified time constraints. To this end, the freight system must be multidimensional and dynamic, consider different modes of transportation (truck, rail, water, air, and pipeline), involve both public and private sectors, and continue to evolve through time. To understand this complex and complicated system, one needs not only reliable data but also sophisticated computational tools.

Over the years, practitioners within transportation research community have collected and created a wealth of information stored in many database systems throughout their organization. Some of these systems are well establish and maintained, while others involve smaller data sets collected by individuals or by an office for a specific purpose. In either case the data are valuable by themselves, and could potentially offer a broader insight if the users were given the ability to integrate these individual data sources

Because of the spatial and temporal components embedded in almost all forms of freight data, the Geographic Information System (GIS) is an ideal computation framework for freight transportation analysis and modelling. GIS is an excellently suitable tool for managing, planning, evaluating and maintaining the transportation system. Huge geographic databases can be created and maintained as well as many forms of spatial data can be integrated into a GIS platform. It provides the means for researchers to evaluate the system and determine the impacts of capacity enhancements, operational improvements, and public transportation investments. The network representation of the freight system in GIS is invaluable for analysts to visualize critical segments, links or nodes, and makes possible large network analysis that would be computationally intense in other platforms. Therefore, GIS was the selected tool to integrate data within the transportation community.

With the aid of a set of analytical and interactive tools, a GIS framework enables the freight transportation analysts to understand the complex and dynamic spatial interactions among commodities (ranging from raw material, semi-finished and finished products), shipper (farmers, mining companies, factories, and manufactories), carriers (trucking, rail and marine companies, freight forwarder, third-party-logistic provider), consignees, and end consumers.

2. Aims of the chapter

This chapter will provide the reader with an understanding of tools and models developed and used by the CTA to understand multimodal freight commodity flow. To this end, we begin with a description of the components of the transportation system: the mode of transportation, the commodities transported, the transportation industry sector, and the network sub-systems. We then address some of modelling techniques and computational tools used by practitioners to estimate current and future freight demand and flows on the network system. Specifically, this chapter focuses on geographic databases and analytical tools that have been developed by the CTA to understand U.S. goods movement. After reading this chapter, the reader will have a general understanding of the current freight transportation system and the several analytical tools that have been developed to estimate freight demand and evaluate the system performance.

3. Understanding the movement of goods

To understand how the movement of goods is shaped we need to define the main agents, or system components, affecting the demand for freight. The resulting freight activity pattern of the interactions between such components can be captured through data surveys. In this section the freight transportation system is introduced and the framework for freight data developed by CTA group is also presented.

3.1. Freight transportation system

The term freight primarily refers to the long-haul component of the supply chain. Freight shipments themselves can move by truck, rail, ocean or air and can be characterized as intercity, port to transport terminal, terminal to terminal, interplant, plant to distribution center, and distribution center to distribution center. The freight transportation system can be seen as a system composed of three main components: decision makers or users with a demand for goods movements (producers, consumers, and transportation companies); the physical system supply (transportation infrastructure); and commodities (all different types of goods to be produced and transported).

The demand for the movement of freight involves multiple decision: *producers* of goods and services who decide how much and how to produce, and where and at what price to sell; *consumers*, either intermediate (production companies) or final (households, business, public agencies, etc.), who decide what to consume and how much; and *transportation companies* (shippers and carriers) who decide how to provide transportation services. These decision makers are responsible for production, logistics, distribution, and marketing. The location of producers and consumers certainly affects the spatial distribution of transportation supply, and therefore the pattern of goods movement.

Commodities are the result of production and represent the entity to be transported between geographic locations. The range of products needed and produced is vast and many final products (finished goods) can require a set of other products (semi-finished or raw materials) to be produced. Therefore, in assessing the freight demand, an analyst must be aware of these commodity matrices. Furthermore, a great variety of vehicle types are necessary to match different commodity types and shipment sizes. Depending on the characteristics (e.g. hazardous materials: toxics and flammable substances; high value; perishable) of a particular commodity, its means of transportation may require specific treatments, in terms of routing and vehicle specifications.

The physical system is an intermodal and multimodal system composed of transportation modes; sub-networks, where commodities move; and the interfaces between sub-networks, where commodities are transferred or handled between different modes. In a basic sense, freight transportation modes can be classified as truck, rail, water, air, multiple-modes, and pipeline. Each mode can include different types of fleet and equipment classifications from different transportation companies. The concept of mode in freight transportation is distinctly different from the case of passenger transportation. In freight transportation, mode encompasses physical (the sequence of transportation modes used for a consignment) and organizational (the sequence of entities responsible for transportation) aspects of movements.

In abstract terms sub-networks are composed of subsets of the single-mode network systems, i.e. highway, railroad, waterway and airway; and each of these subsets represents a different reality whether it is a type of service available, a company ownership, or a different type of vehicle (Southworth and Peterson, 2000). The interfaces connecting two or more sub-networks can be either transfer terminals (e.g. seaports and airports, where commodities are transferred between different modes, or different vehicles), and intermodal points (e.g. rail yards, where commodities transported on the same single-mode system are handled by different companies).

3.2. Framework for geographic freight data and analysis

Transportation engineers and policy makers require data and analytical tools to fully understand the complex movement of goods on the freight transportation system. Although considerable mode and shipper specific data is collected for a variety of purposes, the sheer magnitude of the freight data system as well as industrial confidentiality concerns, limit the freight data which is made available to the public. Typically, a considerable portion of the freight data is not disclosed by public agencies.

The Commodity Flow Survey (CFS) is probably the most comprehensive survey of freight activities in the U.S. The survey is conducted every five years as a part of the Economic Census by the U.S. Census Bureau, in partnership with Bureau of Transportation Statistics (BTS) under the Research and Innovative Technology Administration (RITA) of the Department of Transportation (DOT). The last survey was conducted in 2007 and previous years of study include 1993, 1997, and 2002. For 2007 CFS[1], approximately 102,000 establishments were selected from a universe of about 760,000 "in-scope"[2] establishments.

The BTS consolidates the CFS sample, estimates mileage and ton-miles[3] of shipments and publishes versions of survey for public access. The publicly available CFS data are restricted

---

[1] 2007 Econonic Census – Transportation, 2007 Commodity Flow Survey United States: 2007, EC07TCF-US, issued April 2010

[2] "In-scope" is a term used to refer to industries that are covered by CFS sample. Conversely a "out-scope" industry is not included in the CFS sample.

[3] Ton-miles is a measure of freight movement over the network. It is computed for each link (segment) of the network by simply multiplying the correponding link flow by its length in miles.

to confidentiality and reliability constraints. Reliability refers to the level of belief on the estimates provided by the sampling method (e.g. estimates from small sample with large variance are less reliable), whereas confidentiality corresponds to the impositions on the survey to avoid disclosure of particular industry activities. Consequently many cells in the public database have no data. In addition the data are also aggregated to higher geographic levels. The most detailed data in this database are at the state level.

The CFS data is used a base source for other freight data sources, either proprietary or of public domain. TranSearch, one of the most well-known commercial freight databases, for example, is compiled based on the CFS data and supplemented by private freight data and updated annually, and is available commercially through IHS Global Insight. The TranSearch database estimates data geographically at the county-level.

The Freight Analysis Framework (FAF) is a public database sponsored by the Federal Highway Administration (FHWA), which is composed of U.S. domestic and international (imports and exports) freight flows. FAF integrates CFS data (in-scope to the CFS) from a variety of supplemental sources for industry sectors not in the CFS (out-scope to the CFS) to create a comprehensive picture of freight movement among states and major metropolitan areas by all modes of transportation. FAF also provides forecasts on freight flows up to 30 years in the future as well as providing annual provisional updates for the current year and truck flow assignments for the base year and outlying future year. It is the third database of its kind, with FAF1 providing similar freight data products for calendar year 1997, and FAF2 providing freight data products for calendar year 2002. The CTA was heavily involved in the development and execution of the methodology for FAF2 and FAF3. With data from the 2007 CFS and additional sources, CTA developed estimates for tonnage and value, by commodity type, mode, origin, and destination for 2007 as well as ton-miles by mode.

The movements in FAF are characterized by freight volume (weight in thousand tons and dollar value), geographic dimension, transportation mode, and commodity type. In terms of the geographic dimension, the FAF data provides freight trading between 123 domestic zones and 8 external zones. The geographic level within the national boundaries is based on the CFS geographic strata, as shown in Figure 1, including 74 metropolitan areas, 33 remainder of states, and 16 regions identified as entire states. External zones, or foreign zones, are defined by 8 world regions: Canada, Mexico, and six other regions defined according to United Nations geographic region. Hence a domestic flow is spatially characterized by its origin and destination zones; imports are reported by foreign origin, FAF domestic zone of entry, and FAF destination zone; and exports are reported by FAF domestic origin, FAF domestic zone of exit, and foreign destination.

In terms of commodity classification, FAF reports freight flows using the same 43 2-digit Standard Classification of Transported Goods (SCTG) classes, as reported by the CFS. Figures 2 and 3 illustrate the amount of U.S. foreign trades in 2007 estimated by FAF, by tonnage and value of goods, respectively.

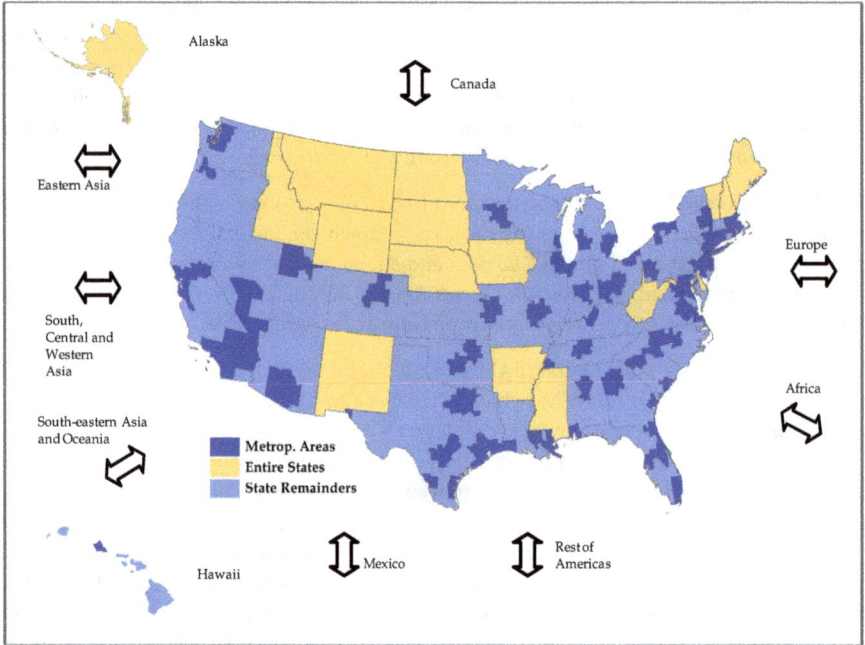

Figure 1. FAF Geographic Zones

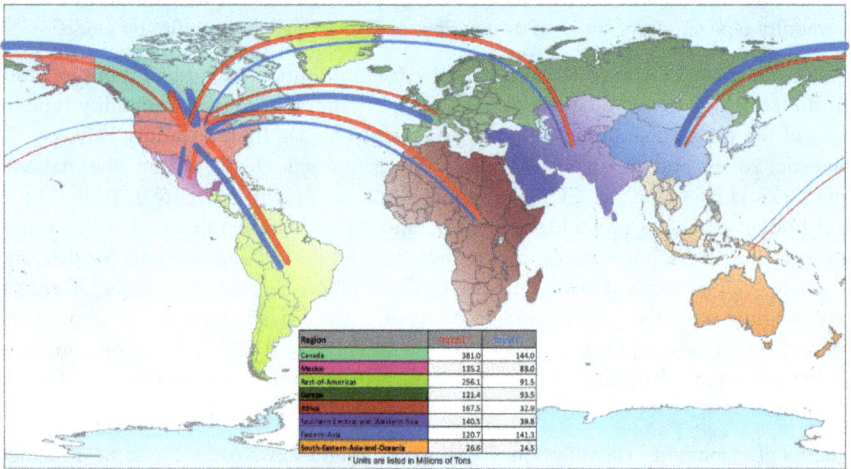

Figure 2. U.S. Foreign Trades, tonnage in 2007

For FAF3, the CTA added the estimations of mileage and ton-miles for freight flows. These ton-miles estimates were derived using models of the network system (see Section 4.1), and freight demand models (see Section 4.2). The CTA team manages detailed geographical data and information of a large multimodal and intermodal freight network, including highways, railroads, waterways, airways, and pipelines, and their associated infrastructure (e.g., intermodal terminals, transfer points, seaports and airports). Section 4.1 describes the construction of geographic representation of the highway-waterway-railway network systems. Among other analysis, this geographic tool makes possible to obtain likely routes for freight movements between geographic zones. The corresponding routing distances in miles are then used as estimates of mileages for freight movements.

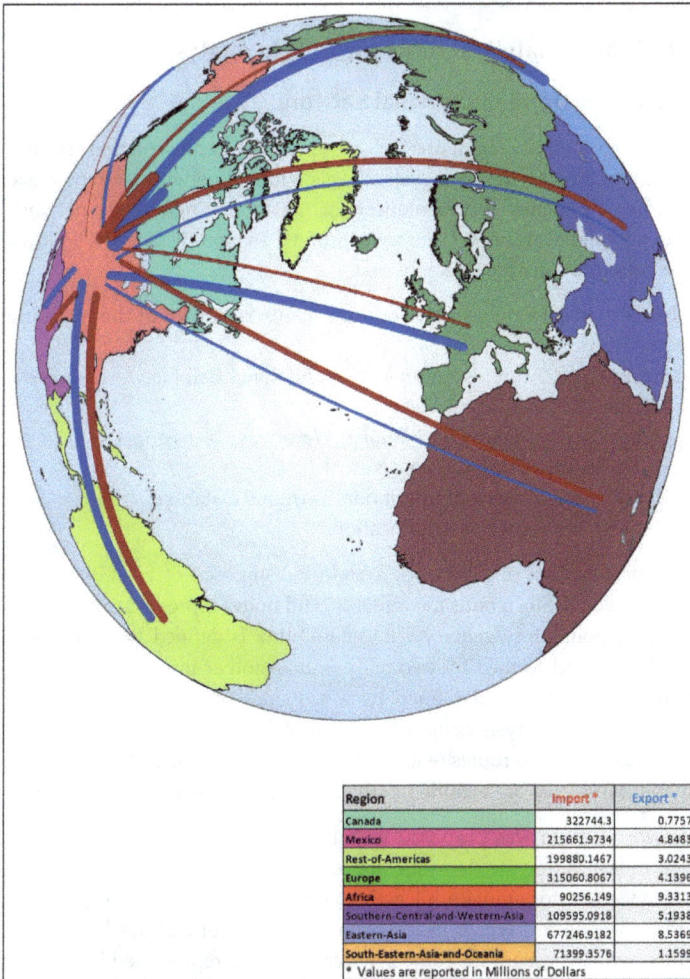

Region	Import*	Export*
Canada	322744.3	0.7757
Mexico	215661.9734	4.8463
Rest-of-Americas	199880.1467	3.0243
Europe	315060.8067	4.1396
Africa	90256.149	9.3313
Southern-Central-and-Western-Asia	109595.0918	5.1938
Eastern-Asia	677246.9182	8.5369
South-Eastern-Asia-and-Oceania	71399.3576	1.1599
* Values are reported in Millions of Dollars		

Figure 3. U.S. Foreign Trades, value in million dollars

Because the geographic detail of the network representation is higher than the geographic level of FAF database, certain extrapolation of freight movements from a higher geographic level (state-, metropolitan-, and remainder-level) to a lower geographic (county-level) is required in order to provide ton-mile estimates. This disaggregation of FAF database is done using freight demand models, which associates freight activity to the exogenous variables related to economic activity and network measurements (e.g. travel times, monetary costs, distances, etc). A discussion on freight demand models is presented in Section 4.2, and some models used by CTA are presented in Section 4.3. Section 4.4 presents how ton-miles were estimated for the 2007 FAF.

4. Geographic and analytical tools for freight analysis

4.1. CTA's intermodal and multimodal network

The CTA team in ORNL maintains a computerized representation of the national intermodal and multimodal network system. This national network system was created by combining earlier digitalized representations of the three single-mode network systems: highway, railway, and waterway (Southworth and Peterson, 2000). The following digital databases were used to construct the network:

- the ORNL National Highway Network and its extensions into the main highways of Canada and Mexico;
- the Federal Railroad Administration's (FRA) National Rail Network and its extensions into the rail lines of Canada and Mexico;
- the U.S. Army Corps of Engineers National Waterways Network;
- the ORNL constructed Trans-Oceanic Network;
- the ORNL constructed National Intermodal Terminal Database;
- a database of 5-Digit Zip-Code area locations.

The abstract representation of a network system is composed of a set of links and nodes, where links represent events (goods movements) and nodes represent connections as well as starting or ending points for events. A "line haul" link is defined by a unidirectional link with positive length and formed by two endpoints. Another important abstract concept is the definition of route which is defined by a sequence of directed connected links. The geographic scope of the analysis defines the detail of network needed. At the national level only main physical links and routes (e.g. interstates, arterials, railroad mainlines and branch lines, ocean and rivers, etc.) are included in the network representation.

The CTA network system was proposed with the main intention of estimating the routes, and therefore mileages, for the domestic and export shipments reported in the 1997 CFS. Therefore, in an effort to simulate all activities reported by the CFS, physical links are represented by logical links which in turn simulate different realities in each single-mode network. In the railway system the same physical links are represented by different links to simulate not only a railroad owner but all different railroad companies that have trackage rights over the link. Similarly in the highway system logical links were included to represent both for-hire and private services. The waterway system was separated into three different sub-networks, each one representing a different type of vessel and/or movement: inland and inter-coastal (largely barge traffic), Great Lakes, and trans-oceanic or "deep sea". This logical separation of the single-mode systems was important to model transfer costs between trucking services, railroad companies, and vessels.

Special links were designed to simulate the interfaces between each of the above logical sub-networks. These logical links are named "terminal links" and "interline links". The former simulate unloading /loading operations within terminals to handle goods between different vehicle types. The latter were specially designed to model the locations where goods movements are switched between two railroad companies. It is worth noting that terminals are represented by nodes in the network and their corresponding transfer links are links of zero-length connecting two logical endpoints at the same location. Interline are also links of zero-length connected at the same physical node.

Points of origination and termination of freight movements are represented by node centroids[4]. Such representation is a way to simplify the network model given that it is infeasible to include all actual locations where freight movements are originated and terminated. At the current static CTA network these nodes represent the county centroids.

The connection between generator points, centroids and terminals, and the network system is made by specific "access/egress links". Such links should therefore represent all movements connecting the real freight generators of a geographic area to the network system. For the CTA network a computation routine has been deployed to create these links "on-the-fly", that is every time there is a request to route a movement from an origin centroid to a destination centroid. Figure 4 illustrates how a shipment with mode sequence truck-rail-truck is routing onto the CTA network system. In simple words, a route is generated using a "shortest-path" routing algorithm that executes the following sequence of searches on the network: initiate the route by accessing the highway network (create access links), search for connection of it via truck-rail terminal to the rail sub-network, and return to the highway network via a second multimodal terminal transfer.

Although each individual layer (sub-network) of the network system can be stored and maintained in a commercial GIS, the combined multimodal network poses impediments for its use in GIS. Such impediments are related to the overlapping of geographic features (logical links of the same physical link), and the representation of many geographic features (i.e., terminal links and interlines) over a single degenerate point. In many standard GIS software zero-length links cannot be accepted and links that meet at the same location must be connected. Therefore, to use the network in standard GIS a special version of the CTA network has been prepared. In this version, the locations of logical nodes, along with the ending vertices of polylines incident to them, falling at the same geographic location are slightly perturbed preventing spurious link connections and allowing slightly positive lengths for the zero-length links. A view of the CTA network in GIS is shown in Figure 5.

---

[4] Centroids are abstract representations of all generators points of a given geographic area.

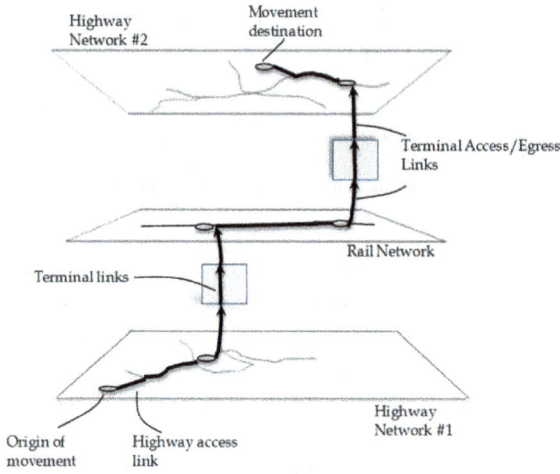

Figure 4. Shipment routing on the CTA network system

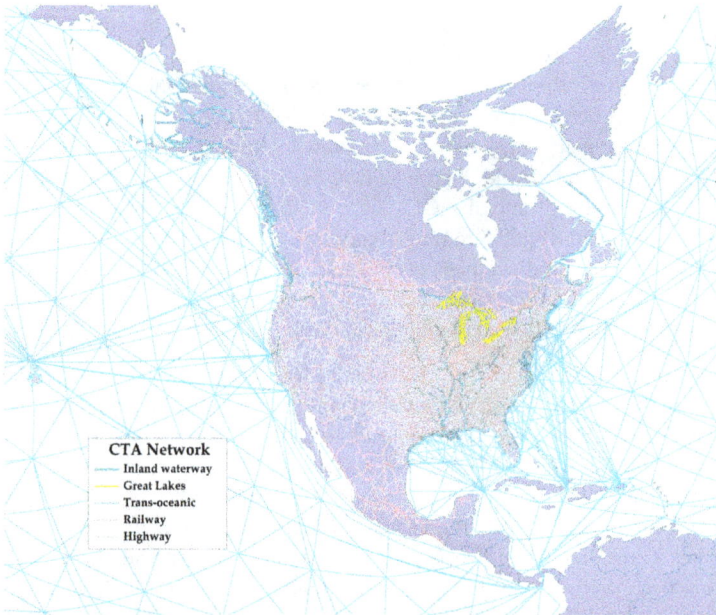

Figure 5. CTA Network System in GIS

A route selection routine was developed to determine likely routes over the network for shipments with known sequence of modes. This required the development of impedance functions to represent the generalized cost of different en-route activities over network facilities (line hauls, terminal links, interlines, access and egress links). This process started with a set of what Southworth and Peterson (2000) termed "native link impedance function". Routes over the highway network are determined according to link operational speed. Therefore, highway impedances are surrogates for link travel time. In the railway network impedances are assigned on the basis of the rail line class, i.e. main lines (long-haul and high capacity lines, thus with lower impedance for movement), and branch-lines (short-haul and less capacity, and therefore with higher impedances). Waterway links received identical native impedance due to the fact that rarely there is more than one choice in routes between any pair of geographic zones.

Native impedance values have been allocated to terminal links and interline links in attempt to simulate transfer costs of unload/loading operations in terminals, and the cost of switching between railroad companies in interlines, respectively. Highway and railway access/egress links received impedance value of 5 times their link length. It is worth noting that the lengths of such links were increased by a circuit factor[5] of 20%. Lengths for water access/egress links were set to zero by assuming that most of the originators for water movements are near to the dock locations alongside the waterway network. However the real link lengths were used to calculate impedance for water access links so that the waterway network could be accessed at points closest to the centroid locations.

Given the native impedances the next step was to determine the relative costs of transports between different modes. Although transportation costs are in reality affected by many factors (related to both cargo and mode characteristics), if we assume that sequence of modes is known for a given shipment, the routing problem may become one of selecting the most likely transfer points between modes. To this end, in the routing algorithm, it has been also assumed that less expensive modes are preferentially used for as large a proportion of the trip as practicable, relegating more expensive modes to access role. Therefore the native impedances defined above were normalized to reflect differences in transportation costs between modes for multimodal movements. This normalization was such that if the water was used it dominated the route miles. Otherwise rail dominated, with highway usually used for terminal access and/or egress.

4.2. Freight demand models

As opposed to demand modelling for passenger transportation, there is not a universal paradigm to model freight demand, only individual examples. However, some of the techniques developed specifically to model inter-city and urban travel demand have been also used to model freight demand. Some of these approaches are presented in this section, most of which are based on the book written by Cascetta (2009).

---

[5] Circuit factor between two points over a network is defined by the ratio between their distance over the network and their straight-line distance.

The objective of estimating freight demand models is to represent the production and distribution of goods, either for intermediate use or final consumption, for a given time period. With aim of freight database and economic variables related to the production and consumption of goods, a system of freight demand models can be formally expressed as

$$d_{od}[p,c,m,k] = d(A, T; \beta) , \qquad (1)$$

where d_{od} represents a demand flow (usually expressed in tons) between a origin zone o and a destination zone d. The characteristics p, c, m and k are associated with *sectors of economic activity, commodity types, transportation modes* and *routes*, respectively. The A variables reflect the economics of production and consumption. T are variables related to attributes of the different transportation modes and services (times, costs, service reliability, etc). Vector β denotes the model coefficients.

Models can be classified according to the assumptions about modelling approach or to the level of data aggregation (see Cascetta, 2009). Based on the model assumptions, models can be of type *descriptive* if they merely describe empirical relationships between the exogenous (explicative variables related to the economic system) variables and the endogenous variables (response variables related to freight demand); or *behavioural* if they explain the behaviour of decision makers in choosing among the universe of choices involved in the production and distribution of goods. According to the unit of variables available for model calibration/estimation models can also be of type aggregate and disaggregate. Aggregate models use average of variables related to aggregate units (e.g. all companies of a given industry sector) aggregated to the geographic zone level, whereas disaggregate models use variables related of small units, within the geographic zone, such as individual companies or individual shipments.

In this section, we will discuss national freight models (aggregate and disaggregate models) for predicting annually domestic and foreign trade in the U.S. In the following discussion, the sector p represents the producer sector (or originated economic sector) which is consistent with the CFS database. In addition, in most of the following discussion, the subscript p is also omitted from the equations for simplicity in the notation. The allocation of freight flows on the transportation system (i.e., interaction between demand and transportation supply) is not represented in the models that will be discussed.

The model in equation (1) represents the freight demand resulting from choices made by decision makers of a given economic sector with respect to production, and spatial and modal distribution of freight demand. Such decisions affecting the freight demand are interrelated. However when the system of models of equation (1) represent the complete universe of choices that decision makers should face, a single model may not be feasible either computationally or analytically. In order to simplify the analytical representation, the single model of equation (1) is usually decomposed into sequence of sub-models each representing a step within a sequential decision process. Partial share models can be estimated by the traditional paradigm of transportation demand modelling (or four-stage model) which consists in separately estimating generation, distribution, mode choice models, and route choice models (see Ortuzar and Willumsen, 2011).

Equations (2) – (3) below present ways of decomposing the global model into sub-models,

$$d_{od}[p,c,m] = d_{o.} [pc](A). p[d/pco](A, T) . p[m/pcod](A, T) , \qquad (2)$$

$$d_{od}[p,c,m] = \alpha d_{o.} [pc](A) . d_{.d} [pc](A) . f(C_{od}) . p[m/pcod](A, T) , \qquad (3)$$

$$d_{od}[p,c,m] = d_{od} [pc](A, T). p[m/pcod](A, T) , \qquad (4)$$

where

$d_{o.}[pc](A)$ is the freight production model, which gives the total production of commodity c of sector p in zone o;

$p[d/pc](A,T)$ is the freight spatial distribution model, which gives the fraction of goods, or the fraction of the total production of c of sector p in zone o, that are transported to zone d;

$p[m/pcod](A,T)$ is the freight mode choice model, which gives the fraction of the total freight flow of commodity c of sector p between zones o and d that is transported by mode m;

$d_{.d}[pc](A)$ is the freight attraction model, which gives the demand for goods in the destination zones;

$d_{od}[pc](A,T)$ is a joint generation and distribution model which gives the actual spatial distribution of freight demand within a study area;

C_{od} is a variable related to the generalized cost of transportation. The generalized cost of transportation is formed by a combination of all attributes of the transportation system (performance variables, monetary costs, and different mode services) that separates zones o and d;

$f(C_{od})$ is an impedance function (sometimes called friction function) that decreases with the C_{od};

α is a constant parameter that should be calibrated to balance out quantities of production and attraction between pair of geographic zones.

In the following sections, we will briefly describe each of the sub-models in equations (3)-(4) and illustrate how some of the sub-models can be formulated as function of aggregate variables provided by the CFS regional database, as well as variables related to the transportation system from the CTA network, and variables related to a given economic pattern provided by the U.S. economic census. Below it is described aggregate descriptive models for generation, descriptive and behavioural models for distribution and mode choice. A descriptive model for the joint generation and distribution of freight demand is also presented.

Generation models, $d_{.d}[pc](A)$ and $d_{.d}[pc](A)$, describe how much is produced (supply) and how much is needed (demand) for a given pattern of economic activity in a geographic area. Descriptive generation models represent the empirical relationship between freight demand (produced or attracted) by geographic area and a given pattern of economic activity. Such models can be applied to short term analysis in which the pattern of economic activity is given. In this case we try to estimate the amount of goods (in tons) generated due to decisions related to what and how much to produce. The freight demand resulting from long term decisions, such as where to produce, are not considered in this simplified descriptive approach. Modelling long term decisions should incorporate behavioural aspects that may require disaggregate data to be modelled.

Equation (5) shows an example of a linear model for production of goods where the unit of aggregation are economic sectors within regions. The parameters of these models are usually obtained with statistical estimation methods, or multiple regression analysis.

$$d_o.[p] = \sum_k \beta_k X_{kpo} \tag{5}$$

Where

X_{kpo} are exogenous variables related to the economic activity of sector p in zone o.

β_k are the model coefficients to be calibrated.

Attraction models are not straightforward to estimate since they should represent the total amount of goods attracted to a zone that are produced by a given sector, from companies located at different zones, and used by multiple sectors in that zone. Therefore, one way to represent aggregate attraction model is:

$$d_d[p] = d(\mathbf{X}, \mathbf{W}; \beta) , \tag{6}$$

where

W is a matrix that containing coefficient factors representing the industry-to-industry trade of goods. These are binary values that are equal to one whenever a sector associated with a row can provide goods to a user associated with a column, and equal to zero otherwise. These assignment coefficients can be obtained from regional or national input-output accounts;

X are variables related to the economic of those industry sectors in the destination zone that use commodities produced by sector p, as well as demographic variables representing final consumption of goods.

Examples of variables related to economic activity are the number of employees or total payroll by industry sector in each geographic zone. Demographic variables are represented by population or number of households within each geographic zone.

Distribution models estimate the trade flow between geographic zones (or regions) of the study area. The product of such models is called an origin-destination matrix of freight flows that satisfies the "trip-end" production and attraction constraints.

The descriptive model, $\circledast d_o.[p]. d_d[p] f(C_{od})$, corresponds to well-known gravity model, where $d_o.[p]$ is the total production in zone o by sector p and $d_d[p]$ is the total freight demand attracted to d that is produced by sector p. The name "gravity" came from the model resemblance to Newton's law of gravity. The impedance function $f(C_{od})$ is a monotonically decreasing function of the generalized transportation cost C_{od}. One typical expression for this function which can be derived from the entropy maximization problem (see Wilson, 1967) is:

$$f(C_{od}) = exp(-\beta\ C_{od}). \tag{7}$$

In the context of passenger transportation, the formulation of equation (7) is obtained by finding the most likely distribution pattern, corresponding to maximizing an *entropy function* (see Subsection 4.3.2) subject to macro constraints on the total number of trips produced and/or attracted as well as on the overall transportation cost. The entropy function is a measure of the number of possible arrangements of individuals that gives rise to a certain distribution pattern. To extend this model to the case of freight, we would have to assume that each trip represents a unit of freight (tons) being transported from an origin to a destination point.

The behavioural model, $p[d/po](A,T)$, estimate a probability of choosing a destination d for a given industry sector and origin of shipment o. Therefore, this model combines the distribution and attraction models into a single formulation. Assuming that the set of alternatives is formed by all zones in the study area a formulation for this model can be estimated on the basis of the random utility theory. Under this paradigm, the most common model form is the *Multinomial Logit (MNL)* model, as expressed below:

$$p[d/po](A,T) = exp(V_d/\theta_d)\ /\ \textstyle\sum_i exp(V_i/\theta_i), \tag{8}$$

where

V_d is the expected utility, or systematic utility, value of choosing a destination d for given industry sector p and origin zone o;

θ_d represents the Gumbel distribution parameter of the perceived utility U_d, such that $U_d = V_d + \varepsilon_d$, where the ε_d values are the random residues, deviations from the mean value V_d, which are assumed to be independently identically distributed as Gumbel random variable with zero mean and scale parameter θ_d.

The systematic utility is formulated as function of the attributes related to characteristics of the alternatives and the decision makers. Equation (9) shows a typically linear specification.

$$V_d = \textstyle\sum_k \beta_k\ X_{kd}. \tag{9}$$

The attributes of the systematic utility are grouped into attributes of the activity system in zone d, or attractiveness attributes; and attributes that quantify the accessibility or cost of travel between zones o and d. Attractiveness attributes are variables that measure the attractiveness of a zone as destination. As mentioned before, they might be a function of the number of employees or the total payroll for a given consumer industry, or client industry. Demographic variables such as population are also measures of attractiveness. Cost or accessibility attributes are variables reflecting the generalized cost of moving goods between o and d. They can be a straight-line distance connecting the centroids of zones o and d, or generalized cost variables that take into account the different contributions for each of the modes available between zones o and d. By explicitly showing the measures of attractiveness and transportation cost, equation (9) becomes

$$V_d = \sum_h \beta_h \, A_{hd} - \beta \, C_{od}, \tag{10}$$

where

A_{hd} are the measures of attractiveness in zone d.

Mode choice model, $p[m/pod](A,T)$, predict the fraction or choice probability that decision makers of a given sector p select mode or service m to ship goods from zone o to zone d. The first random utility models were formulated to analyze transportation mode choice. The MNL formulation can be then applied to predict these mode choice probabilities.

The definition of the mode choice alternatives constitutes the first step in the modelling process. In the context of freight transportation the alternatives of a mode choice model are the individual transportation modes (truck, rail, water, air, pipeline, etc) and combinations of single modes (truck and rail, truck and water, truck and air) as described in the CFS database. Different services provided by carriers – related to delivery time, security, levels of priority, type and capacity of vehicles, suitability of vehicles to certain types of commodities, etc. – can also be used to represent elementary alternatives of transportation.

The next step corresponds to the definition of the choice set, which is the set of mutually exclusively alternatives or group of alternatives available for a given decision maker. In this case it is possible that some alternatives are unavailable for a given decision maker and therefore should be excluded from choice set. It may also happen that the analyst has not exact knowledge of the alternatives available. To handle the mode availability special treatments have been proposed in the literature (see Cascetta, 2009). By assuming that all decision makers face the same set of alternatives, the MNL formulation for the probability choice is

$$p[m/pod](A,T) = exp(V_m/\theta_m) \, / \, \sum_i exp(V_i/\theta_i). \tag{11}$$

An alternative to deal with mode availability would be to force the systematic utility to minus infinity, whenever the mode is not available, which would result in a choice probability of zero.

The joint generation and distribution model, $d_{od}[p](A,T)$, predicts the actual demand from a zone o to a zone d for a given pattern of economic activity and transportation system. This model combines the generation and distribution models into a single functional form. In some sense, such models are analogous to the gravity model, presented before, in which the number of trips (or zone mass) produced or attracted to a zone are replaced by exogenous measures of production and attractiveness of that zone. The descriptive formulation presented bellow is the traditional gravity model formulation in economics (Brocker *et al.*, 2011):

$$d_{od}[p](A,T) = \alpha_p \, A_o{}^{\xi p} A_d{}^{\zeta p} f_{pod}, \tag{12}$$

where

α_p is a multiplier representing the overall scale of industry p;

A_o and A_d are the measures of production and attractiveness of zones o and d, respectively. The Gross Domestic Product (GDP) can be used as an indicator for variables A_o and A_d;

f_{pod} is a distance decay function representing the trade impeding effect or transport cost as well as other barriers;

ξ_p and ζ_p are the elasticities to be estimated.

4.3. CTA's freight models

The CTA group developed descriptive freight generation models (production and attraction models) to predict freight demand by U.S. state resultant from domestic interstate commerce as a function of economic activity (Oliveira Neto *et al.*, 2012). Fundamentally, the industry activities, translated into capital and labour by industry sector, should be the main base to explain the economic activity within a geographic area. In addition, the CTA team uses the structure of the national freight database in GIS and network analysis tools to estimate descriptive models for freight distribution. Specifically the process of employing the gravity model for a given origin-destination matrix of freight demand and transportation supply system is presented here.

4.3.1. Generation models

This section presents the estimation of static freight demand model due to the domestic trade in the U.S. for the year 2007. Two major data sources were used in this effort: 2007 CFS tabulations and the 2007 County Business Pattern (CBP). Specific CFS tabulations requested as supplement information for FAF estimation process were used to obtain sample estimates for freight productions and attractions in weight units. These tables contain the information of domestic movements of goods between U.S. states for 28 industry sectors in 2007, classified by the 3-digit North American Industry Classification System (NAICS), which are responsible for the production freight transportation. The list of these NAICS codes can be found in CFS document referenced above[6] and BTS website[7]. It is worth noting that in this special tabulation only a small number of cells were suppressed due to likely small sample sizes and their resultant large sample variances. These cells were treated as missing information during the modelling process.

---

[6] 2007 Econonic Census – Transportation, 2007 Commodity Flow Survey United States: 2007, EC07TCF-US, issued April 2010.

[7] Source: 2007 Commodity Flow Survey – Survey Overview and Methodology – Industry Coverage, Accessed July, Available from: 2011http://www.bts.gov/publications/commodity_flow_survey/methodology/index.html#industry _cove

The CBP is an annual series that provides sub-national economic data by industry. The series is useful for studying the economic activity of small areas and for analyzing economic changes over time. In addition, it can be used as a benchmark for statistical series, surveys, or other databases between economic censuses. The survey covers most of the U.S. economic activity, except self-employed individuals, employees of private households, railroad employees, agricultural production employees, and most government employees. The database provides information on payroll, number of establishments and number of employees for industries classified, since 1998, according to NAICS. The variable chosen to represent the economic activity of a given geographic area was the annual payroll by industry sector.

In the modelling process, it was assumed that the origin zones (U.S. states) are producers, where the products or commodities (raw materials or final products) are obtained or produced. In contrast, the destination zones are considered the "users" where the products or commodities are used/assembled/modified by intermediate industry sectors. Both production and attraction models by industry sector were specified as a power model with single explanatory variable by the following equation:

$$y_{pi} = \alpha \, x_{pi}^{\,\beta} \, , \tag{13}$$

where,

y_{pi} denotes the response variable (shipment tons by industry sector p and state i);

x_{pi} denotes the exogenous variable (annual payroll by industry sector p and state i);

α and β are the model parameters, to be estimated.

Production and attraction equations have been estimated for each of the 28 industry sectors. With respect to production equations, the response variables in CFS represent total demand produced by a given industry sector and state. In this case the exogenous variable was the corresponding payroll by industry and state. As for the attraction equations, the response variable represent the total demand attracted by a state that was originally produced by one of the industry sectors. Therefore, the total demand attracted is not separated by the industries that use the goods, rather is classified by industry sectors responsible for production. To be consistent with the CFS data, the single exogenous variable, or explanatory variable, in the attraction equations is composed of the payroll by state for those industries that use the commodities produced by the originated sectors. Figure 6 shows a scatter plot of freight produced versus total payroll for industry sector 311 – Food Manufacturing. The data points are sample observations for the 50 U.S. states and the District of Columbia. The fitted curve and estimated production equation are also presented in the chart.

It is worth noting that Oliveira Neto et al. (2012) compared the structure of the 2007 models with models estimated based on 2002 CFS similar tabulations. The results of their empirical analysis indicated some structural change in the production and attraction models between 2002 and 2007 due to a possible increase in productivity and a significant economic growth over this short time period. In addition, even when no significant change in model structure was detected, the modelling process did not result in reasonable predictions of freight for a future year. After applying 2002 models to predict freight volumes in 2007, it was found that there may be other factors, besides payroll, ought to be included in the modelling process.

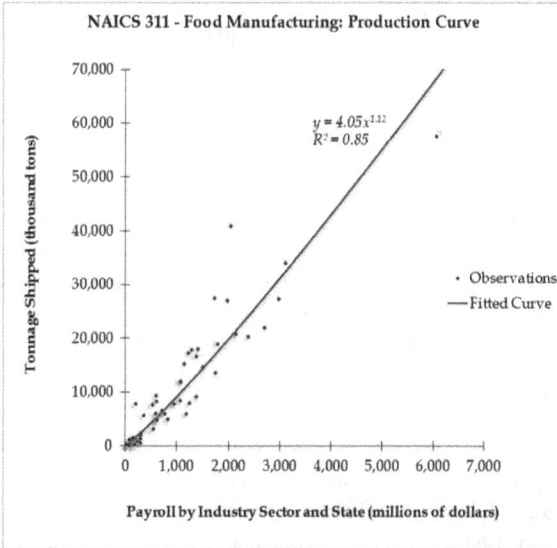

Figure 6. Fitted production curve for industry sector 311 – Food Manufacturing

4.3.2. Distribution models

In this section we illustrate how a gravity model can be estimated employing the same 2007 CFS tabulation used for estimating freight generation models earlier. As seen in Section 4.2, in transportation planning gravity model is used to balance an origin-destination trip matrix so that the zone-to-zone flows are consistent with the total trips generated at each origin zone and the total demand terminating at each destination zone. The model form that will be estimated for this exercise is

$$T_{pij} = \alpha_{ij} P_{pi} A_{pi} \exp(-\beta_p c_{ij}), \tag{14}$$

where T_{pij} is the annual freight flow in thousand tons to be estimated; P_{pi} is the total amount of freight (thousand tons) produced in state i by sector p; A_{pj} is the total amount of freight (thousand tons) attracted to state j from the originated industry p; c_{ij} is a measure of the distance that separates the states i and j; β_p is the specified parameter, the value of which says how important is the distance variable in explaining the trade between zones i and j for a given producer p; and α_{ij} are adjustment factors that should be calibrated to balance the freight flows by assuring that $\sum_j T_{pij} = P_{pi}$ and $\sum_i T_{pij} = A_{pj}$ for all i and j.

The formulation in Equation (14) is known as the classical gravity model. As demonstrated by Wilson (1967), it is derived by applying the Lagrange multipliers on the following optimization problem:

$$Maximize \; \xi = - \sum_{ij}(T_{pij}\log T_{pij} - T_{pij}) , \qquad (15)$$

$$\sum_j T_{pij} = P_{pi} , \text{ for all } i , \qquad (16)$$

$$\sum_i T_{pij} = A_{pj} , \text{ for all } j , \qquad (17)$$

$$\sum_i \sum_j T_{pij}c_{ij} = C_p , \text{ for all } i \text{ and } j , \qquad (18)$$

where the objective function in equation (15) is a monotonic function, often referred to as *entropy function*; Equations (16) and (17) are constraints representing our knowledge about the total productions and attractions per zone; Equation (18) is a constraint corresponding to our knowledge about the total expense, denoted by C_p, in using the network system by industry sector p. If we measure all c_{ij} in miles, C_p is therefore a measure of total annual ton-miles loaded on the network system for a given industry sector.

It is important to mention that if the parameter β_p is known in Equation (14) the values $exp(-\beta_p c_{ij})$ become reference factors, or the elements of a priory trip matrix, and the model reduces to what is called in the literature as ordinary gravity model. In this case the adjustment factors a_{ij} can be estimated by a bi-proportional matrix balancing method, also known as "iterative proportional fitting". This balancing method was apparently first described by Kruithof (1937), who used the model for prediction of telephone traffic distribution (see Lamond and Stewart, 1981). In transportation planning this method is referred to "Fratar Method" in the U.S. or "Furness Method" (see Furness, 1965) elsewhere.

Since we do not have complete information about the total expenditure C_p, the estimation of β_p cannot be done by directly solving the problem (15)-(18). If we had C_p it can be shown that a unique solution of the problem could be found by solving a system of linear equations of the flows T_{pij} and the Lagrange multipliers, one for each constraint. Approximation methods have been proposed for estimating β_p when the knowledge about the overall system expenditure is unknown. The method devised by Hyman (1969) was used to obtain estimations of the parameters β_p for each industry sector listed in Table 2. Hyman's method is an iterative procedure based on successive applications of matrix balancing techniques for a given sequence of estimates for β_p, that are appropriately readjusted at each iteration (see Ortúzar and Willumsen, 2011).

To estimate the model of Equation (14), it is first necessary to obtain a matrix of distances c_{ij} for travelling between origin and destination zones. In this exercise we assumed that the impedance to travel between a pair of U.S. states is determined by a function of the average distance in miles of the set of paths on the CTA network system between all corresponding pair of contiguous counties, as expressed by

$$c_{ij} = \sum_{rs}d_{rs} / n \text{ for all } r \in i \text{ and } s \in j , \qquad (19)$$

where

r denotes a county centroid within origin state i and s denotes a county centroid within destination state j;

n is the number of possible pair-wise combinations of counties that exists between zones i and j;

d_{rs} is the average distance in miles to travel on the CTA's network between the county centroids r and s.

The routes between county centroids are determined in terms of the impedance functions defined in Section 4.1. For the highway and the waterway system, distances between counties are determined on the basis on the minimum shortest paths so that one single route is found for each pair of counties. When railway is available the distance is estimated as an average distance calculated from the set of shortest routes obtained from all possible combinations of available railroad carriers. Recall that since the sequence of modes is unknown in the CFS tabulation used for calibration, the resulting distance between counties will be predominantly on water modes, whenever available, followed by railway distances, with highway used for short-hauls.

After applying the Hyman's methods using the 2007 CFS tabulation and the average distance matrix described above, we obtained the parameters listed in Table 1. A statistic for comparing the set of estimated flows $\{T_{pij}\}$ and the set of observed flows $\{N_{pij}\}$ is also provided in Table 1. This measure, Equation (20), dubbed standardized root mean square (s.r.m.s), was proposed by Pitfield (1978) as an alternative to deal with sparse matrices and also to consider the scale of the variables involved. However, it does not have any statistical property. It is merely a descriptive measure of goodness-of-fit.

$$\text{s.r.m.s}_p = [\; \Sigma_i \Sigma_j (T_{pij} - N_{pij})^2 \,/\, M] ^{\frac{1}{2}} \,/\, \Sigma_i \Sigma_j N_{pij} \,/\, M \qquad (20)$$

where M is the number of cells in the estimated set $\{T_{pij}\}$.

Table 1 also shows the R^2 statistics of the models and the total freight flows in thousands of tons for each industry sector. Note that the conclusion about the model's goodness-of-fit should not be made solely based on the R^2 statistics, as systematic errors and difference in variances cannot be captured by the R^2 statistic.

4.4. Ton-miles for U.S. freight movements

This section presents two methods for estimating the ton-miles of freight movement over the U.S. network system (within the U.S. boundaries). Based on the freight network and demand models described so far, two procedures can be identified to estimate ton-miles for freight movements:

- Shipment allocation for a given sequence of modes;
- Prediction of freight flows and average mileage between counties using CTA's network system.

2002 NAICS Industry	Total shipment (thousand tons)	β_p	R^2	s.r.m.s$_p$
212	3,638,114	0.0081	0.82	3.08
311	568,907	0.0027	0.94	2.12
312	143,523	0.0043	0.97	1.75
313	8,937	0.0020	0.81	4.64
314	6,939	0.0016	0.99	2.52
315	1,421	0.0007	0.78	5.57
316	584	0.0002	0.13	11.02
321	218,808	0.0039	0.92	1.74
322	166,443	0.0024	0.87	2.42
323	33,588	0.0022	0.85	2.67
324	1,415,085	0.0072	0.98	2.02
325	594,228	0.0028	0.97	2.42
326	66,709	0.0018	0.90	2.86
327	1,060,895	0.0103	0.97	2.19
331	201,312	0.0024	0.82	3.39
332	118,304	0.0032	0.95	1.87
333	40,484	0.0011	0.62	3.90
334	5,307	0.0009	0.78	4.94
335	18,736	0.0008	0.52	4.18
336	93,962	0.0023	0.92	3.10
337	18,635	0.0017	0.87	3.41
339	10,857	0.0008	0.21	5.38
423	1,361,118	0.0057	0.95	2.02
424	2,244,332	0.0061	0.94	2.15
454	55,654	0.0074	0.77	5.62
493	187,175	0.0049	0.98	1.43
511	11,825	0.0055	0.86	3.38
551	250,216	0.0038	0.89	3.07

Table 1. Gravity model parameters and goodness-of-fit statistics

The first alternative was the one proposed by Southworth and Peterson (2000) to allocate CFS shipments on the network system when the detailed sequence of modes is given. This method can be used to allocate freight movements onto the U.S. network system resultant from both U.S. domestic and foreign trade as long as the sequence of modes and the main geographic references (i.e. origin, destination and U.S. ports of entry/exit) are provided. Note that in CFS a shipment is characterized by its volume (weight and monetary value), the Zip Codes for origin and destination, and the sequence of modes used, as well as port of exit and foreign cities for exports. As discussed in Section 4.1 routing procedures were developed to find the most likely route and transfer points for a shipment with given sequence of modes. Such a computational tool can be used to allocate shipper-based databases onto the CTA network system to provide estimates of link ton-miles and, subsequently, the overall system ton-miles, with Equation (21), by transportation mode. As an illustration, Figure 7 shows an example of freight flows for shipments on the main railroad lines (i.e. Class I railroads), on the highway system, and on the inland waterway sub-network.

$$C_m = \sum_a T_{ma} \, d_{ma} , \qquad (21)$$

where

C_m denotes the overall ton-miles by mode of transportation m, as listed in Table 1;

T_{ma} denotes the total freight tons through a link a;

d_{ma} is the distance in miles over a link a;

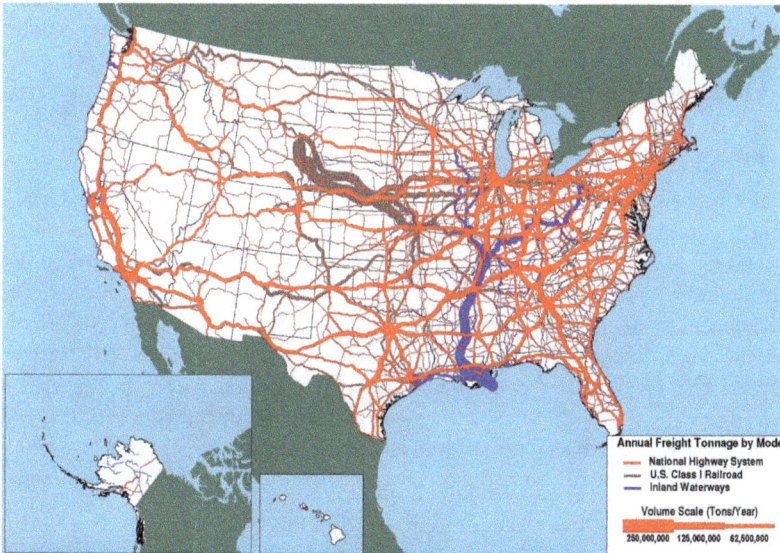

Figure 7. 2007 annual freight flow map by mode in U.S.

When a detailed description of shipments is not available, which is the case for the public CFS, the second procedure may be used. In general, the method is designed to estimate freight flows between U.S. counties as well as the mileages over the most likely routes connecting county centroids. With respect to the estimation of freight flows, freight models based on aggregated data (freight movements, and economic activity) are projected and applied to estimate freight movements in more disaggregated geographic level, based on economic disaggregated data. This process is called disaggregation of freight data. In our case, such disaggregation procedure relied on the estimation of separated nationwide generation and distribution models by industry sector as discussed in Section 4.3. Note that those models in Section 4.3 only represent the U.S. domestic trade and may not be used for estimating freight movements on the U.S. network resultant form foreign trade between U.S. and external zones (trading countries). Nevertheless, as we will see, such modes have been used to estimate ton-miles on the U.S. system for the 2007 FAF database resultant from both domestic and foreign trade.

As seen in Subsection 4.3.1, a single variable (i.e. state payroll by industry sector) was used to explain the effect of economic activity on internal freight demand. In that process a set of 28 production and attraction equations were estimated. With these models, freight productions and attractions were estimated at the county level (using the county payroll by industry sector). Such estimates were then used as weights to disaggregate each of the total FAF origin-destination flows to the corresponding county productions and attractions. These final estimates were then used as marginal totals for the distribution models. Regarding to distribution models, gravity models with exponential deterrence function have been estimated for each industry sector. The variable (argument of the deterrence function) used to explain the resistance for travelling between zones was an estimate of the average distance to travel between states on the CTA's network system (see Subsection 4.3.2). Using the estimated model parameters, freight movements between counties by the 7 classes of modes listed in Table 1 can then be obtained.

The following steps describe the FAF disaggregation procedure: a) FAF database is organized by industry sector and mode, and its flows classified by SCTG are grouped to the corresponding 28 NAICS groups; b) for each FAF origin-destination pair (classified by NAICS and mode), productions and attractions by industry sector were estimated (using the freight generation models) for the counties within the corresponding FAF origin and destination zones, respectively; c) using the estimated values from b) as weights, the FAF flow is then disaggregated to generate the productions and attractions by the corresponding counties; d) a matrix balancing procedure (i.e. gravity model by industry sector) is applied to distribute the estimated total productions and attractions from d) in order to generate the freight flows between the corresponding counties. The set of Equations (22)-(26) summarizes the disaggregation process.

$$w_{pur} = x_{pr}{}^{\mu_p} / \sum_r (x_{pr}{}^{\mu_p}) \text{, for all } r \text{ within } u \,, \tag{22}$$

$$w_{pvs} = x_{ps}{}^{\gamma_p} / \sum_s (x_{ps}{}^{\gamma_p}) \text{, for all } s \text{ within } v \,, \tag{23}$$

$$P_{pmr} = T_{pmuv}\, w_{pur} \,, \tag{24}$$

$$A_{pms} = T_{pmuv}\, w_{pus} \,, \tag{25}$$

$$T_{pmrs} = \alpha_{prs}\, P_{pmr}\, A_{pms}\, exp(-\theta_p\, d_{mrs}) \,, \tag{26}$$

where

T_{pmuv} is the FAF freight flow between FAF domestic zones u and v by mode m to be disaggregated;

w_{pur} is the estimated fraction of the total freight produced by sector p in FAF zone u that is generated by county r;

w_{pur} is the estimated fraction of total freight produced by sector p and terminated in FAF zone v that is attracted to county s;

P_{pmr} and A_{pms} are the estimates for freight production and attractions in the counties r and s, respectively, shipped and delivered by mode m;

μ_p, γ_p and θ_p are the estimated model parameters by sector p;

α_{prs} are the county adjustment factors calibrated for each industry sector p;

d_{mrs} is the distance in miles to travel over the modal network denoted by m.

The set of modal distances to travel over the U.S. network system is the second piece of information necessary for estimating freight ton-miles. To this end, the CTA network system is used for estimating the mileages d_{mrs} for likely freight movements between county centroids. In this case, distances over the U.S. network are estimated for each mode class (see FAF3, 2007, for a description of mode categories), as follows:

- For *truck*, *rail* and *water* modes, the distance between a pair of county centroids is determined based on the most likely route over the corresponding single-mode systems (highway, railway and waterway networks);
- Distances for *air* are estimated on the basis of the Great Circle Distances (GCD), which is the distance along the earth circumference between any two geographic points;
- The distances over the highway are used as surrogates for *pipeline* distances;
- For the *multiple mode & mail* category, distances are estimated by finding the most likely routes for a given *truck-rail-truck* mode sequence;
- For the *Other & Unknown modes* category, distances are determined by finding the most likely routes over the multimodal and intermodal network system, similarly to that described in Subsection 4.3.2;
- The *no domestic mode* does not use the domestic U.S. network and therefore is ignored in the ton-miles calculation.

With the disaggregated estimation of flows between counties and the county distances by mode, ton-miles for freight movements by mode can be calculated as follows

$$C_m = \sum_p \sum_r \sum_s T_{pmrs}\, d_{mrs}. \tag{27}$$

Table 2 shows comparisons of the ton-miles estimates based on FAF3 and other data sources over the major U.S. freight transportation systems: highway, railway, and waterway. In addition to the 2007 CFS, other sources of information include: a) *2007 U.S. freight railroad statistics*[8] from Association of American Railroads (AAR), and b) *2007 waterborne commerce*[9]

---

[8] For more information see AAR's report at: http://www.aar.org/~/media/aar/Industry%20Info/AAR%20Stats%202010%200524.ashx

[9] More information about the waterborne commerce can be found at:
http://www.ndc.iwr.usace.army.mil/wcsc/pdf/wcusnatl08.pdf

of the U.S. Army Corp of Engineers (USACE) Navigation Data Center. Note that, the *2007 U.S. freight railroad statistics* reports all freight activities for all North American railroad companies using the U.S. system. The *2007 waterborne commerce* data is based a complete census of all cargo via the U.S. waterway system (except military cargo carried on military ships). Besides CFS, there is no other current source for comparison of highway movements.

As expected all FAF3 estimates are larger than the corresponding CFS estimates due to the inclusion of freight activities for out-scope industries and import movements in the FAF3 database. Both FAF3 estimates of ton-miles for rail and water are consistent with estimates reported by modal data programs, AAR and USACE, respectively; with 2% difference in waterway movements and about 5% for railway movements. The discrepancy observed for railway movements is likely due to the following reasons. The FAF3 database does not account for transhipments from Canada (e.g., Canada-Mexico), which is estimated to be about 15 billion ton-miles annually. Furthermore, AAR report includes some movements of empty containers and the weight of containers for mixed freight (estimated to be about 60 billion ton-miles), which are not considered in the FAF3 database. With these two considerations, the gap between FAF3 and the *AAR* can be reduced to less than 1%.

U.S. Network Sub-system	Data source / Modes	Ton-miles (billions)
Highway	*FAF3 (Truck single mode only)*	2,420
	2007 CFS (Truck single mode only)	1,342
Railway	*FAF3 (Rail single mode plus rail portion of multiple modes)*	1,732
	2007 CFS (Rail single mode and portion of mutiple modes which includes rail)	1,530
	2007 AAR report (all rail activities)	1,820
Waterway	*FAF3 (water and the water portion of multiple modes)*	495
	2007 CFS (water and the portion of multiple modes which indlues water)	348
	2007 USACE waterborne commerce (all water activities)	506

Table 2. 2007 ton-miles estimates by truck, rail and water modes

5. Concluding remarks

This chapter presented a framework for national freight data suitable for national decisions with respect to movements of goods in the U.S. The CTA team in the ORNL developed and maintain a comprehensive database with all necessary dimensions (geographic, commodity nature, mode of transportations) to understand how goods move on the U.S. transportation system. To construct such database, geographic representation of the network system and analytical tools were extensively used. The chapter provided a number of analytical examples for modelling the demand for freight, and specifically presents the demand models developed by the CTA in the recent years. Although it is mostly descriptive, the chapter presents a simple application of how to estimate ton-miles over the freight transportation system.

Estimation of ton-miles is an import application of such tools to gauge the freight system usage and provide insightful information for national political decisions. The methodology for estimating ton-miles over the transportation system was based of prediction of freight flows and distances between U.S. counties. In the demand side, the disaggregation of freight flows was a necessary step to reconcile the freight database with the detailed representation of network system and derive more accurate estimates of ton-miles. As for the transportation system, the geographic representation of multimodal and intermodal interfaces over the transportation network system allows the analyst to predict likely routes that are more representative of real world activities involved in the transportation of goods. In sum, for a given transportation mode ton-miles are estimated by an element-by-element multiplication between a matrix of freight volumes and a matrix of average distances for the set of likely routes.

Models of the freight demand and the supply transportation system can also be applied in several other transportation related problems. Besides ton-miles estimation, the analytical and geographic tools described in this paper can help perform impact analysis at macro level such as *energy use and environmental impacts of the transportation activity, as well as effects of external events*. In future work CTA will apply the proposed models to investigate alternatives (e.g. modal shifting, vehicle technologies, etc) for alleviating some of negative effects due to the high use of fossil fuels. In addition, the network models will be used for identifying the system vulnerability and resilience to damage and disruptions caused by natural and manmade events (e.g. hurricane, flooding, terrorist attacks, chemical spills, etc).

Author details

Shih-Miao Chin, Francisco M. Oliveira-Neto, Ho-Ling Hwang,
Diane Davidson and Bruce Peterson
Oak Ridge National Laboratory, USA

Lee D. Han
The University of Tennessee, Knoxville, TN, USA

Acknowledgement

The authors are highly indebted to Michael Sprung and Rolf Schmitt of the Federal Highway Administration (FHWA) Office of Freight Operations and Management for their sponsorship of our freight system data and analysis work and their deep commitment to improving the quality and availability of freight data and analytical tools for the transportation community. We also express thanks to our colleagues who comprise the Bureau of Transportation Statistics (BTS) Commodity Flow Survey (CFS) team as well as the staff at the Bureau of the Census who worked on the CFS. Last, we appreciate the excellent maps that were prepared for this Chapter by Rob Taylor and Ryan Parten (University of Tennessee) and Frank Xu (ORNL).

The contents of this document reflect the views of the authors, who are responsible for the facts and accuracy of the information presented herein. The contents do not necessarily reflect the official views or policies of the Department of Transportation State or the Federal Highway Administration. This report does not constitute a standard, specification or regulation.

6. References

Brocker, J., Korzhenevych, A. & Riekhof, M. C. (2011). Predicting freight flows in a globalising world. *Research in Transportation Economics*, Vol. 31, pp. 37-44.

Cascetta, E. (2009). *Transportation Systems Analysis: Models and Applications*, 2nd Edition, Springer, ISBN 978-0-387-75856-5, New York, United States.

CBP (2007). County Business Pattern. U. S. Census Bureau, Accessed July 2011, Available from: http://www.census.gov/econ/cbp/index.html

CFS (2007). Commodity Flow Survey – Survey Overview and Methodology, In: *Research and Innovative Technology Administration. Bureau of Transportation Statistics*, Accessed July 2011, Available from: http://www.bts.gov/publications/commodity_flow_survey/methodology

FAF3 (2010). The Freight Analysis Framework, Version 3: Overview of the FAF3 National Freight Flow Tables, In: *FHWA's Office of Freight Management and Operations. Federal Highway Administration, U.S. Department of Transportation*, Washington, D. C., 2010, Accessed July 2011, Available from: http://faf.ornl.gov/fafweb/Documentation.aspx

Furness, K. P. (1965). Time function iteration. *Traffic Engineering and Control*, Vol. 7, pp. 458-460.

Hyman, G. M. (1969). The calibration of trip distribution models. *Environment and Planning*, Vol. 1, pp. 105-112.

Kruithof, J. (1937). Calculation of telephone traffic. *Der Ingenieur*, Vol. 52, E15-E25.

Lamond, B. & Stewart, N. F. (1981). Bregman's balancing method. *Transportation Research*, Vol. 15B, No. 4, pp. 239-248.

Oliveira-Neto, F. M., Chin, S. M. & Hwang, H. L. (2011). Aggregate Freight Generation Modelling: Assessing the Temporal Effect of Economic Activity on Freight Volumes using a Two-Period Cross-Sectional Data. Submitted to *Transportation Research Record: Journal of the Transportation Research Board.*

Ortuzar, J. D. & Willumsen, L. G. (2011). *Modelling Transport*, 4th Edition, John Wiley & Sons, Ltd, ISBN 978-0-470-76039-0, Chichester, United Kingdom.

Pitfield, D. E. (1978). Sub-optimality of freight distribution. *Transportation Research*, Vol. 12, pp. 403-409.

Southworth, F. & Peterson, B. E. (2000). Intermodal and international freight network modelling. *Transportation Research Part C*, Vol. 8, pp. 147-266.

Wilson, A. G. (1967). A statistical Theory of Spatial Distribution Models. *Transportation Research*, Vol. 1, pp. 253-269.

Chapter 17

Developing Web Geographic Information System with the NDT Methodology

J. Ponce, A.H. Torres, M.J. Escalona, M. Mejías and F.J. Domínguez-Mayo

Additional information is available at the end of the chapter

1. Introduction

The increasing popularity of the Internet has led to the development of Web applications, known as Web GIS. A geographic information system (called GIS from now) is a software system that manages geo-referenced information. GIS systems are an automated system used for storing, analysing and manipulating geographical information. Geographical information represents objects and actions where geographical location is indispensable information (Aronof, 1989; Bull, 1994). In this context, a Web GIS system offers different GIS services for analysis and visualization of geographical information on the Web (Kim, 2002).

Thus, there is a growing need for tools and methodologies that support the rapid development of this kind of Web applications and their modification to meet the ever changing business needs. Recently, some frameworks have been developed Autodesk MapGuide, ESRI ArcIMS, gvSIG, Quantum GIS, PostgreSQL + PostGIS that enable users to develop and deploy Web GIS applications selecting functionalities a user needs.

In order to develop such systems, it is necessary to follow a proper development process. This process should cover the inherent characteristics of geographic information processing and also is available on Web platform. With this motivation, Escalona (Escalona et al, 2008) made a first approximation of such processes. They introduced a process for developing Web GIS (Geographic Information Systems) applications. This process integrated the NDT (Navigational Development Techniques) (Escalona & Aragon, 2008) approach with some of the Organizational Semiotic models (Liu, 2000). The use of the proposed development process is illustrated for a real application: the construction of the WebMaps system. WebMaps is a Web GIS system whose main goal is to support harvest planning in Brazil (Macário et al., 2007). The process can be seen on the next figure:

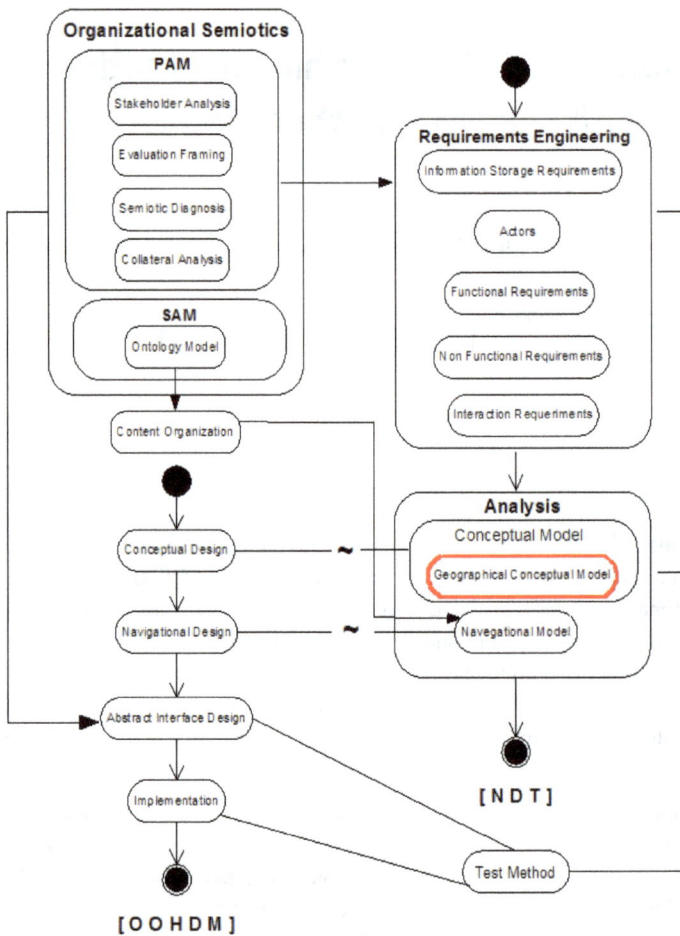

Figure 1. Development process for Web GIS applications (Escalona et al, 2008)

In the first activity, the process begins with a Requirement Engineering phase. Reinforce this phase with improving the process of capturing requirements GIS environments by defining a template for all those requirements. Next, an Analysis phase is performed. Later, the Abstract Interface model is built. Finally, the target system is implemented. Figure 1 shows the development process. Shows how models from NDT and from Organizational Semiotics are built following this process. The OOHDM (Oriented-Object Hypermedia Design Method) (Rossi, 1996) models are presented in the Figure 1 in order to reference the classic models. OOHDM is one of the most studied approaches. This approach for web developing is based on Hypermedia Design Model (HDM) (Garzotto et al., 1993) a structured approach to model hypermedia system.

The objective of the analysis phase is to detail, define and validate the requirements. In the NDT methodology, the analysis phase involves the definition of two main models: conceptual model and navigational model. A relational model might be generated from the conceptual model. The relational model is useful for the implementation of the databases of the system under development. NDT is helpful in the generation of navigational model for the pages that include information processing. However, NDT is not helpful for organizing static information. The Organizational Content approach is, then, used to complement the content and models generated by NDT. The Organization content used has been proposed and adapted from the ontology diagram by the research team at Institute of Computing Unicamp (Baranauskas et al., 2005). This process is left open and pending the definition and detailed specification Geographical Conceptual Model (highlighted in Figure 1). Therefore, the purpose of this paper is the formal definition of Conceptual Model inside the process for developing Web GIS systems.

This work is organized as follows. Section 2 introduces preliminary concepts needed to understand the process. Section 3 describes the formal definition of Geographic Conceptual Model inside development for Web GIS with NDT. Then, section 4 presents the case study based on formal definitions. Finally, section 5 outlines conclusions and future works.

2. Preliminary concepts

Then, will show some preliminary concepts. The first section will focus on Web GIS systems and the second; will outline briefly some concepts of NDT methods.

2.1. Web GIS

Web GIS are a special case of Web applications, meant to deal with complex geographic data and share them across several users for different business goals. Geographic information is usually distributed across different layers, which a Web GIS user should be able to handle separately or in overlay modality. Thus, besides common Web navigation and composition tasks, far more complex functionalities are needed in Web GIS for visualization and content management. As a matter of fact, with respect to traditional Web applications, this kind of systems require special focus on spatial data which may be acquired from different sources and stored in different formats, for all of which the user should be offered direct support. From a design point of view, Web GIS present many specific characteristics, making them different from traditional data intensive Web applications. Among them, two are of fundamental relevance: one is related to the data model and the other is concerned with the navigational model. As for the first aspect, the complex nature of geographic data, where the two components, descriptive and spatial, should be analyzed and managed in a joint manner requires the use of a different modelling approach, named Geodata and Metadata Conceptual Model. In our proposed methodology we will present the formal definition of this type of model with NDT. However, it is important to note the popularity of the E-R graphical notation, as well the intuitiveness of its extension to represent geographic data, led us to exploit the Spatial E-R model as core of the Geodata and Metadata Modeling phase. According to this model, each set of geodata is described as a spatial entity characterized by a set of attributes, geometry and a couple of coordinates. Moreover, a similar extension has also been defined for relationships, so as to model entities which are spatially related, in terms of topological relationships. The expected output from this geodata design task consists of a set of logical schemas, which organize data in a relational way and associate them with layers, according to their meaning and the underlying data structure, which can be raster- or vector-model based. In addition, since a Web GIS application may also be aimed at supporting data sharing, metadata management is a crucial issue to be faced. Making interoperable data, which are heterogeneous in terms of formats and/or sources, may guarantee better performances during data exchanging and retrieving.

As for the navigational model a set of functionalities meant to dynamically navigate a map, commonly known as Web mapping. The generation of the Web Mapping model is not the aim of this work.

2.2. Navigational Development Techniques (NDT)

NDT (Navigational Development Techniques) is a model-driven Web methodology that was initially defined to deal with requirements on Web development. NDT starts with a goal-oriented phase of requirements and defines a set of transformations to generate analysis models.

NDT has evolved in the last years and offers a complete support for the whole life cycle. Figure x, represents processes supported by NDT. It covers viability study, requirements treatment, analysis, design, construction, implementation as well as maintenance and test phases as software development phases. Additionally, it supports a set of processes to bear out project management and quality assurance.

In the last year, it evolved to support different life cycles: sequential, iterative, agile processes, constitute some examples, although in Figure 2, the sequential life cycle was only shown to make the representation easier.

One advantage is that NDT can be used in the enterprise environment. Nowadays, a high number of companies in Spain work with NDT and the associated tools to develop software. This is possible due to the fact that NDT is completely supported by a set of free tools, grouped in NDT-Suite. This suite enables the definition and use of every process and task supported by NDT and offers relevant resources for quality assurance, management and metrics to develop software projects. NDT is based on the model-driven paradigm. It selects a set of metamodels for each development phase (requirements, analysis, design, implementation, construction, test and maintenance) in order to support each artifact defined in the methodology.

All concepts in every phase of NDT are metamodeled and formally related to other concepts by means of associations and/or OCL constraints. For instance, Figure 3, presents the metamodel for the requirements phase.

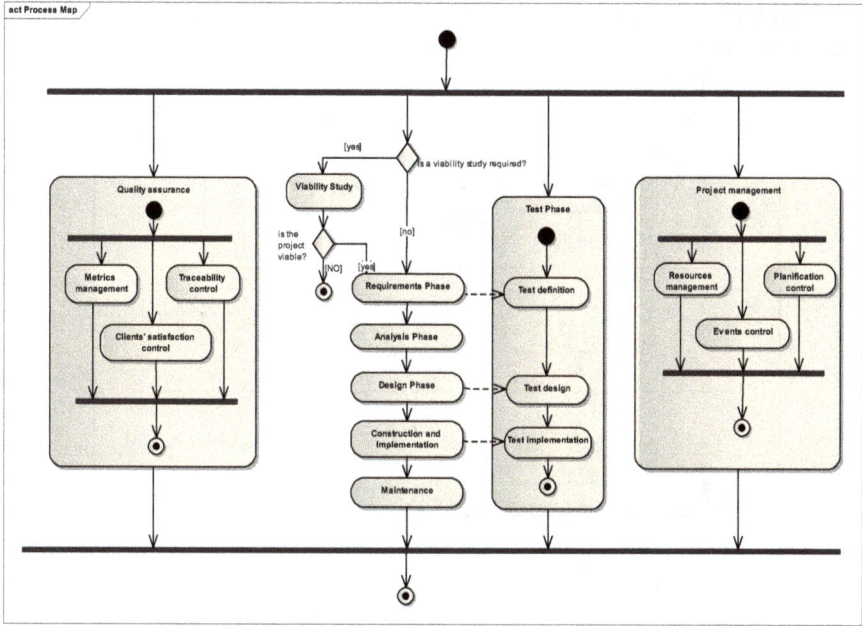

Figure 2. A general view of NDT Processes

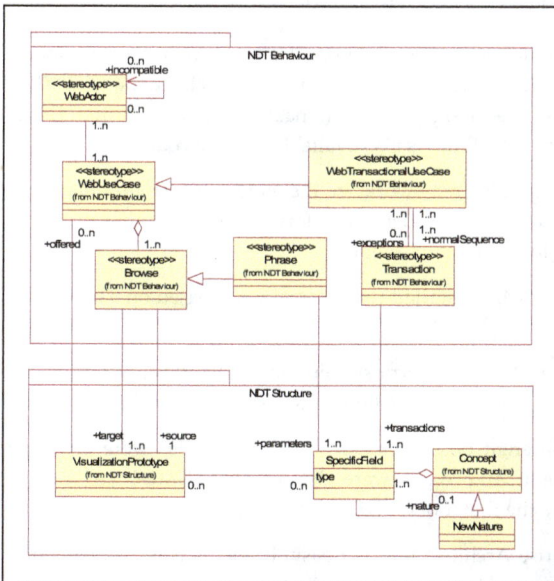

Figure 3. Requirements Metamodel in NDT

NDT proposes a set of QVT (Query/View/Transformation) among each metamodel for every phase, that enables to get one phase results from the previous one. Figure 4, shows this idea in a sequential development life cycle.

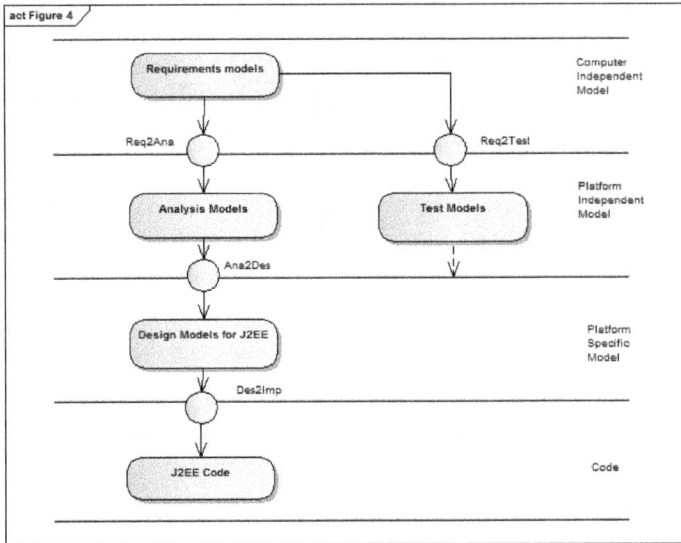

Figure 4. Transformations in a sequential life cycle

In this case, NDT defines transformations from the requirements model to the analysis model (Req2Ana), from the requirements model to Functional Test (Req2Test), from the analysis model to the design model (Ana2Des) and from the Design model to the implementation one (Des2Imp) is also available for Java code.

Nevertheless, using this framework in the enterprise environment is not possible. In companies, the use of metamodels, transformations and other elements is not real and technology seems too abstract for them.

For this reason, in 2004, we started two lines of work based on a new set of tools to support these NDT ideas:

1.   Enterprise line: Offer an enterprise solution for the use of NDT.
2.   Research line: Offer a way to validate transformations rules and metamodels.

After assessing different possibilities, some UML-profiles were developed for each NDT metamodel. These UML-profiles were defined in a UML-based tool named Enterprise Architect and then, the first tool for NDT-Suite was developed and called NDT-Profile.

To choose Enterprise Architect was not easy. In fact, a comparative study was carried out together by our research group and the Andalusian Government. The result was the tool that offered the best position in price/quality.

We selected UML profiles since, after some empirical studies, we realized that it was the easiest device for people in companies; on one hand, UML is commonly used in software companies, thus development teams already know its notation. On the other hand, they usually work with UML-based tools.

Other suitable possibility was using Eclipse and EMF technologies. However, as NDT is mainly oriented to requirements and end users' work, it does not offer as good interfaces as UML models, for instance, with use cases.

The problem of this election dealt with defining transformations. Enterprise Architect or other UML-based tools analyzed do not support QVT transformations. Thus, NDT transformations were translated into Java and implemented in a new tool that was included in NDT-Suite, named NDT-Driver.

This solution, NDT-Profile and NDT-Driver, provides development teams with a tool environment to define models as well as an engine to execute transformations.

These seeds prepare the environment for using NDT both for research and practical use.

Nowadays, NDT-Suite has been improved with some tools: NDT-Prototypes, NDT-Checker, NDT-Glossary, NDT-Report, NDT-Quality, NDT-Tracer and NDTQ-Framework. All of them bear out different aspects in NDT: quality assurance, exit generation or code checking, among others. Likewise, all of them are available on the Web.

This paper focuses on analyzing in depth tools that support, among other things, requirements validation and model-driven paradigm to reduce the development time and cost.

3. Developing web geographical information systems with NDT

The purpose of this section is to present the formal definition necessary to achieve the geographic conceptual model using NDT. This requires starting from the definitions of requirements engineering phase, then set the derivation to get the models of analysis. Therefore, the next section explains the relationship of the models in the existing methodology and is followed by extension for Web GIS.

3.1. Relationship models

NDT (Escalona et al., 2004) is a methodological process based on the navigation of web and hypermedia systems. NDT defines the Requirements Engineering and Analysis phases of a software development process. In the Requirements Engineering phase, NDT defines four models: information storage requirement model, functional requirements model, actors model, and interaction requirements model.

The information storage requirements model specifies the information needs of the system under development. This model answers the following questions: what information must the system store? What information does the system need?

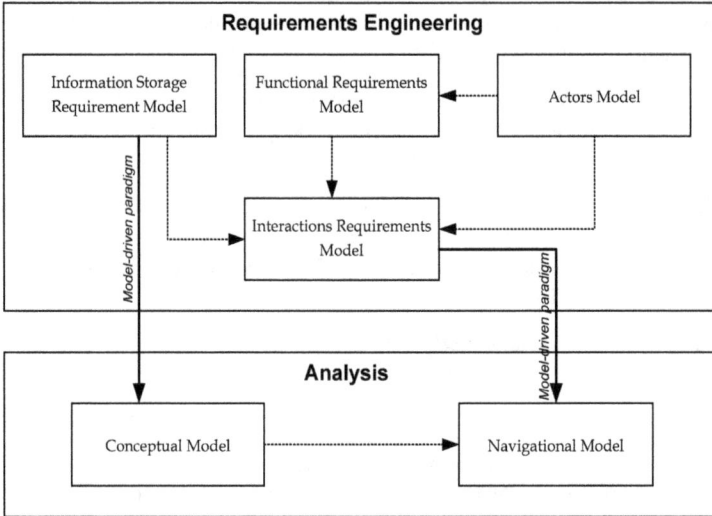

Figure 5. Relationship models

Functional requirements model specifies the operation of the system; this model answers the following question: what can the system do? The actors model specifies the roles of actors that interact with the system, their incompatibilities and generalization among them.

The interaction requirements model is a relevant model for a navigational system. The interaction requirement model provides the information and the functionality asked by the user.

In the Analysis phase, NDT defines two models. The conceptual model is a set of conceptual classes. These classes represent the static structure of the system. The navigational model is a class diagram with special classes that offers a view of the conceptual model and shows how to navigate through the information managed by the system.

The main documents obtained in NDT are: the System Requirement Document (SRD) in the Requirements Engineering phase and the System Analysis Document (SAD) in the Analysis phase.

3.2. Relationship models (extended to GIS)

Figure 6 shows the relationship extended for the treatment of Web GIS. In the phase of Requirements Engineering introduces the Geographical Data Storage Requirement Model, which is the basis for the derivation of Geographical Conceptual Model. The derivation of models is based on model-driven paradigm.

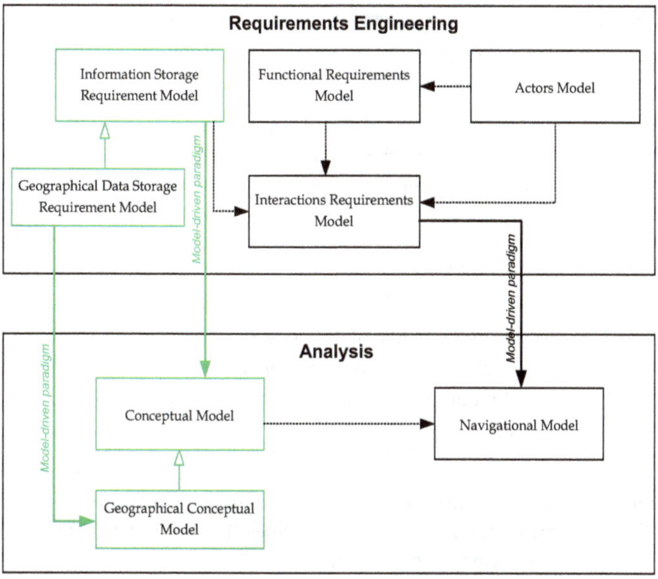

Figure 6. Relationship models (extended to GIS)

3.3. Geographical data storage requirement model

One of the keys for navigation in a navigational system is the information it handles. Some authors have defined the navigation system (Rossi, 1996)(Cachero, 2003) as a set of views of the information it stores and manages the system.

Being, therefore, the information displayed and handled one of the elements that are relevant for navigation during requirements engineering, is necessary to generate the model of storage requirements. This model represents the needs of storage that has the system and defines the characteristics of the information that will manage and display in the navigational system.

The structure of this model, including specification of geographic data, is presented in this section, by a corresponding metamodel, as well as more accurate description technique that is based on the use of patterns.

3.3.1. Model definition

The model of storage requirements contains a description of the information handled by the system and specifies its structure and meaning.

The items that appear in the model of storage requirements and the relationships are represented by the metamodel in Figure 7. We used the notation of the UML class diagram to describe it.

In this model, StorageRequirement class represents the set of information needs of the system. Instances of this class are used to define what information is handled in the system and its structure and meaning.

Each storage requirement is a relevant concept for the need to store information in the system. Each is described by an identifier, which turns out to be a unique code for every storage requirement, a name that represents it, and a description, which can define its meaning.

In addition, each storage requirement has associated a specific data set are represented in the diagram in Figure 7 by EspecificData class.

A specific piece of information is each of the specific items to be stored for a storage requirement. The storage requirement defines the general concept of information to manage the system, while the specific data concretely describes each of the items of information to be stored in a storage requirement.

The specific data are defined by a name that should be intuitive enough to describe the concept it represents, and a description detailing its meaning. In addition, each has specific data cardinality. The cardinality is a range that defines the minimum and maximum values of specific data that can be found in the requirement.

Another important concept is the nature of specific data. Nature defines the domain of that specific data. Let define the set of values and structural details that has the specific data. The concept of nature, although very close, does not match the data type concept. Nature represents a domain as a set of values that have a specific meaning within the system without going into details of low level, is the view that the user has over the domain and structure information.

Each nature, in turn, is defined by a short name so it must express its meaning. In the metamodel, it is possible to define three types of natures:

1. Predefined natures, which represent a set of predefined domains that are presupposed in any system and that the model of Figure 7 is represented by the class PredefinedNature. There are several predefined natures represented as child classes in the metamodel. Keep in mind that nature, as has been said, is not specific data type in the sense that is understood in the programming language, but not the broad sense of the programming language, since there is no indication that the type of this attribute is implemented later as a string, but in requirements specification, the user sees it as such. The final type of specific information or even the decision of whether this specific data is implemented as just a particular field is task of the designers and implementers of the system.
2. New natures, new domains that are defined in a concrete way for the system being modeled. In the class diagram in Figure 7 is represented by the class NewNature. Each new nature is described by an identifier, a name, which inherits from nature, and description, whose meanings are the same as for the storage requirements seen before. But it also has other attributes that do not have to take value in all cases, and they are:

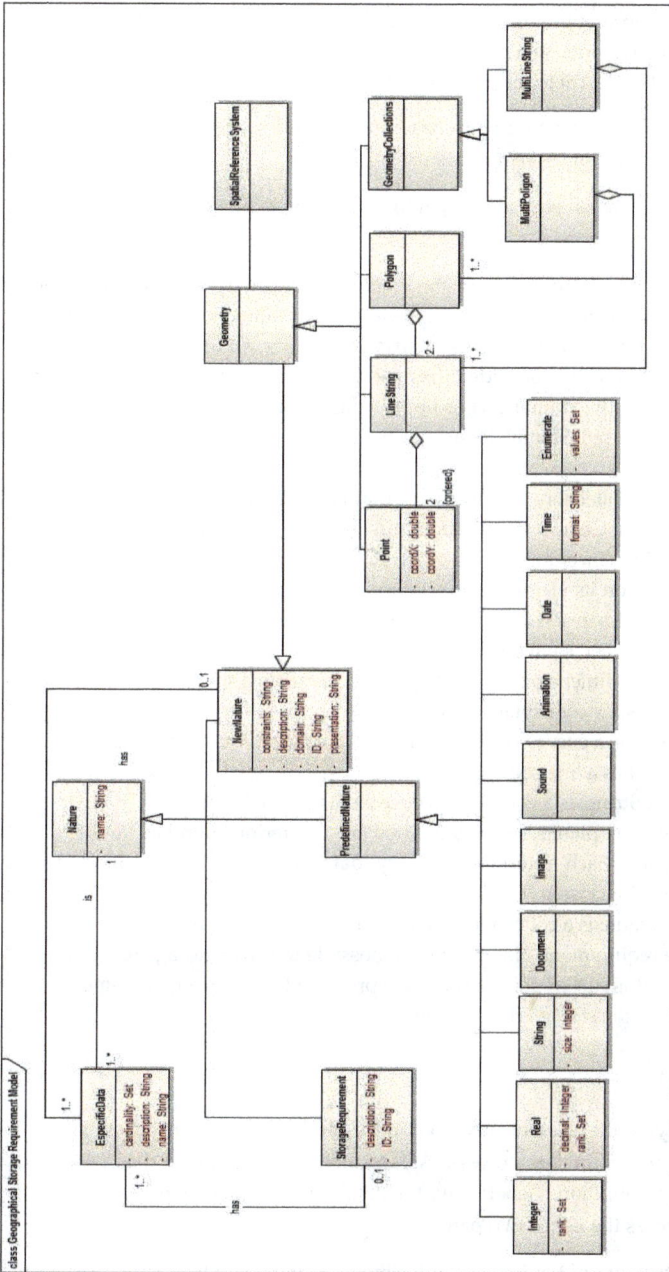

Figure 7. Geographical Data Storage Requirement Model

a. domain, representing the set of possible values that nature can take.
b. a set of constraints, which express constraints that must be fulfilled by the nature.
c. presentation restricts specific ways of how to represent.

Besides this, each new nature has an associated set of specific data whose meaning is similar to the storage requirements.

To own extension and adaptation to GIS requirements define the following elements of the previous figure.

- Geometry: is the root class of the hierarchy. Geometry is an abstract (non-instantiable) class. The instantiable subclasses of Geometry defined in this specification are restricted to 0, 1 and two dimensional geometric objects.
- SpatialReferenceSystem: identifies the coordinate system for all geometries, and gives meaning to the numeric coordinate values for any geometry instance. Examples of commonly used Spatial Reference Systems include 'Latitude Longitude', and 'UTM Zone 10'.
- Geometry Collection: is a geometry that is a collection of 1 or more geometries. All the elements in a GeometryCollection must be in the same Spatial Reference. This is also the Spatial Reference for the GeometryCollection. GeometryCollection places no other constraints on its elements. Subclasses of GeometryCollection may restrict membership based on dimension and may also place other constraints on the degree of spatial overlap between elements.
- Point: is a 0-dimensional geometry and represents a single location in coordinate space. A Point has a x-coordinate value and a y-coordinate value.
- LineString: is a curve with linear interpolation between points. Each consecutive pair of points defines a line segment.
- MultiLineString: is a collection whose elements are LineStrings.
- Polygon: is a planar Surface, defined by 1 exterior boundary and 0 or more interior boundaries. Each interior boundary defines a hole in the Polygon. Polygons are topologically closed.
- MultiPolygon: is a collection whose elements are Polygons
3. Storage requirement. Nature may be possible to turn a storage requirement. This means that the domain of this nature is represented by the set of elements that abstractly represents the storage requirement.

3.3.2. Model description

After specifying the elements involved in the model of storage requirements, we must study how to represent in the process. As mentioned, we propose the use of patterns as a technique of definition. Generically, for defining the elements of the storage requirements model proposes the use of two patterns.

The basic structure of the first, includes the aspects related to storage requirements that are described in the metamodel. This pattern, shown in Table 1 is an evolution of the proposed staffing Environment Requirements Engineering Methodology for Information Systems (Durán, 1999).

It is assumed the structure of this proposal, but we have added new concepts such as cardinality or natures which are necessary to capture for the treatment of navigation as presented in the following chapters.

The meaning of the fields of this pattern is easily deduced from the definition of the metamodel.

The identifier is a code that identifies uniquely the requirement. Coding is proposed beginning with the letters RA, indicating that it is a requirement of storage, and continues for a unique sequence number. The use identifiers with a certain meaning and not, for example, numbers, is very convenient for quick identification of the requirement.

The descriptive name and description attributes match the name and description of the metamodel. As for the specific data field contains all the information regarding the specifics of the requirement. Each instance contains the attributes and relationships represented in the metamodel: the name and description, nature, and its cardinality and properties.

<ID>	<descriptive name of the requirement>	
Description	The system will store the information on the <concept relevant>. Specifically:	
Specifics data	**Name and description**	**Nature**
	<data name>:< brief description of the data>	<nature of the data> [Cardinality:<cardinality>]
	...	
	<data name>:< brief description of the data>	<nature of the data> [Cardinality:<cardinality>]

Table 1. Pattern for the definition of storage requirement

The second pattern is proposed to describe quite similar and the information concerning new natures. It is described in Table 2.

<ID>	<descriptive name of the nature that is being defined>	
Description	This nature represents <description of the nature>	
Specifics data	**Name and description**	**Nature**
	<data name>:< brief description of the data>	<nature of the data> [Cardinality:<cardinality>]
	...	
	<data name>:< brief description of the data>	<nature of the data> [Cardinality:<cardinality>]
Domain	<value domains of nature>	
Constraints	<constraints that need to have the fields of nature>	
Presentation	<description of the way it presents the fields of nature>	

Table 2. Pattern for the definition of new natures

3.4. Geographical conceptual model

The conceptual model represents the static structure of the system. Can model and describe the information handled by the system, and its structure. Classically been called static model, and although the concept is the same, in the surroundings of web engineering and navigational systems has extended the terminology of the conceptual model. It is in the web world of engineering where there has been more emphasis on the aspect of navigation, so it is assumed in this paper the terminology of the conceptual model.

Within the development of navigation, the conceptual model represents and describes the type of information will be treated, made or modified during the navigation process and within the navigational system and the relationships established between the items of this information . In this way, the navigation of a system is defined on the basis of the information handled.

3.4.1. Model definition

To describe the conceptual model is proposed using a class diagram for definition. As the authors of UML indicate, when using a class model to represent the static model of a system is used with one of these three objectives (Booch et al., 1999):

1. To model the vocabulary of a system, this involves making decisions about the limits of the system.
2. To model simple collaborations, or whatever it is, to represent the relationships and collaborations that are established in the society of classes, interfaces and other system elements.
3. To model a logical scheme of the database. The class model is the first model the system on the final design of the database.

The conceptual model is oriented mainly to the two first points. Allows to define the universe of discourse of the problem and the relationships that occur between the abstract elements defined in the model.

Figure 8 shows the metamodel that describes the conceptual model. This metamodel is based on the proposed by UML v.2.0. Note that this metamodel is a view of the UML metamodel proposed v.2.0. Actually, the class model includes many more aspects than those included in the metamodel of Figure 8. Concepts allowed in the class model as the invariant or stereotypes or other means of expansion are not included in our meta-model, but could be added in the conceptual model if desired. The metamodel in Figure 8 deals only with aspects of the relevant class model for defining the navigation system.

The main element of this model is the UML Classifier class v.2.0. Under this kind are shown the two main elements of the model classes: classes, included in Class, and partnerships, contained in the Association class.

All classifier is represented by an identifier, a name and description. These attributes (attributes) do not appear under this name in the UML metamodel, however, have remained in figure 8 for the metamodel classes follow the structure of previous models.

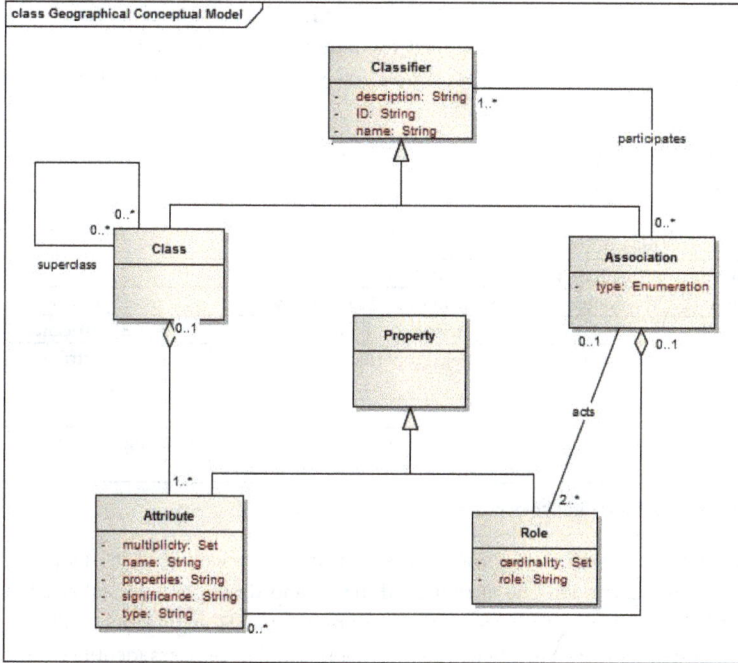

Figure 8. Conceptual model

Specifically entering the first class that inherits from the classifiers, the Class, say their class instances represent any conceptual system.

These may establish hierarchical structures that are represented by the association it has with herself. The school also has an associated set of attributes that describe them. These attributes are represented in the Attribute class, which according to the UML metamodel is a daughter v.2.0 Property class in UML v. 2.0.

With respect to Class Association, their instances represent the relationships can be established between different classifications of the model. Furthermore, in a partnership at least two roles, represented in the class Role, each described by a role name, contained in the attribute role, and a cardinality. These concepts are consistent with those proposed in UML and are not objective of this work to define them in detail.

3.4.2. Model description

To represent the conceptual class model proposes two techniques basically:

1. The class diagram, it is advisable to follow the standard UML notation.
2. The data dictionary. The data dictionary to describe the graphical model class diagram in a more concrete and detailed.

There are many possibilities to develop a data dictionary. This proposal recommends the use of patterns similar to those described for requirements engineering since, as already mentioned, offer a structured and simple description of the model and facilitates automation.

For the data dictionary conceptual class model, we propose two patterns. The first one is aimed to describe the system classes and is described in Table 3.

<ID>	<descriptive name of the class>	
Inherited from	<ID and name of the parent class>	
Description	The system will store the information on the <concept relevant>. Specifically:	
Attributes	**Description**	**Significance**
	name [multiplicity][:type][{Properties}]	<Meaning of attribute>
	...	
	name [multiplicity][:type][{Properties}]	<Meaning of attribute>

Table 3. Pattern for the definition of classes

This pattern is quite similar to those used so far and the meaning of its fields can easily guess from the metamodel. The attributes id, name and description are collected in fields that are labeled as his name. It is advised, as requirements engineering, follow specific criteria for assigning identifiers. In this case, it is advised that the class identifier begins with CL or CLn and will be followed by a unique number. By studying the criteria for referral in the next chapter, it indicates when you have to opt for a code that starts with CL or one that starts with CLn. The field 'inherits from', reflects the hierarchical relationships in which it participates as a child class and the attributes field gives the description of the attributes associated with the class.

The second proposed standard aims to provide a tool to describe the associations. The structure is displayed in Table 4.

<ID>	<descriptive name of the association>		
Description	Classes <identifiers of the classes that are related> are related by the association that represents <significance of the association>		
Type	{unidirectional, bidirectional}		
Classes	**Class name**	**Role name**	**Cardinality**
	<ID and name of the class>	<role of the association for the class>	<cardinality of the association in the class>
	...		
Attributes	**Description**		**Significance**
	name [multiplicity][:type][{Properties}]		<Meaning of the attribute>
	...		

Table 4. Pattern for the definition of associations

It's a fairly intuitive pattern also taking into account the above definitions. This pattern is intended only for associations, since the hierarchical relationships are reflected in the previous pattern. The pattern reflects associations in the fields ID, name, description and type attributes with the same name which appear in the metamodel. In the case of partnerships it is advisable to use an identified beginning with AS and will be followed by a unique sequence number.

Class field contains the reference to classes participating in the association keeping for each code and name of the class, the role it assumes in the association and the cardinality of the same. There is also the attributes field which gathers the attributes of the association in the event that you are defining what is a class association.

3.5. Basic conceptual model derivation

As mentioned, the basic conceptual model can be generated from the information provided by the elements of the model storage requirements engineering requirements.

There are different relationships between elements of the model storage requirements and the basic conceptual model. In Figure 9 are represented by a diagram of these classes. The classes participating in this diagram are classes that come from the metamodel presented in the previous chapter, specifically the metamodel storage requirements and conceptual metamodel. In the next figure has only had relationships that are reflected throughout this section and have ignored other metamodels shown in the previous chapter that are not relevant to the referral process.

Keep in mind that the added relationships, generates four labeled in the figure, only hold for the case of the basic conceptual model need not be met in the final model. In the figure, the thick line is divided for part of the diagram that comes from the storage requirements metamodel (which is at the top) and comes from the conceptual metamodel (which is at the bottom). The Objective of the analysis phase is to detail, define and validate the requirements. In the NDT Methodology, the analysis phase Involves the definition of two main models: conceptual model and navigational model. A relational model generated from Might Be the conceptual model. The relational model is Useful for the Implementation of the databases of the system under Development.

3.5.1. Derivation of classes from the storage requirements and natures

The first relationship can be studied in Figure 9 is what allows us to derive lessons from the natures and storage requirements. From it, one might conclude that, in the basic model, each storage requirement and every new nature generates a class that has the same name and same description or nature of the requirement that generates it.

In the previous section are advised to give IDs format requirements beginning with RA and natures starting with NA. It is also advisable to identify classes with identifiers that begin with CL and CLn.

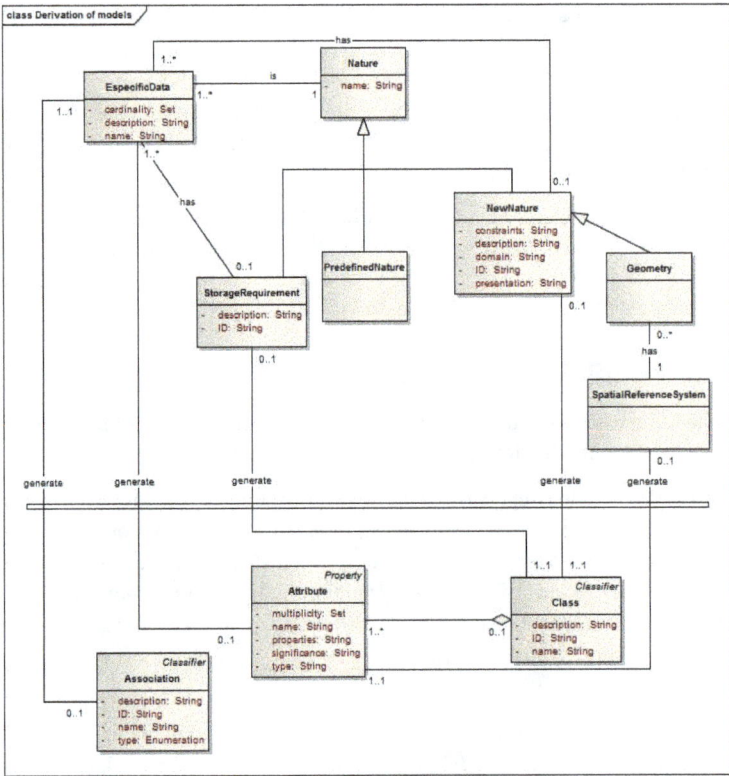

Figure 9. Relations between the model of storage requirements and conceptual model

3.5.2. Deriving attributes from the specific data

Another relationship that will allow to establish referral systems is that between specific data and attributes. Each class has an associated set of attributes, in the same way that every nature and every storage requirement is related to a specific data set.

Figure 9 shows a relationship which indicates that each specific data generated from 0 to 1 attribute. The specific data storage requirement or a new nature will serve to generate the attributes of the class that is generated from that specific data or from that nature. However, not all the specifics of the requirement or derived nature of the class attribute, taking into account only those whose nature is either a default or nature a new nature.

3.5.3. Derivation associations from specific data

The latest model of relationship between storage requirements and the basic conceptual model of Figure 9 to study is made between specific data and associations. This time, specific data are taken into account are those which are naturally a storage requirement.

Each of them will cause a binary association between two classes. So, is restricted to the beginning than the number of classifiers involved in the association are two and that its type is Class.

4. Case study

This will model the patterns described above, the requirements model and the conceptual model that includes the following case. Just try to store the geographic information world's rivers that pass through cities. The cities are in countries.

That is why we get information from both cities by passing a river crossing countries, etc. In each of the features shown in the example data is stored minimum and necessary.

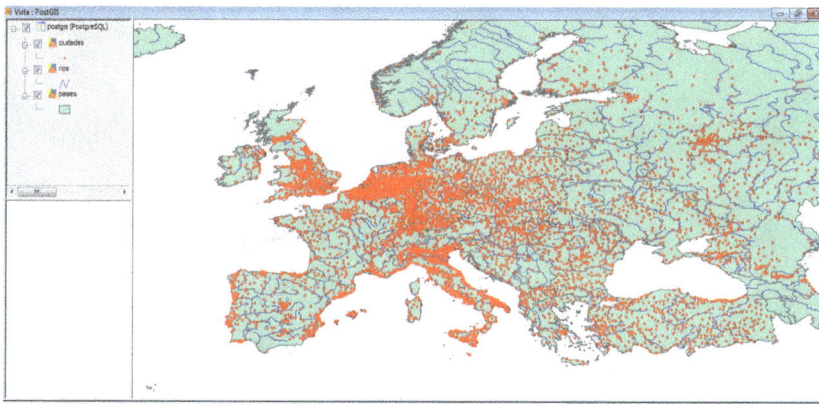

Figure 10. GIS Data Representation

Then define the patterns corresponding to the storage requirements.

RA-01	Country	
Objectives associated	OBJ-01: provide full view of geographic information layers	
Description	The system will store data for countries. Specifically:	
Especific Data	**Name and Description**	**Nature**
	codeCountry: stores information about the code that uniquely identifies each country	NA-01
	Name: stores information about the country's name	String
	Area: stores information about the country's area	Real
	Population: stores information about the country's population	Integer
	Capital: stores information about the country's capital	RA-02 Cardinality: 1..1
	layerCountry: layer stores the type which represents information	NA-04
	Cities: stores a list of cities	RA-02 Cardinality: 1..*

Table 5. Requirement pattern RA-01

RA-02	City	
Objectives associated	OBJ-01: provide full view of geographic information layers	
Description	The system will store data for cities. Specifically:	
Especific Data	**Name and Description**	**Nature**
	codeCity: stores information about the code that uniquely identifies each city	NA-02
	Name: stores information about the city's name	String
	Area: stores information about the city's area	Real
	Population: stores information about the city's population	Integer
	Country: stores the country that owns the city	RA-01
	layerCity: layer stores the type which represents information	NA-05
	Rivers: stores a list of rivers	RA-03 Cardinality: 0..*

Table 6. Requirement pattern RA-02

RA-03	River	
Objectives associated	OBJ-01: provide full view of geographic information layers	
Description	The system will store data for rivers. Specifically:	
Especific Data	**Name and Description**	**Nature**
	codeRiver: stores information about the code that uniquely identifies each river	NA-03
	Name: stores information about the river's name	String
	Length: stores information about the river's length. Be stored in kilometers	Real
	layerRiver: layer stores the type which represents information	NA-06
	Cities: stores a list of river crossing towns	RA-02 Cardinality: 1..*

Table 7. Requirement pattern RA-03

Then define the patterns corresponding to the natures.

NA-01	codeCountry	
Objectives associated	OBJ-01: provide full view of geographic information layers	
Description	This nature is the structure describing the unique code identifying each country.	
Especific Data	**Name and Description**	**Nature**
	country: store, through a code of three numbers, the country	String Size: 3
Presentation	Data are presented using a 3-digit numeric code.	

Table 8. Nature pattern NA-01

NA-02	codeCity	
Objectives associated	OBJ-01: provide full view of geographic information layers	
Description	This nature is the structure describing the unique code identifying each city within the country.	
Especific Data	**Name and Description**	**Nature**
	country: store, through a code of three numbers, the country	NA-01
	city: is a three-digit saves the particular number that has this city in the country.	String size: 3
Presentation	Data are presented using a 6-digit numeric code. The first three allude to the country and three from the city. Are separated by a point. Example: CCC.ccc	

Table 9. Nature pattern NA-02

NA-03	codeRiver	
Objectives associated	OBJ-01: provide full view of geographic information layers	
Description	This nature is the structure describing the unique code identifying each river.	
Especific Data	**Name and Description**	**Nature**
	river: store, through a code of six numbers, the river	String Size: 6
Presentation	Data are presented using a 6-digit numeric code.	

Table 10. Nature pattern NA-03

NA-04	layerCountry	
Objectives associated	OBJ-01: provide full view of geographic information layers	
Description	This nature is the structure describing the unique code identifying each country.	
Especific Data	**Name and Description**	**Nature**
	typeGeometric: defines the type of geometry of the layer	MultiPolygon
	dataGeometric:	NA-07
Presentation	Data are presented using a 3-digit numeric code.	

Table 11. Nature pattern NA-04

NA-05	layerCity	
Objectives associated	OBJ-01: provide full view of geographic information layers	
Description	This nature is the structure describing the unique code identifying each city.	
Especific Data	**Name and Description**	**Nature**
	typeGeometric: defines the type of geometry of the layer	Point
	dataGeometric:	NA-07
Presentation	Data are presented using a 3-digit numeric code.	

Table 12. Nature pattern NA-05

NA-06	layerRiver	
Objectives associated	OBJ-01: provide full view of geographic information layers	
Description	This nature is the structure describing the unique code identifying each river.	
Especific Data	**Name and Description**	**Nature**
	typeGeometric: defines the type of geometry of the layer	MultiLineString
	dataGeometric:	NA-07
Presentation	Data are presented using a 3-digit numeric code.	

Table 13. Nature pattern NA-06

NA-07	dataGeometric	
Objectives associated	OBJ-01: provide full view of geographic information layers	
Description	Defines the number of dimensions and spatial reference system	
Especific Data	**Name and Description**	**Nature**
	nDimensions	Integer
	spatialSystemReference	NA-08

Table 14. Nature pattern NA-07

NA-08	spatialSystemReference	
Objectives associated	OBJ-01: provide full view of geographic information layers	
Description	Define spatial reference system	
Especific Data	**Name and Description**	**Nature**
	code	Integer

Table 15. Nature pattern NA-08

5. Conclusion

This paper has presented the formal definition of geographic data models with NDT methods. It was based on the extension of the definition of existing models of complex navigation systems. He presented the definition of the model Engineering Requirement phase and the derivation of Analysis models. However, we are still working on the specifications of constraints of the models using OCL.

Finally, a Web GIS model is necessary to store information, which is discussed in this work, but it is also important to define the navigation-related models of this type of geo-referenced data. Figure 11 shows a future work, which is the generation of the Web Mapping Model.

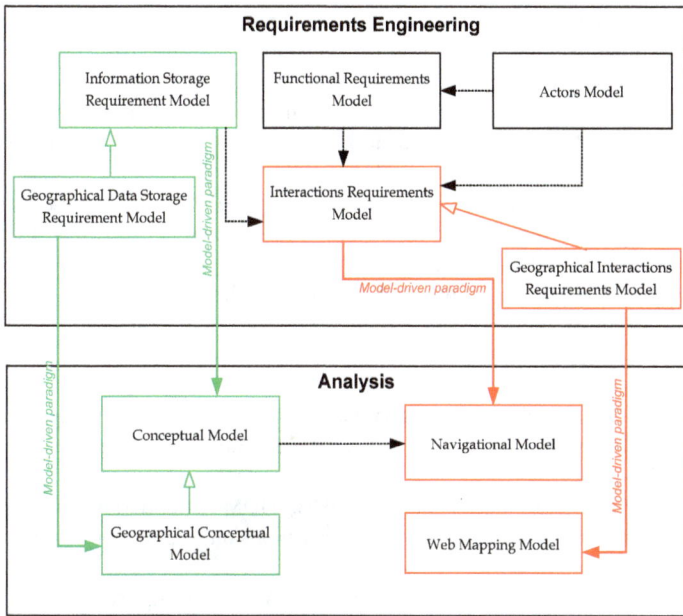

Figure 11. Generating Web Mapping Model (Future works)

Author details

J. Ponce, A.H. Torres, M.J. Escalona, M. Mejías and F.J. Domínguez-Mayo
Department of Computer Languages and Systems, University of Seville, Seville, Spain

Acknowledgement

This research has been supported by the project Tempros project (TIN2010-20057-C03-02) and Red CaSA (TIN 2010-12312-E) of the Ministerio de Ciencia e Innovación, Spain and NDTQ-Framework project of the Junta de Andalucia, Spain (TIC-5789).

6. References

Aronoff, S., 1989. Geographic Information Systems. WDL Publications, Canada.

Baranauskas, M. Cecilia C., Schimiguel, Juliano, Simoni, Carlos A., Medeiros, Cláudia B., 2005. Guiding the Process of Requirements Elicitation with a Semiotic-based Approach – A Case Study, HCI – International USA.

Booch, G., Rumbaugh, J., Jacobson, I. Unified Modelling Language User Guide. Ed. Addison-Wesley, 1999.

Bull, G., 1994. Ecosystem Modelling with GIS. Environmental Management. 18(3):345-349.– 438.

Cachero, C. Una extensión a los métodos OO para el modelado y generación automática de interfaces hipermediales. PhD Thesis. Universidad de Alicante. Alicante, España. Enero 2003.

Durán A., Bernárdez, B., Ruiz, A., Toro M. A Requirements Elicitation Approach Based in Templates and Patterns. Workshop de Engenharia de Requisitos. pp.17-29 . Buenos Aires, Argentina. 1999

Durán A. Un Entorno Metodológico de Ingeniería de Requisitos para Sistemas de Información. Ph. Tesis. Departamento de Lenguajes y Sistemas Informáticos. Universidad de Sevilla. Sevilla, 1999.

EMF Technologies. Available in www.eclipse.org/modeling/emft/. Accessed in February 2011.

Escalona, M. J., Aragón, G. 2008. NDT. A Model-Driven approach for Web requirements. IEEE Transaction on Software Engineering, 34(3), 370-390.

Escalona, M. J., 2004. Modelos y técnicas para la especificación y el análisis de la navegación en sistemas software. PHD Thesis. Escuela Técnica Superior de Ingeniería Informática. Universidad de Sevilla.

Garzotto, F., Schwabe, D. and Paolini P., 1993. HDM-A Model Based Approach to Hypermedia Application Design. ACM Transactions on Information System, 11 (1), pp 1-26.

Kim, Do-Hyun, Kim, Min-Soo, 2002. Web GIS service component based on open environment. Geoscience and Remote Sensing Symposium, IGARSS '02. IEEE International Volume 6, Page(s):3346 - 3348 vol.6.

Liu, K., 2000. Semiotics in Information Systems Engineering. Cambridge University Press, Cambridge.

Macário, Carla, Medeiros, Claudia, Senra, Rodrigo, 2007. IX Brazilian Symposium on GeoInformatics - Geoinfo 2007 p.239-250.

M.J. Escalona, A.H. Torres, J.J. Gutiérrez, E. Martins, R.S. Torres, M. Cecilia, C. Baranauskas. A Development Process for Web Geographic Information System A Case of Study. ICEIS 2008International conference on enterprise information system; España (2008). Vol. HCI, pp. 112-117, ISBN/ISSN: 978-989-8111-40-1

Rossi, G., 1996. An Object Oriented Method for Designing Hipermedia Applications. PHD Thesis, Departamento de Informática, PUC-Rio, Brazil.

UML, Unified Modeling Language. http://www.uml.org/.

Chapter 18

Using Geographic Information Systems for Health Research

Alka Patel and Nigel Waters

Additional information is available at the end of the chapter

1. Introduction

The connection between public health and geography can be traced back to Hippocrates (c. 400 BC) who deduced that spatially varying factors such as climate, elevation, environmental toxins, ethnicity and race contributed to the spatial patterns of illness (Parchman *et al.*, 2002). The observations of Hippocrates still hold true today and these relationships between geography and disease have allowed geospatial methods to become valuable within the field of public health. Maps have long been a useful tool for visualizing patterns in health care. One of the earliest and most commonly cited examples is from the mid 1800s when Dr. John Snow deduced the source of a cholera outbreak in London based on a simple visualization of the incidents of cholera in relation to water pumps (Johnson, 2007). Although visualizing data geographically is still very valuable for uncovering patterns and associations over space, geospatial analysis has become more sophisticated over time.

One tool that can be used to apply advanced geospatial methods to health care problems is a geographic information system (GIS). The power of GIS lies in its ability to analyze, store and display large amounts of spatially referenced data. In a field where manual data analysis can become overwhelming, GIS is a valuable tool. The medical research applications of GIS are numerous and include finding disease clusters and their possible causes (Murray *et al.*, 2009; Srivastava *et al.*, 2009), improving deployment for emergency services (Ong *et al.*, 2009; Peleg & Pliskin, 2004) and determining if an area is being served adequately by health services (Cinnamon *et al.*, 2009; Schuurman *et al.*, 2008). There have been several reviews and textbooks published in the past decade that focus on the application of GIS to different areas of health research (Cromley & McLafferty, 2002; Kurland & Gorr, 2009; McLafferty, 2003; Parchman *et al.*, 2002; Rushton, 2003).

While the use of GIS is gaining favor with health researchers, barriers remain in the uptake of more advanced geospatial methods by health care decision-makers. A recent qualitative study conducted in the UK found that although health care decision-makers see the value of using GIS in the decision making process, many (especially those in the community setting) still view GIS primarily as a visualization tool (Joyce, 2009). This chapter aims to go beyond a simple discussion of GIS and its standard capabilities (e.g. mapping, buffering, clipping, intersecting, etc.) in order to focus more broadly on geospatial methods (including spatial statistics) that can be used to describe and study the distribution of disease and health care delivery. It adds to current literature by taking an approach that compares and contrasts different geospatial methods with the goal of informing health care decision-makers about the strengths and weaknesses of each.

This chapter is written for an audience that has a basic knowledge of statistics and GIS. It does not give a comprehensive overview of all geospatial methods that are available for health research but instead highlights ten different geospatial methods that can be used to address problems posed when studying the distribution of disease and health service delivery. Five topic areas will be covered: representing populations, identifying disease clusters, identifying associations in a spatial context, identifying areas with access to health care and identifying the best location for a new health service. Although each method is discussed briefly, references will be provided where health researchers and decision-makers can find details. Each of these five sections will be further subdivided into two sections. 'Representing populations' will discuss the topics of *Geographically linked census data* and *Dasymetric modeling*. 'Identifying disease clusters' will discuss the topics of *Measures of spatial autocorrelation* and *Kernel density estimation*. 'Identifying associations in a spatial context' will cover *Spatial regression* and *Geographically weighted regression*. 'Identifying areas with access to a health care service' will discuss the topics of *Road network analysis* and *Interpolation*. Finally, 'Identifying the best location for a new service' will discuss the topics of *Coverage problems* and *Distance problems*.

2. Representing populations

Studying relationships between disease or health service delivery and environmental or population characteristics, requires researchers to represent populations in terms of their location in space. Two geospatial methods that can be used to represent population groups will be discussed in this section: geographically linked census data and dasymetric modeling.

2.1. Geographically linked census data

Studies in health research may sometimes be focused on the general population that is at risk for a disease or that has access to a specific health service in an area. This group can be represented by population counts in the form of census data. Census data can be represented at many different levels. For example, in Canada census data may be represented at the census block level, the dissemination area level, and at the census tract level (Apparicio *et al.*, 2008). This method of representing population links census data tables to geographic boundary files using unique area level identifiers. The aggregation of populations into larger areal units can affect the results obtained from a spatial analysis. For example, when studying geographic access to urban amenities, Canadian studies have found that when accessibility was measured for smaller spatial units, the measures were less subject to aggregation error than when larger spatial units were used to represent populations (Apparicio *et al.*, 2008; Hewko *et al.*, 2002).

There are limitations to geospatial analysis when populations are represented in this way. As discussed above, the aggregation errors differ depending on the spatial unit used. One study showed that for some amenities the aggregation to larger areas for abundant resources (e.g. playgrounds) should use smaller areal units more representative of where the population is distributed (Hewko *et al.*, 2002). Another limitation occurs because census data are standardized based on population numbers. This causes rural areas to be represented by larger areal units that do not precisely represent locations of where the population is distributed. Finally, the modifiable areal unit problem (MAUP) should be considered when using different census boundaries to represent populations. The MAUP occurs because the results of any geographic analysis can be sensitive to the areal units that are being used for the study (Fotheringham & Wong, 1991; Oppenshaw, 1996). For example, results can differ depending on the level of census data used in analysis. Notable strengths of using census data to represent populations are: 1) in most developed countries these data are largely publically available and 2) these data can be grouped by age categories and linked with other census variables of interest to study associations (with, for example, income, employment, education, ethnicity and other socio-economic indicators).

2.2. Dasymetric modeling

Census data are often used by researchers to represent population numbers evenly within a specified area. For example, for analytical purposes the population within a census tract is often represented by a single point at the geographic centre of the polygon, known as the centroid of the area. A method used to represent population distribution more accurately is dasymetric modeling. Dasymetric modeling uses additional data sources to estimate the distribution of aggregated population data within a specified unit of analysis (Briggs *et al.*, 2007a; Mennis, 2009; Poulsen & Kennedy, 2004). This technique essentially uses a combination of variables to represent where population is truly located within an area. Dasymetric models have used light emission and remotely sensed land cover data to represent populations at the small area level in Europe (Briggs *et al.*, 2007b). One study has shown how population estimates in residential areas can be more accurately represented using grid based land cover data in combination with a mailing database to represent where residential populations are located within a census area (Langford *et al.*, 2008). Dasymetric modeling produces population numbers that are identical to those within a census area; it is the spatial distribution of the population that is modified within these spatial units. A major limitation of using this method for representing population distribution is that it is more time intensive and requires the use of additional data sources when compared to simply using the raw census data. The strength of this method is that it more accurately represents population distribution in large areas (e.g. rural census areas). One study found lower access to primary health care services using dasymetric modeling when compared to using point census data to represent populations in rural areas (Langford & Higgs, 2006). This shows that census methods of representing population data could be overestimating access in rural areas.

Both methods for representing population distribution have been shown to have strengths and limitations. In the absence of patient data or when the general population is of interest, it is possible to use census data at varying levels of aggregation. Using this method, populations can be represented as a centroid within the areal unit or as being evenly distributed throughout the area. Dasymetric mapping allows for a more accurate representation of the population distribution within areal units but at the cost of increased time and data collection effort. However, this method can be valuable when population data are represented for rural areas. Both methods suffer from the problem of the MAUP and this should be considered when applying any level of population data to analysis. Figure 1 highlights the differences in representing populations using two different representations of census data and dasymetric modeling.

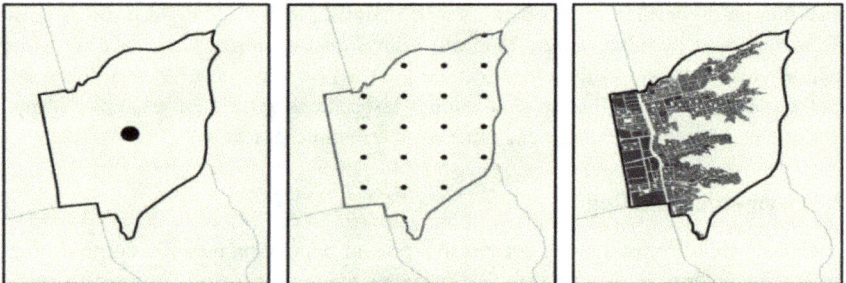

Figure 1. Comparison of methods to represent populations (from left to right) by the centroid, evenly distributed points, and dasymetric modeling methods (Wielki, 2009)

3. Identifying disease clusters

The identification of clusters of disease can help to target health care delivery and identify where resource allocation efforts should be focused. Point pattern analysis is a broad category of geospatial methods that can be used to identify disease clusters. In this section we will discuss two specific methods: measures of spatial autocorrelation and kernel density estimates.

3.1. Measures of spatial autocorrelation

Spatial autocorrelation can generally be thought of as the association of a variable with itself over space. Statistical measures of spatial autocorrelation can be used to determine if the clustering of variables with similar attributes occurs over space. Moran's I is a global measure for evaluating spatial autocorrelation. It is measured using the location of a given set of points and their associated attributes over an entire study area. The values of Moran's I range from -1 to +1 (dispersed to clustered, respectively, with values close to 0 representing no spatial autocorrelation as might be observed in a random pattern) (Rogerson, 2010). One study conducted to evaluate the survival from Cardiac arrest in a US county used Moran's I to determine that there was no clustering of survival rates for cardiac arrest in the study area (Warden et al., 2007). The Moran's I statistic can also be used to define the extent of spatial bands used for other spatial statistics as exemplified in a study that determined hot spots using the Getis-Ord Gi* statistic (McEntee & Ogneva-Himmelberger, 2008). The value of the Moran's I measure is that it allows for an overall assessment of whether spatial autocorrelation exists over a study area. The limitation of the Moran's I measure is that it does not take into account variation at the local level and reports only one measure for the entire study area.

In order to deal with the limitation of the global measure of Moran's I, a local measure of spatial association was introduced by Anselin (Anselin, 1995). This local indicator of spatial association (LISA) breaks down the global indicator to allow for an understanding of how each observation contributes to the overall measure. There are three main purposes that the LISA statistic can serve: 1) to indicate local areas of spatial non-stationarity, 2) to evaluate how different locations within an area are contributing to the overall global spatial autocorrelation statistic and 3) to evaluate if spatial outliers exist (Anselin, 1995). A positive LISA value for an area indicates that the feature is surrounded by features with similar values (a cluster). A negative LISA value indicates that the feature is surrounded by features with dissimilar values (an outlier) (Anselin, 1995). LISA has been used to uncover clusters of overweight and obesity in Canada (Pouliou & Elliott, 2009) and clusters of chronic ischemic heart disease in relation to aerosol air pollution in the Eastern United States (Hu & Rao, 2009). As with most geospatial methods, the above methods of cluster analysis are limited by the quality and scale of the data that are available (Shi, 2009).

3.2. Kernel density estimation

Quadrat analysis is a method that evaluates the points within polygon areas to assess whether a pattern is random, dispersed or clustered in space. Standard quadrat analysis is limited by edge effects, where the density of points between adjoining areas can change abruptly from one value to another. Kernel density estimates address the limitations of simple quadrat analysis though the use of a kernel function that smoothes the values of interest over space by forming a count of events per unit area within a moving quadrat or 'window'. This produces a more spatially 'smooth' estimate of variation in spatial autocorrelation than can be obtained by using a fixed grid of quadrats (Gatrell et al., 1996). A detailed description of the kernel function is outlined in several resources (de Smith et al., 2007a; Gatrell et al., 1996).

One important consideration when using kernel density estimation is that the degree of smoothing produced is dependent on the size of the bandwidth. It is often useful to examine several density plots of the same samples, all smoothed by different amounts, in order to gain greater insight into the data (Farwell, 1999). Guidelines for choosing bandwidths vary from simple visual inspection, to specific parameters such as specified resolution limits or nearest neighbor distances (Levine, 2007). One study by Lerner *et al.* used kernel density estimation in ArcGIS with default bandwidth settings to identify out of hospital cardiac arrests in Rochester, NY. By integrating census data with the data from a cluster analysis using kernel density estimation, they were able to discover that out of hospital cardiac arrest clusters were associated with communities with lower income, higher number of African-Americans residents and more residents without a high school diploma than the population of the city in general (Lerner *et al.*, 2005). The identification of clusters of events could lead to an increased understanding of where health care resources should be allocated. One of the major limitations of kernel density estimation is that the choice of size of the 'moving window' (bandwidth of the kernel function) can alter the results of the cluster analysis (de Smith *et al.*, 2007a). This method is superior to a simple quadrat analysis because it takes into account edge effects and creates a more realistic representation of spatial variation through a smoothing process (Gatrell *et al.*, 1996).

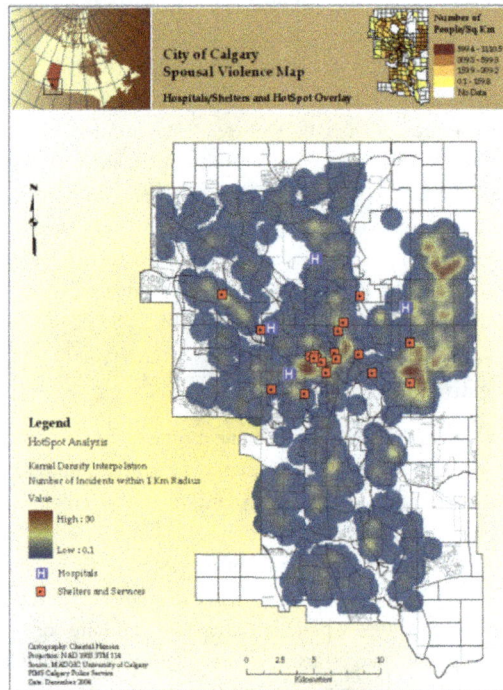

Figure 2. Using kernel density estimation to identify hot spots (Hansen, 2005)

The two methods discussed to identify disease clusters are similar in that they both allow for a statistical measure of spatial association within areas. Using the global and local measures of spatial association can give a quantitative measure of whether global or local spatial autocorrelation (reflecting clustering) exists in a dataset while the kernel density estimates are a way of using data locally within windows to create a grid surface of where clustering is occurring. The spatial measures of association that show whether spatial autocorrelation exists in a dataset can indicate what type of further analysis should be undertaken. The presence of spatial autocorrelation suggests that spatial regression techniques should be used in place of simple linear regression. While the LISA method can be used to measure spatial association within a defined area, the method of kernel density estimation allows for minimizing edge effects between areal units with sharp contrasts. One recent study has shown how LISA and kernel density estimation can be used together to uncover hot spots of diarrhea in Thailand (Chaikaew *et al.*, 2009). Figure 2 shows how kernel density estimation has been used to identify areas of high spousal violence incidents in Calgary, Canada (Hansen, 2005).

4. Identifying associations in a spatial context

The purpose of many studies of the distribution of health services is to determine if inequity exists. These studies strive to determine if those who are considered disadvantaged are served as equally by health services as those who are not (Meade & Emch, 2010). Studies may also seek to understand relationships between disease prevalence and various environmental variables. Spatial regression and geographically weighted regression are two spatial statistical methods that can be used to study these types of associations between dependent and independent variables.

4.1. Spatial regression

When exploring variables over space it is often found that spatial autocorrelation exists. Spatial dependence violates the assumption of independence needed for standard linear regression (Legendre, 1993). Spatial regression is a valuable method in situations where spatial autocorrelation exists because it considers the spatial dependence between variables by giving weights to them based on distance. The weighting matrix used in the spatial regression process accounts for the spatial dependency in the dependent variable (Fotheringham *et al.*, 2005b). In order to determine accurately the weights used within this matrix it is necessary to first determine the range (distance) of the spatial dependence on the dependent variable using a semivariogram, a graph that visually provides a description of how the data are related (correlated) with distance. Beyond the range defined by the semivariogram there is no longer spatial dependence (Fotheringham *et al.*, 2005b). This incorporation of spatial dependence is important because spatial autocorrelation can reduce the efficiency of the Ordinary Least Squares estimator used in linear regression. Since most features over space have an element of spatial dependence associated with them, it suggests that spatial regression may be a better choice than simple linear regression when modeling features over space, although spatial dependence can be removed using a variety of methods including detrending procedures where spatial trends are evident.

The benefit of using the spatial regression technique is its ability to create a model that can be used to make generalized inferences in an area while accounting for the effects of spatial dependence. A study by Zenk *et al.* examined the association between neighborhood racial composition, neighborhood poverty and spatial accessibility to supermarkets using spatial regression and found that racial segregation places African-Americans in more disadvantaged neighborhoods and at further distances from supermarkets (Zenk *et al.*, 2005). Although spatial regression can account for spatial dependence in a dataset, a recent review states that there is always the possibility of ecological bias when area level data are used to represent individual characteristics within studies (Wakefield, 2007).

4.2. Geographically weighted regression

Geographically weighted regression (GWR) is a local regression model that creates unique local coefficients in order to describe the variation in the dependent variable. This creates local models that are well suited to explore the variation in the data graphically (Jetz *et al.*, 2005). Instead of incorporating the spatial dependence into the model as is done through the process of spatial regression, GWR "measures the relationships inherent in the model at each point" (Fotheringham *et al.*, 2005a). The weighting matrix used in GWR is dependent on the location of individual points and thus must be recalculated at each point. This is in contrast to the spatial regression method where a single contiguity weighting matrix is used for the entire study area (based on the extent of spatial dependence on the dependent variable). In GWR, kernels can be used which associate aspects of density of data into the process (Fotheringham *et al.*, 2005a). GWR also has the benefit of being able to evaluate changes in strength of relationships over different scales, which is difficult to do using spatial regression (Jetz et al., 2005). Previous publications have summarized GWR and local analysis in further detail (Brunsdon *et al.*, 1996; Fotheringham *et al.*, 1998).

GWR has been used in a recent study to evaluate the relationship between cold surges and cardiovascular disease (CVD) mortality. This study found that the CVD mortality rates increased significantly after cold surges and that the tolerance of different populations to these weather changes differed over space (Yang *et al.*, 2009). Another study compared the findings from linear regression and GWR when studying access to parks and physical activity sites and found that while OLS regression found weak relationships between park density and physical activity site density, GWR indicated disparities in accessibility that varied over space (Maroko *et al.*, 2009). This study shows the importance of using the appropriate method of analysis for the data that is under study. Yet another study used GWR as a tool to indicate the need for spatial coefficients when linking health and environmental data in geographical analysis (Young *et al.*, 2009).

The main difference between spatial regression and GWR is that one is a global model and one is a local model. Overall, GWR should be used when there is spatial non-stationarity or when relationships between variables differ over space (Fotheringham *et al.*, 2005a). A simple map of the residuals of a model indicates whether a spatial regression method should be employed over a traditional linear regression method; when the residuals are clustered into high and low values over space it is likely necessary to use a spatial regression method. The residuals of a spatial regression or GWR can also be mapped to assess the areas where the models are under or over estimating (Fotheringham *et al.*, 2005b).

5. Identifying areas with access to a health care service

The concept of access to health care is a central focus of many health research studies. Spatial accessibility is one aspect of access to health care that can be measured using geospatial methods. Geographic access to a service can be represented as straight line distance, network distance and travel time based on travel along a road network (Apparicio *et al.*, 2008); travel effort can also be represented as a continuous travel time surface over a specified area. The two methods that we will focus our discussion on in this section are road network analysis and interpolation.

5.1. Road network analysis

Network analysis has been considered by many researchers to be a more accurate method for evaluating accessibility in terms of travel time and distance when the focus is on travel via roads (Brabyn & Skelly, 2002; Christie & Fone, 2003). These researchers argue that although straight line Euclidean distance and the evaluation of catchments/service areas using Thiessen polygons are commonly used in accessibility analysis, these methods only allow for a crude measure of accessibility (Brabyn & Skelly, 2002). Network analysis allows for a more realistic representation of travel times along a road network by accounting for the different travel speeds along the various road links. Network analysis has been used in health care research to study access to video lottery terminals (Robitaille & Herjean, 2008), access to parks (Comber *et al.*, 2008) and access to palliative care (Cinnamon *et al.*, 2008). The measures of access obtained through network analysis can then be used in combination with census data to assess the population numbers with access to a facility within a specified distance or travel time (Christie & Fone, 2003; Nallamothu *et al.*, 2006). The travel times and distances can also be used to evaluate associations between measures of access and various health indicator (socio-economic) variables (Hare & Barcus, 2007). One UK study found that travel time estimates based on network analysis may be preferred to reported travel times for modeling purposes because the reported times can reflect unusual circumstances and reporter error (Haynes *et al.*, 2006). The strength of road network analysis is that it allows for the estimation of travel times between different point locations along a road network. It does not allow for continuous travel time surfaces to be mapped. In order to do this we can use a method referred to as interpolation.

5.2. Interpolation

The method of interpolation is based on Tobler's first law of geography: things that are near are more similar than things that are further apart (Tobler, 2004). The process of interpolation uses mathematical modeling to 'fill in the gaps' between point data values in order to create a continuous surface of values. There are a variety of resources available to researchers that discuss in detail the theoretical basis of the different interpolation methods (Lam, 1983; Mitas & Mitasova, 1999; Waters, 1997a, 1997b). The advantage of interpolation over simple network analysis is that it can allow for a comparison between different modes of transportation to health care facilities. To do this it would be necessary to have point data representing travel times via different modes for this type of analysis. For example, using interpolation, a study conducted by Lerner *et al.* considered how GIS could be used as a tool to help make transport decisions for trauma patients (Lerner *et al.*, 1999). The goal of that study was to create a map that would show where air or ground ambulance should be preferred to transport patients to a trauma center from a given patient starting location. One Canadian study used this method to compare where ground, helicopter or fixed-wing transport was faster when transporting patients to a cardiac catheterization facility (Patel *et al.*, 2007). Figure 3 highlights how interpolation can be used to model travel times via different modes of emergency transport over an area.

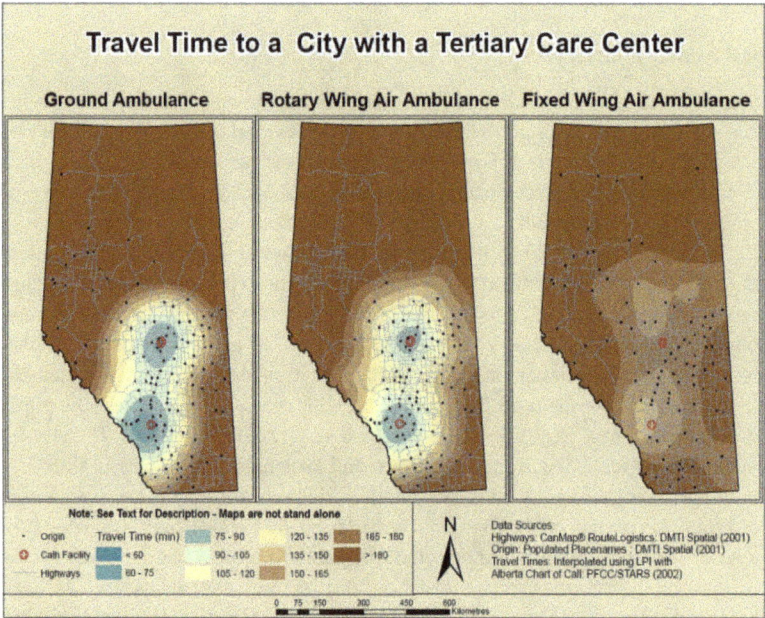

Figure 3. Using interpolation to model emergency transport (Patel *et al.*, 2007)

There are strengths and limitations to using interpolation as part of a geospatial analysis. One limitation of using this method is that an accurate interpolation requires many points that are evenly distributed over the study area. Another limitation is that depending on the type of interpolation used (e.g. global or local, exact or inexact) the results could vary (Lam, 1983; Mitas & Mitasova, 1999). In spite of these limitations, interpolation is a powerful method to compare access via different modes of travel. This is achieved by evaluating those areas that are served faster by different modes of travel using grid cell comparisons. Both network analysis and interpolation can be used to estimate travel time values that can be used as a measures of geographic access. These measures can then be used to evaluate the associations between spatial access to services and other variables (e.g. socio-economic variables). These access measures can also be linked with census or patient data to study the populations with access to certain services (Nallamothu et al., 2006; Schuurman et al., 2008). It is important to note that because the method of interpolation focuses on creating continuous surfaces from point data, it can also be used to model the concentration of environmental variables from a point source or the spread of an infectious agent over boundaries (Kistemann et al., 2002).

6. Identifying the best location for a new service

The spatial analysis of health services is based on the principle that populations and their need for healthcare vary across space. People are not located randomly about the earth and it is often observed that different areas are populated by groups with differing characteristics (e.g. socio-economic status, race, age, income, education, etc.); these varying characteristics of a group often influence their need for health services (McLafferty, 2003). Recent studies have used GIS and the principles of location analysis to evaluate optimal locations for pre-hospital helicopter emergency medical services (Schuurman et al., 2009) and to evaluate how patient access can change depending on where cardiac services are located (Pereira et al., 2007). The operations research literature is a strong source of methods papers and includes literature on how models can be used to maximize service area (Mahmud & Indriasari, 2009), how models can simulate different starting point locations for ambulance (Ingolfsson et al., 2003) and how stochastic approaches can be superior to deterministic approaches when planning health service delivery (Harper et al., 2005). It is beyond the scope of this chapter to discuss the mathematical models that are used for location analysis; this information is available elsewhere (de Smith et al., 2007b; ReVelle & Eiselt, 2005). This section will discuss two different approaches for identifying the best location for a new service: coverage and distance problems.

6.1. Coverage problems

There are two types of coverage problems that are commonly referred to in the operations research literature: the location set covering problem (LSCP) and the maximal covering location problem (MCLP). The goal of LSCP is to "find the minimum number of facilities and their locations such that all neighborhoods are covered within the maximal distance or time standard" (Church & Murray, 2009a). In this model, each demand point is covered at least once. The goal of MCLP is to "find a prespecified number of facilities such that coverage within a maximal service distance or time is maximized" (Church & Murray, 2009a). Given these constraints this model seeks to provide the best possible coverage with the limited available resources and does not guarantee service. Geospatial tools available within GIS software can be used to assemble the needed data to address both types of coverage problems. The three essential data components that need to be represented in a GIS in order to solve this problem are the demand points that are to be served by the facilities, the set of possible facility locations and the service area capabilities of these facilities (Church & Murray, 2009a). Ambulance and fire department response are two examples of services that require complete coverage of an area and thus a LSCP approach to coverage. Other services may strive for as much coverage as possible based on the resources available. One example is the placement of public health care facilities. By creating catchment areas within GIS, the best possible locations for new facilities can be evaluated (based on the most area and population covered) (Murad, 2008). The strength of using these types of coverage models is that they provide a spatial-standard based framework for siting new health service facilities that considers population needs and available resources. While these frameworks can be used to study the best locations based on the coverage of a facility over a given area, in other instances a distance based approach may be required.

6.2. Distance problems

There are two common approaches to finding an optimal location based on distance for a new facility: p-median and p-centre approaches. The p-median approach focuses on minimizing the average distance/cost to or from patients, while the p-centre approach focuses on minimizing the maximum distance/cost to or from patients. The limitation of solving these problems discretely is that when the number of facilities to be placed increases, the computational intensity of the calculations quickly increases (Church & Murray, 2009b). In order to deal with this limitation, heuristic algorithms are employed that use systematic procedures to trade off the quality of the solution with processing speed (de Smith *et al.*, 2007b). These algorithms strive to find an optimal solution based on specified search strategies (Church & Murray, 2009b). While these methods are founded theoretically in the mathematical modeling and operations research literature, there are a number of components to these strategies that can be evaluated using GIS (Church, 2002). For example, previous studies have used geospatial methods to evaluate the suitability of selected sites as possible locations for a new facility (Cinnamon et al., 2009; Schuurman et al., 2008) and to measure the distance between populations and proposed services (Pereira et al., 2007).

Both the coverage approach and the distance based approach to the placement of new facilities are important. They provide a framework for optimizing the solution to the best location for a new facility. The coverage model focuses on ensuring the population is best served in terms of the placement of a new facility. The location-allocation approach is more focused on the tradeoffs in terms of minimizing the average or maximum distance that patients have to travel in order to access a service. The incorporation of patient service utilization data for both coverage and distance models will add value to the optimal solutions; these data appear to be one component that is lacking in current research that uses GIS and location-allocation approaches to recommend the placement of new services.

7. Conclusions

This chapter has shown that the geospatial methods that can be used with GIS extend beyond the simple visualization of patterns. There are powerful geospatial methods available to address problems that are commonly posed by epidemiologists and health services researchers. This chapter goes beyond a simple discussion of the basic methods of data creation and mapping with GIS, to introduce researchers having a basic understanding of GIS and statistical methods to more robust methods for spatial data analysis. It is hoped that a broader understanding of the tools available for spatial analysis will ensure that the most appropriate method to address a spatial problem is used, while a consideration is made of the strengths and limitations of the geospatial methods employed in health research. While sophisticated tools have been introduced to deal with spatial data issues such as spatial autocorrelation, the mapping of spatial data remains a useful way of conveying the results of any spatial analysis. In future, health researchers should aim to work in collaborative teams with geographers and operations research specialists to ensure that the most appropriate geospatial methods are being used to study the distribution of disease and health service delivery.

Author details

Alka Patel
University of Calgary, Canada

Nigel Waters
George Mason University, USA

Acknowledgement

We would like to thank Jeff Wielki and Chantal Hansen for giving permission to include Figure 1 and Figure 2, respectively.

8. References

Anselin, L. (1995). Local indicators of spatial association - LISA. *Geographical Analysis, 27*(2), 93-115.

Apparicio, P, Abdelmajid, M, Riva, M, *et al.* (2008). Comparing alternative approaches to measuring the geographical accessibility of urban health services: Distance types and aggregation-error issues. *International Journal of Health Geographics, 7*:7.

Brabyn, L, & Skelly, C. (2002). Modelling population access to New Zealand public hospitals, *International Journal of Health Geographics,* 1:3.

Briggs, D, Fecht, D, & De Hoogh, K. (2007a). Census data issues for epidemiology and health risk assessment: Experiences from the small area health statistics unit. *Journal of the Royal Statistical Society: Series A (Statistics in Society), 170*(2), 355-378.

Briggs, DJ, Gulliver, J, Fecht, D, *et al.* (2007b). Dasymetric modelling of small-area population distribution using land cover and light emissions data. *Remote Sensing of Environment, 108*(4), 451-466.

Brunsdon, C, Fotheringham, AS, & Charlton, ME. (1996). Geographically weighted regression: A method for exploring spatial nonstationarity. *Geographical Analysis, 28*(4), 281-298.

Chaikaew, N, Tripathi, NK, & Souris, M. (2009). Exploring spatial patterns and hotspots of diarrhea in Chiang Mai, Thailand. *International Journal of Health Geographics,* 8:1.

Christie, S, & Fone, D. (2003). Equity of access to tertiary hospitals in Wales: A travel time analysis. *Journal of Public Health Medicine, 25*(4), 344-350.

Church, RL. (2002). Geographical information systems and location science. *Computers & Operations Research, 29*(6), 541-562.

Church, RL, & Murray, AT. (2009a). Chapter 9: Coverage. In *Business site selection, location analysis, and GIS* (pp. 209-233). Hoboken, NJ: John Wiley & Sons, Inc.

Church, RL, & Murray, AT. (2009b). Chapter 11: Location-allocation. In *Business site selection, location analysis, and GIS* (pp. 259-280). Hoboken, NJ: John Wiley & Sons, Inc.

Cinnamon, J, Schuurman, N, & Crooks, VA. (2008). A method to determine spatial access to specialized palliative care services using GIS. *BMC Health Services Research,* 8:140.

Cinnamon, J, Schuurman, N, & Crooks, VA. (2009). Assessing the suitability of host communities for secondary palliative care hubs: A location analysis model. *Health & Place, 15*(3), 792-800.

Comber, A, Brunsdon, C, & Green, E. (2008). Using a GIS-based network analysis to determine urban greenspace accessibility for different ethnic and religious groups. *Landscape and Urban Planning, 86*(1), 103-114.

Cromley, EK, & McLafferty, SL. (2002). *GIS and public health.* USA: The Guilford Press.

de Smith, M, Goodchild, M, & Longley, P. (2007a). Chapter 4: Building blocks of spatial analysis. In *Geospatial analysis: A comprehensive guide to principles, techniques and software tools* (2 ed., pp. 71-180). Leicester, UK: Matador.

de Smith, M, Goodchild, M, & Longley, P. (2007b). Chapter 7: Network and location analysis. In *Geospatial analysis: A comprehensive guide to principles, techniques and software tools* (2 ed., pp. 71-180). Leicester, UK: Matador.

Farwell, D. (1999). Specifying the bandwidth function for the kernel density estimator., from http://www.mrc-bsu.cam.ac.uk/bugs/documentation/coda03/node44.html

Fotheringham, AS, Brunsdon, C, & Charlton, ME. (2005a). Chapter 5: Local analysis. In *Quantitative geography: Perspectives on spatial data analysis* (pp. 93-129). London, UK: SAGE Publications.

Fotheringham, AS, Brunsdon, C, & Charlton, ME. (2005b). Chapter 7: Spatial regression and geostatistical models. In *Quantitative geography: Perspectives on spatial data analysis* (pp. 162-183). London, UK: SAGE Publications.

Fotheringham, AS, Charlton, ME, & Brunsdon, C. (1998). Geographically weighted regression: A natural evolution of the expansion method for spatial data analysis. *Environment and Planning A, 30*(11), 1905-1927.

Fotheringham, AS, & Wong, DWS. (1991). The modifiable areal unit problem in multivariate statistical-analysis. *Environment and Planning A, 23*(7), 1025-1044.

Gatrell, AC, Bailey, TC, Diggle, PJ, *et al.* (1996). Spatial point pattern analysis and its application in geographical epidemiology. *Transactions of the Institute of British Geographers, 21*(1), 256-274.

Hansen, C. (2005). *A GIS analysis of reported spousal violence in Calgary, AB.* University of Calgary, Calgary.

Hare, TS, & Barcus, HR. (2007). Geographical accessibility and Kentucky's heart-related hospital services. *Applied Geography, 27*(3-4), 181-205.

Harper, PR, Shahani, AK, Gallagher, JE, *et al.* (2005). Planning health services with explicit geographical considerations: A stochastic location-allocation approach. *Omega-International Journal of Management Science, 33*(2), 141-152.

Haynes, R, Jones, AP, Sauerzapf, V, *et al.* (2006). Validation of travel times to hospital estimated by GIS. *International Journal of Health Geographics, 5*:40.

Hewko, J, Smoyer-Tomic, KE, & Hodgson, MJ. (2002). Measuring neighbourhood spatial accessibility to urban amenities: Does aggregation error matter? *Environment and Planning A, 34*(7), 1185-1206.

Hu, ZY, & Rao, KR. (2009). Particulate air pollution and chronic ischemic heart disease in the eastern United States: A county level ecological study using satellite aerosol data. *Environmental Health, 8.*

Ingolfsson, A, Erkut, E, & Budge, S. (2003). Simulation of single start station for Edmonton EMS. *The Journal of the Operational Research Society, 54*(7), 736.

Jetz, W, Rahbek, C, & Lichstein, JW. (2005). Local and global approaches to spatial data analysis in ecology. *Global Ecology and Biogeography, 14*(1), 97-98.

Johnson, S. (2007). *The ghost map: The story of London's most terrifying epidemic--and how it changed science, cities, and the modern world.* London: Riverhead Trade, Penguin Group.

Joyce, K. (2009). "To me it's just another tool to help understand the evidence": Public health decision-makers' perceptions of the value of geographical information systems (GIS). *Health & Place, 15*(3), 831-840.

Kistemann, T, Dangendorf, F, & Schweikart, J. (2002). New perspectives on the use of geographical information systems (GIS) in environmental health sciences. *International Journal of Hygiene and Environmental Health, 205*(3), 169-181.

Kurland, KS, & Gorr, WL. (2009). *GIS tutorial for health* (3rd ed.). Redlands, CA: ESRI Press.

Lam, NSN. (1983). Spatial interpolation methods - a review. *American Cartographer, 10*(2), 129-149.

Langford, M, & Higgs, G. (2006). Measuring potential access to primary healthcare services: The influence of alternative spatial representations of population. *Professional Geographer, 58*(3), 294-306.

Langford, M, Higgs, G, Radcliffe, J, *et al.* (2008). Urban population distribution models and service accessibility estimation. *Computers Environment and Urban Systems, 32*(1), 66-80.

Legendre, P. (1993). Spatial autocorrelation - trouble or new paradigm. *Ecology, 74*(6), 1659-1673.

Lerner, EB, Billittier, AJ, Sikora, J, *et al.* (1999). Use of a geographic information system to determine appropriate means of trauma patient transport. *Academic Emergency Medicine,* 6(11), 1127-1133.

Lerner, EB, Fairbanks, RJ, & Shah, MN. (2005). Identification of out-of-hospital cardiac arrest clusters using a geographic information system. *Academic Emergency Medicine, 12*(1), 81-84.

Levine, N. (2007). CrimeStat: A spatial statistics program for the analysis of crime incident locations. Washington, DC: National Institute of Justice.

Mahmud, A, & Indriasari, V. (2009). Facility location models development to maximize total service area. *Theoretical and Empirical Researches in Urban Management,* 87.

Maroko, AR, Maantay, JA, Sohler, NL, *et al.* (2009). The complexities of measuring access to parks and physical activity sites in New York City: A quantitative and qualitative approach. *International Journal of Health Geographics,* 8:34.

McEntee, JC, & Ogneva-Himmelberger, Y. (2008). Diesel particulate matter, lung cancer, and asthma incidences along major traffic corridors in MA, USA: A GIS analysis. *Health & Place, 14*(4), 817-828.

McLafferty, SL. (2003). GIS and health care. *Annual Review of Public Health, 24,* 25-42.

Meade, MS, & Emch, M. (2010). *Medical geography* (Third ed.). New York: The Guilford Press.

Mennis, J. (2009). Dasymetric mapping for estimating population in small areas. *Geography Compass, 3*(2), 727-745.

Mitas, L, & Mitasova, H. (1999). Chapter 34: Spatial interpolation. In P. Longley, Goodchild, M. F., Maguire, D. J. and Rhind, D. W. (Ed.), *Geographical information systems: Principles, techniques, applications and management* (Second ed., pp. Pp. 481-492): Wiley, New York.

Murad, AA. (2008). Defining health catchment areas in Jeddah city, Saudi Arabia: An example demonstrating the utility of geographical information systems. *Geospatial Health, 2*(2), 151-160.

Murray, EJ, Marais, BJ, Mans, G, *et al.* (2009). A multidisciplinary method to map potential tuberculosis transmission 'hot spots' in high-burden communities. *International Journal of Tuberculosis & Lung Disease, 13*(6), 767-774.

Nallamothu, BK, Bates, ER, Wang, YF, *et al.* (2006). Driving times and distances to hospitals with percutaneous coronary intervention in the United States - implications for prehospital triage of patients with ST-elevation myocardial infarction. *Circulation, 113*(9), 1189-1195.

Ong, ME, Ng, FS, Overton, J, *et al.* (2009). Geographic-time distribution of ambulance calls in Singapore: Utility of geographic information system in ambulance deployment (CARE 3). *Annals of the Academy of Medicine, Singapore, 38*(3), 184-191.

Oppenshaw, S. (1996). Chapter 4: Developing GIS-relevent zone-based spatial analysis methods. In Longley & Batty (Eds.), *Spatial analysis: Modelling in a GIS environment*: John Wiley & Sons, Inc.

Parchman, ML, Ferrer, RL, & Blanchard, KS. (2002). Geography and geographic information systems in family medicine research. *Family Medicine, 34*(2), 132-137.

Patel, AB, Waters, NM, & Ghali, WA. (2007). Determining geographic areas and populations with timely access to cardiac catheterization facilities for acute myocardial infarction care in Alberta, Canada. *International Journal of Health Geographics, 6*:47.

Peleg, K, & Pliskin, JS. (2004). A geographic information system simulation model of EMS: Reducing ambulance response time. *American Journal of Emergency Medicine, 22*(3), 164-170.

Pereira, A, Niggebrugge, A, Powels, J, *et al.* (2007). Potential generation of geographical inequities by the introduction of primary percutaneous coronary intervention for the management of st segment elevation myocardial infarction. *International Journal of Health Geographics, 6*:43.

Pouliou, T, & Elliott, SJ. (2009). An exploratory spatial analysis of overweight and obesity in Canada. *Preventive Medicine, 48*(4), 362-367.

Poulsen, E, & Kennedy, LW. (2004). Using dasymetric mapping for spatially aggregated crime data. *Journal of Quantitative Criminology, 20*(3), 243-262.

ReVelle, CS, & Eiselt, HA. (2005). Location analysis: A synthesis and survey - invited review. *European Journal of Operational Research, 165*(1), 1-19.

Robitaille, E, & Herjean, P. (2008). An analysis of the accessibility of video lottery terminals: The case of Montreal. *International Journal of Health Geographics, 7*:2.

Rogerson, PA. (2010). *Statistical methods for geography* (Third ed.). Los Angeles, CA: Sage.

Rushton, G. (2003). Public health, GIS, and spatial analytic tools. *Annual Review of Public Health, 24*, 43-56.

Schuurman, N, Bell, N, Hameed, MS, *et al.* (2008). A model for identifying and ranking need for trauma service in nonmetropolitan regions based on injury risk and access to services. *Journal of Trauma-Injury Infection & Critical Care, 65*(1), 54-62.

Schuurman, N, Bell, NJ, L'Heureux, R, *et al.* (2009). Modelling optimal location for pre-hospital helicopter emergency medical services. *BMC Emergency Medicine, 9*, 6.

Shi, X. (2009). A geocomputational process for characterizing the spatial pattern of lung cancer incidence in New Hampshire. *Annals of the Association of American Geographers, 99*(3), 521-533.

Srivastava, A, Nagpal, BN, Joshi, PL, *et al.* (2009). Identification of malaria hot spots for focused intervention in tribal state of India: A GIS based approach. *International Journal of Health Geographics, 8*:30.

Tobler, W. (2004). On the first law of geography: A reply. *Annals of the Association of American Geographers, 94*(2), 304-310.

Wakefield, J. (2007). Disease mapping and spatial regression with count data. *Biostatistics, 8*(2), 158-183.

Warden, CR, Daya, M, & LeGrady, LA. (2007). Using geographic information systems to evaluate cardiac arrest survival. *Prehospital Emergency Care, 11*(1), 19-24.

Waters, NM. (1997a). Unit 40 - spatial interpolation I. from
http://www.geog.ubc.ca/courses/klink/gis.notes/ncgia/u40.html

Waters, NM. (1997b). Unit 41 - spatial interpolation II. from
http://www.geog.ubc.ca/courses/klink/gis.notes/ncgia/u41.html

Wielki, J. (2009). The development of barriadas & access to medical services in Lima, Peru. Retrieved November, 2009, from
http://www.focal.ca/pdf/media_Peru_Lima%20Hospital%20Access%20Wielki.pdf

Yang, TC, Wu, PC, Chen, VYJ, *et al.* (2009). Cold surge: A sudden and spatially varying threat to health? *Science of the Total Environment, 407*(10), 3421-3424.

Young, LJ, Gotway, CA, Yang, J, *et al.* (2009). Linking health and environmental data in geographical analysis: It's so much more than centroids. *Spatial and Spatio-temporal Epidemiology, 1*(1), 73-84.

Zenk, SN, Schulz, AJ, Israel, BA, *et al.* (2005). Neighborhood racial composition, neighborhood poverty, and the spatial accessibility of supermarkets in metropolitan Detroit. *American Journal of Public Health, 95*(4), 660-667.

Permissions

The contributors of this book come from diverse backgrounds, making this book a truly international effort. This book will bring forth new frontiers with its revolutionizing research information and detailed analysis of the nascent developments around the world.

We would like to thank Bhuiyan Monwar Alam, for lending his expertise to make the book truly unique. He has played a crucial role in the development of this book. Without his invaluable contribution this book wouldn't have been possible. He has made vital efforts to compile up to date information on the varied aspects of this subject to make this book a valuable addition to the collection of many professionals and students.

This book was conceptualized with the vision of imparting up-to-date information and advanced data in this field. To ensure the same, a matchless editorial board was set up. Every individual on the board went through rigorous rounds of assessment to prove their worth. After which they invested a large part of their time researching and compiling the most relevant data for our readers. Conferences and sessions were held from time to time between the editorial board and the contributing authors to present the data in the most comprehensible form. The editorial team has worked tirelessly to provide valuable and valid information to help people across the globe.

Every chapter published in this book has been scrutinized by our experts. Their significance has been extensively debated. The topics covered herein carry significant findings which will fuel the growth of the discipline. They may even be implemented as practical applications or may be referred to as a beginning point for another development. Chapters in this book were first published by InTech; hereby published with permission under the Creative Commons Attribution License or equivalent.

The editorial board has been involved in producing this book since its inception. They have spent rigorous hours researching and exploring the diverse topics which have resulted in the successful publishing of this book. They have passed on their knowledge of decades through this book. To expedite this challenging task, the publisher supported the team at every step. A small team of assistant editors was also appointed to further simplify the editing procedure and attain best results for the readers.

Our editorial team has been hand-picked from every corner of the world. Their multi-ethnicity adds dynamic inputs to the discussions which result in innovative outcomes. These outcomes are then further discussed with the researchers and contributors who give their valuable feedback and opinion regarding the same. The feedback is then collaborated with the researches and they are edited in a comprehensive manner to aid the understanding of the subject.

Apart from the editorial board, the designing team has also invested a significant amount of their time in understanding the subject and creating the most relevant covers. They scrutinized every image to scout for the most suitable representation of the subject and create an appropriate cover for the book.

The publishing team has been involved in this book since its early stages. They were actively engaged in every process, be it collecting the data, connecting with the contributors or procuring relevant information. The team has been an ardent support to the editorial, designing and production team. Their endless efforts to recruit the best for this project, has resulted in the accomplishment of this book. They are a veteran in the field of academics and their pool of knowledge is as vast as their experience in printing. Their expertise and guidance has proved useful at every step. Their uncompromising quality standards have made this book an exceptional effort. Their encouragement from time to time has been an inspiration for everyone.

The publisher and the editorial board hope that this book will prove to be a valuable piece of knowledge for researchers, students, practitioners and scholars across the globe.

List of Contributors

Rodrigo Nobrega, Charles O'Hara and Bethany Stich
Geosystems Research Institute, Mississippi State University, USA

Colin Brooks
Environmental Science Laboratory, Michigan Tech Research Institute, USA

Enguerran Grandchamp
University of Antilles and Guyana, France, Guadeloupe

Süleyman İncekara
Fatih University, Department of Geography, Turkey

Darka Mioc and François Anton
Technical University of Denmark, Denmark

Christopher M. Gold
University of Glamrogan, United Kingdom

Bernard Moulin
Université Laval, Canada

Jung-Sup Um
Kyungpook National University, South Korea

Barbara Goličnik Marušić
Urban Planning Institute of the Republic of Slovenia, Slovenia

Damjan Marušić
The Surveying and Mapping Authority of the Republic of Slovenia, Slovenia

Marek Kachnic
Nicolas Copernicus University in Toruń, Departure of Geology and Hydrogeology, Poland

Xinshen Diao, Liangzhi You, Vida Alpuerto and Renato Folledo
International Food Policy Research Institute (IFPRI), Washington DC, USA

Nikos Krigas
Laboratory of Systematic Botany & Phytogeography, Department of Botany, School of Biology, Aristotle University of Thessaloniki, Greece

Kimon Papadimitriou
Office for Sustainability, Aristotle University of Thessaloniki, Greece

Nikos Krigas and Antonios D. Mazaris
Department of Ecology, School of Biology, Aristotle University of Thessaloniki, Greece

Rolando Rodríguez and Pedro Real
Forest Science Faculty, Universidad de Concepción, Chile

Tatiana S. da Silva, Maria Luiza Rosa and Flávia Farina
Federal University of Rio Grande do Sul, Institute of Geoscience, Brazil

David Sanz, Santiago Castaño and Juan José Gómez-Alday
University of Castilla - La Mancha / Remote Sensing and GIS Group, Albacete, Spain

Fazel Amiri
Spatial and Numerical Modeling Laboratory, Institute of Advance Technology (ITMA), Faculty of Engineering, Universiti Putra Malaysia, Malaysia

Abdul Rashid B. Mohamed Shariff
Spatial and Numerical Modeling Laboratory, Faculty of Engineering, Universiti Putra Malaysia, Malaysia

Taybeh Tabatabaie
Environnemental Science, Faculty of Engineering, Islamic Azad University Bushehr Branch, Iran

Ilham S. M. Elsayed
University of Dammam, College of Engineering, Saudi Arabia

Nicolai Guth and Philipp Klingel
Karlsruhe Institute of Technology (KIT), Institute for Water and River Basin Management, Germany

Shih-Miao Chin, Francisco M. Oliveira-Neto, Ho-Ling Hwang, Diane Davidson and Bruce Peterson
Oak Ridge National Laboratory, USA

Lee D. Han
The University of Tennessee, Knoxville, TN, USA

J. Ponce, A.H. Torres, M.J. Escalona, M. Mejías and F.J. Domínguez-Mayo
Department of Computer Languages and Systems, University of Seville, Seville, Spain

Alka Patel
University of Calgary, Canada

Nigel Waters
George Mason University, USA